Advances in Design by Metallic Materials

Advances in Design by Metallic Materials: Synthesis, Characterization, Simulation and Applications

Editors

Cristiano Fragassa
Jeremy Epp
Grzegorz Lesiuk
Miroslav Zivkovic

MDPI • Basel • Beijing • Wuhan • Barcelona • Belgrade • Manchester • Tokyo • Cluj • Tianjin

Editors

Cristiano Fragassa
University of Bologna
Italy

Jeremy Epp
Leibniz Institute for Materials Engineering
Germany

Grzegorz Lesiuk
Wroclaw University of Science and Technology
Poland

Miroslav Zivkovic
University of Kragujevac
Serbia

Editorial Office
MDPI
St. Alban-Anlage 66
4052 Basel, Switzerland

This is a reprint of articles from the Special Issue published online in the open access journal *Metals* (ISSN 2075-4701) (available at: https://www.mdpi.com/journal/metals/special_issues/metallic_synthesis_characterization_simulation_applications).

For citation purposes, cite each article independently as indicated on the article page online and as indicated below:

LastName, A.A.; LastName, B.B.; LastName, C.C. Article Title. *Journal Name* **Year**, *Volume Number*, Page Range.

ISBN 978-3-0365-0746-0 (Hbk)
ISBN 978-3-0365-0747-7 (PDF)

© 2021 by the authors. Articles in this book are Open Access and distributed under the Creative Commons Attribution (CC BY) license, which allows users to download, copy and build upon published articles, as long as the author and publisher are properly credited, which ensures maximum dissemination and a wider impact of our publications.

The book as a whole is distributed by MDPI under the terms and conditions of the Creative Commons license CC BY-NC-ND.

Contents

About the Editors vii

Cristiano Fragassa
Advances in Design by Metallic Materials: Synthesis, Characterization, Simulation and Applications
Reprinted from: *Metals* **2021**, *11*, 272, doi:10.3390/met11020272 1

Ivana O. Mladenović, Nebojša D. Nikolić, Jelena S. Lamovec, Dana Vasiljević-Radović and Vesna Radojević
Application of the Composite Hardness Models in the Analysis of Mechanical Characteristics of Electrolytically Deposited Copper Coatings: The Effect of the Type of Substrate
Reprinted from: *Metals* **2021**, *11*, 111, doi:10.3390/met11010111 5

Jelena Živković, Vladimir Dunić, Vladimir Milovanović, Ana Pavlović and Miroslav Živković
A Modified Phase-Field Damage Model for Metal Plasticity at Finite Strains: Numerical Development and Experimental Validation
Reprinted from: *Metals* **2021**, *11*, 47, doi:10.3390/met11010047 23

Rahel Jedamski and Jérémy Epp
Non-Destructive Micromagnetic Determination of Hardness and Case Hardening Depth Using Linear Regression Analysis and Artificial Neural Networks
Reprinted from: *Metals* **2021**, *11*, 18, doi:10.3390/met11010018 51

Grzegorz Lesiuk, Bruno A.S. Pedrosa, Anna Zięty, Wojciech Błażejewski, Jose A.F.O. Correia, Abilio M.P. De Jesus and Cristiano Fragassa
Minimal Invasive Diagnostic Capabilities and Effectiveness of CFRP-Patches Repairs in Long-Term Operated Metals
Reprinted from: *Metals* **2020**, *10*, 984, doi:10.3390/met10070984 73

Antonio Carlos de Figueiredo Silveira, William Lemos Bevilaqua, Vinicius Waechter Dias, Pedro José de Castro, Jeremy Epp and Alexandre da Silva Rocha
Influence of Hot Forging Parameters on a Low Carbon Continuous Cooling Bainitic Steel Microstructure
Reprinted from: *Metals* **2020**, *10*, 601, doi:10.3390/met10050601 87

Rodoljub Vujanac, Nenad Miloradović, Snežana Vulović and Ana Pavlović
A Comprehensive Study into the Boltless Connections of Racking Systems
Reprinted from: *Metals* **2020**, *10*, 276, doi:10.3390/met10020276 99

Yuxuan Wang, Xuebang Wu, Xiangyan Li, Zhuoming Xie, Rui Liu, Wei Liu, Yange Zhang, Yichun Xu and Changsong Liu
Prediction and Analysis of Tensile Properties of Austenitic Stainless Steel Using Artificial Neural Network
Reprinted from: *Metals* **2020**, *10*, 234, doi:10.3390/met10020234 117

Grégori Troina, Marcelo Cunha, Vinícius Pinto, Luiz Rocha, Elizaldo dos Santos, Cristiano Fragassa and Liércio Isoldi
Computational Modeling and Constructal Design Theory Applied to the Geometric Optimization of Thin Steel Plates with Stiffeners Subjected to Uniform Transverse Load
Reprinted from: *Metals* **2020**, *10*, 220, doi:10.3390/met10020220 133

Nenad Miloradović, Rodoljub Vujanac, Slobodan Mitrović and Danijela Miloradović
Dry Sliding Wear Performance of ZA27/SiC/GraphiteComposites
Reprinted from: *Metals* **2019**, *9*, 717, doi:10.3390/met9070717 . . . . **161**

Xing Wang, Yongzhe Fan, Xue Zhao, An Du, Ruina Ma and Xiaoming Cao
Process and High-Temperature Oxidation Resistance of Pack-Aluminized Layers on Cast Iron
Reprinted from: *Metals* **2019**, *9*, 648, doi:10.3390/met9060648 . . . . **175**

Galileo Santacruz, Antonio Shigueaki Takimi, Felipe Vannucchi de Camargo,
Carlos Pérez Bergmann and Cristiano Fragassa
Comparative Study of Jet Slurry Erosion of Martensitic Stainless Steel with Tungsten Carbide
HVOF Coating
Reprinted from: *Metals* **2019**, *9*, 600, doi:10.3390/met9050600 . . . . **185**

Cristiano Fragassa, Matej Babic, Carlos Perez Bergmann and Giangiacomo Minak
Predicting the Tensile Behaviour of Cast Alloys by a Pattern Recognition Analysis on
Experimental Data
Reprinted from: *Metals* **2019**, *9*, 557, doi:10.3390/met9050557 . . . . **201**

About the Editors

Cristiano Fragassa is adjunct professor in the design of machines in the Department of Industrial Engineering, University of Bologna, Italy. He received his PhD in Mechanics of Materials in 2004 and habilitation as associate professor in Mechanical Engineering in 2019. He has over 20 years of experience in research and teaching at the university level on aspects such as the design of machines, structural materials, experimental mechanics, reliability and safety. He is the author of 200 publications, with more than half of them appearing internationally. His main fields of investigation have involved structural and safe design, advanced materials including metals and their alloys and process technologies such as metal casting. These topics have been approached both at the level of numerical modeling and experimental mechanics but always saving a special attention to their addressability to real applications. Recently, his research directions have evolved to include sustainable design and circular economy. He is also very active in terms of R&D or T&T, amassing a long list of internationally and nationally funded projects where he has been the coordinator or principal investigator. Finally, he is a member of different scientific committees including the Metals Editorial Board.

Jeremy Epp is head of the Department of Physical Analysis at the Leibniz Institute for Materials Engineering-IWT in Bremen, Germany, and is a member of the MAPEX Center for Materials and Processes at the University of Bremen. He received his PhD (Dr.-Ing.) in Metallurgical Science in 2016. He is heading one the world's largest X-ray diffraction labs with 21 modern diffractometers and is working in the field of high-strength materials, including heat treatment and manufacturing processes, characterization with advanced methods and evaluation and improvement of mechanical properties. His speciality is the development of methods for in situ X-ray diffraction experiments in lab and neutron/synchrotron facilities. He has published over 60 papers in reviewed national and international journals as well as in conference proceedings. He has been a member of the review panel of "Materials Science" at PETRA III Synchrotron, DESY, Hamburg, since 2020.

Grzegorz Lesiuk is associate professor at the Faculty of Mechanical Engineering, Wrocław University of Technology, Poland. He received his PhD in Technical Sciences in 2013 and habilitation in Mechanical Engineering in 2020. His scientific activities are related to mechanics, fracture mechanics and fatigue of materials and structures as well as new materials and modeling their durability under cyclic and static loads, lightweight structures and dimensional analysis. He has over 90 scientific papers published in national and international scientific and technical journals and over 70 conference papers, monographs and patents in this field. He is a Polish Society of Theoretical and Applied Mechanics and ESIS (European Structural Integrity Society) member.

Miroslav Zivkovic is full professor at the Faculty of Engineering, University of Kragujevac, Serbia, attaining this position in 2007. In 2018, he became head of the Department of Applied Mechanics and Automatic Control. He has more than 300 scientific papers published in international and national journals and scientific conference proceedings. He was the manager of three international projects and four scientific research projects financed by the Serbian government. He is the leader of the research team at the Center for Engineering Software and Dynamic Testing, developing the PAK software package for structural analysis (in the areas of linear and nonlinear static and implicit and explicit dynamic analysis of structures, heat conduction, coupled problems, fracture mechanics, damage mechanics, material fatigue, etc.). He is a member of the Serbian Chamber of Engineers, Serbian Society of Mechanics, Serbian Society of Computational Mechanics, Society for Integrity and Life of Structures and the Danubia-Adria-Society on Experimental Methods.

Editorial

Advances in Design by Metallic Materials: Synthesis, Characterization, Simulation and Applications

Cristiano Fragassa

Department of Industrial Engineering, Alma Mater Studiorum University of Bologna, Viale del Risorgimento 2, 40136 Bologna, Italy; cristiano.fragassa@unibo.it; Tel.: +39-347-6974-046

Received: 3 February 2021; Accepted: 3 February 2021; Published: 5 February 2021

1. Introduction and Scope

Metals have exerted a significant influence throughout the history of mankind, so much so that the different periods of development have often been marked with the name of some material: bronze age and iron age. And all these centuries are studded with continuous discoveries and improvements that have involved materials. However, perhaps in a fairly recent period something has changed in the relationship between Man and Metals: even if until today the growth of humanity remains substantially based on the full exploitation of metals, their convenience compared to other emerging families of unconventional materials is increasingly questioned.

In particular, the central argument for this collection was chosen considering that very recently, a great deal of attention has been paid by researchers and technologists to trying to eliminate metal materials in the design of products and processes in favor of plastics and reinforced composites. After a few years, it is possible to state that metal materials are even more present in our lives and this especially is thanks to their ability to evolve. This Special Issue is focused on that and on the recent evolution of metals and alloys with the scope of presenting the state-of-the-art of solutions where metallic materials have become established, without a doubt, as a successful design solution thanks to their unique properties. The Special Issue also intends to outline the fundamental development trends in the field, together with the most recent advances in the use of the metallic materials.

2. Contributions

The collection includes papers regarding the most multifaced aspects of metals as synthesis [1–3], treatments [2–4], experimental characterization [4–7], material models [7–9] and engineering applications [10–12] providing a clear cross-section of the wide variety of topics and research arguments under investigation in the scientific community now.

It is the case of [1], for instance, where the authors propose a way for predicting the effects of changes in metal process parameters in terms of metal materials properties. The focus was addressed to a conventional cast iron foundry, with the aim of monitoring the process phases, but it also permitted to investigate a rather uncommon cast iron (i.e., compacted) in respect to others (i.e., spheroidal). With such a scope, standard mechanical experiments (i.e., tensile tests) were combined with an advanced approach based on pattern recognition and machine learning able to find physical recurrencies where a human eye cannot discover anything. Similar artificial intelligence tools, based on artificial neural networks, were also adopted in other papers such as [7] on the tensile behavior of an austenitic stainless steel or [5] on the determination of hardness and other surface properties.

Regarding metal processing, in [2] the authors investigated the process of hot forging and how its changes can influence the metal microstructures in the case of a carbon continuous cooling bainitic steel. Other unconventional processes and treatments are proposed, too, as, for instance, in [4] where the protective effect of tungsten carbide provided by HVOF coating on martensitic stainless-steel with respect to jet slurry erosion is discussed. This kind of erosion is rather uncommon in terms of present contributions to the state of the art, while it is quite frequent in practical applications (such as marine engineering, oil and gas applications and so on). However, surface hardening and wear continues to represent a relevant aspect to take care on in the use of metals, as demonstrated in [6] where metal matrix composites are experimentally investigated or in [8] where numerical models are developed in a way that predicts the final effect of hardening treatments. Numerical models, sometimes powered by finite elements, are also present in other contributions with the scope to predict plasticity [9] or failures [10] or, even, to support design actions with the scope to optimize the use of metal structures [11]. Finally, the last paper in the list [12] is maybe able to provide a synthesis in the proposed concept, regarding the opportunity of metallic materials, often considered as belonging to a quite old past, in an epoch where a large assortment of new materials is emerging, as in the case of polymer composites. This paper, not even doing it on purpose, deals with an intrinsic connection that our near future could bring out, merging two worlds only apparently strangers, when it proposes to use composite patches to repair metal bridges in order to extend their life for the benefit of society, not replace them.

3. Conclusions and Outlook

The realization of a Special Issue dealing with metallic materials represents quite a complex task for anyone, whatever the special topics it is focused on. This collection, concerning the use of metallic materials in advanced applications, is certainly no exception. This is the reason why there is no claim to completeness here, but only the desire to attract relevant articles coming from new and promising fields of investigation. In this, a proper result was achieved for sure.

Topics includes a large assortment of metals and alloys, such as steel, cast iron, aluminum, copper and metal matrix composites, together with their advanced use. In terms of processes the collection includes traditional processes such as casting, deformation or material removal, but special attention is also addressed to the most recent processes, such as additive manufacturing, metal deposition and so on. Contributions were selected with the scope to represent an element of novelty in the world of metallic materials as well as in the advanced characterization and use of metals for effective design solutions.

Conflicts of Interest: The author declares no conflict of interest.

References

1. Fragassa, C.; Babic, M.; Bergmann, C.; Minak, G. Predicting the Tensile Behaviour of Cast Alloys by a Pattern Recognition Analysis on Experimental Data. *Metals* **2019**, *9*, 557. [CrossRef]
2. Silveira, A.; Bevilaqua, W.; Dias, V.; de Castro, P.; Epp, J.; Rocha, A. Influence of Hot Forging Parameters on a Low Carbon Continuous Cooling Bainitic Steel Microstructure. *Metals* **2020**, *10*, 601. [CrossRef]
3. Wang, X.; Fan, Y.; Zhao, X.; Du, A.; Ma, R.; Cao, X. Process and High-Temperature Oxidation Resistance of Pack-Aluminized Layers on Cast Iron. *Metals* **2019**, *9*, 648. [CrossRef]
4. Santacruz, G.; Shigueaki Takimi, A.; Vannucchi de Camargo, F.; Pérez Bergmann, C.; Fragassa, C. Comparative Study of Jet Slurry Erosion of Martensitic Stainless Steel with Tungsten Carbide HVOF Coating. *Metals* **2019**, *9*, 600. [CrossRef]
5. Jedamski, R.; Epp, J. Non-Destructive Micromagnetic Determination of Hardness and Case Hardening Depth Using Linear Regression Analysis and Artificial Neural Networks. *Metals* **2021**, *11*, 18. [CrossRef]
6. Miloradović, N.; Vujanac, R.; Mitrović, S.; Miloradović, D. Dry Sliding Wear Performance of ZA27/SiC/Graphite Composites. *Metals* **2019**, *9*, 717. [CrossRef]

7. Wang, Y.; Wu, X.; Li, X.; Xie, Z.; Liu, R.; Liu, W.; Zhang, Y.; Xu, Y.; Liu, C. Prediction and Analysis of Tensile Properties of Austenitic Stainless Steel Using Artificial Neural Network. *Metals* **2020**, *10*, 234. [CrossRef]
8. Mladenović, I.; Nikolić, N.; Lamovec, J.; Vasiljević-Radović, D.; Radojević, V. Application of the Composite Hardness Models in the Analysis of Mechanical Characteristics of Electrolytically Deposited Copper Coatings: The Effect of the Type of Substrate. *Metals* **2021**, *11*, 111. [CrossRef]
9. Živković, J.; Dunić, V.; Milovanović, V.; Pavlović, A.; Živković, M. A Modified Phase-Field Damage Model for Metal Plasticity at Finite Strains: Numerical Development and Experimental Validation. *Metals* **2021**, *11*, 47. [CrossRef]
10. Vujanac, R.; Miloradović, N.; Vulović, S.; Pavlović, A. A Comprehensive Study into the Boltless Connections of Racking Systems. *Metals* **2020**, *10*, 276. [CrossRef]
11. Troina, G.; Cunha, M.; Pinto, V.; Rocha, L.; dos Santos, E.; Fragassa, C.; Isoldi, L. Computational Modeling and Constructal Design Theory Applied to the Geometric Optimization of Thin Steel Plates with Stiffeners Subjected to Uniform Transverse Load. *Metals* **2020**, *10*, 220. [CrossRef]
12. Lesiuk, G.; Pedrosa, B.; Zięty, A.; Błażejewski, W.; Correia, J.; De Jesus, A.; Fragassa, C. Minimal Invasive Diagnostic Capabilities and Effectiveness of CFRP-Patches Repairs in Long-Term Operated Metals. *Metals* **2020**, *10*, 984. [CrossRef]

© 2021 by the author. Licensee MDPI, Basel, Switzerland. This article is an open access article distributed under the terms and conditions of the Creative Commons Attribution (CC BY) license (http://creativecommons.org/licenses/by/4.0/).

Article

Application of the Composite Hardness Models in the Analysis of Mechanical Characteristics of Electrolytically Deposited Copper Coatings: The Effect of the Type of Substrate

Ivana O. Mladenović [1,*], Nebojša D. Nikolić [1], Jelena S. Lamovec [2], Dana Vasiljević-Radović [1] and Vesna Radojević [3]

1. Institute of Chemistry, Technology and Metallurgy, University of Belgrade, Njegoševa 12, 11 000 Belgrade, Serbia; nnikolic@ihtm.bg.ac.rs (N.D.N.); dana@nanosys.ihtm.bg.ac.rs (D.V.-R.)
2. University of Criminal Investigation and Police Studies, Cara Dušana 196, Zemun, 11 000 Belgrade, Serbia; jelena.lamovec@kpu.edu.rs
3. Faculty of Technology and Metallurgy, University of Belgrade, Karnegijeva 4, 11 000 Belgrade, Serbia; vesnar@tmf.bg.ac.rs
* Correspondence: ivana@nanosys.ihtm.bg.ac.rs; Tel.: +381-11-262-8587

Received: 15 December 2020; Accepted: 4 January 2021; Published: 8 January 2021

Abstract: The mechanical characteristics of electrochemically deposited copper coatings have been examined by application of two hardness composite models: the Chicot-Lesage (C-L) and the Cheng-Gao (C-G) models. The 10, 20, 40 and 60 µm thick fine-grained Cu coatings were electrodeposited on the brass by the regime of pulsating current (PC) at an average current density of 50 mA cm^{-2}, and were characterized by scanning electron (SEM), atomic force (AFM) and optical (OM) microscopes. By application of the C-L model we determined a limiting relative indentation depth (RID) value that separates the area of the coating hardness from that with a strong effect of the substrate on the measured composite hardness. The coating hardness values in the 0.9418–1.1399 GPa range, obtained by the C-G model, confirmed the assumption that the Cu coatings on the brass belongs to the "soft film on hard substrate" composite hardness system. The obtained stress exponents in the 4.35–7.69 range at an applied load of 0.49 N indicated that the dominant creep mechanism is the dislocation creep and the dislocation climb. The obtained mechanical characteristics were compared with those recently obtained on the Si(111) substrate, and the effects of substrate characteristics such as hardness and roughness on the mechanical characteristics of the electrodeposited Cu coatings were discussed and explained.

Keywords: copper coatings; pulsating current (PC); composite hardness models; hardness; creep resistance

1. Introduction

Copper electrodeposition is of high significance, attracting much attention in both the scientific and technological sectors, with numerous applications in many industrial branches. The main industrial branches using the copper coatings are the electrical, electronic, automotive, defense industries, etc. [1]. Application of this metal is based on its excellent electrical and thermal conductivity, as well as its corrosion-resistant characteristics, enabling the application of electrolytically formed Cu thin films and coatings for the interconnection of printed circuit boards (PCBs) and ultra large scale integration (ULSI), wiring so-called damascene process, etc. [2,3]. The electrolytically synthesized Cu coatings possess a good adhesiveness, making Cu a highly effective undercoat before applying other coatings such as tin or nickel. Sometimes a harder surface may be required, and then copper electroplating can be used to increase surface strength [1].

The good mechanical characteristics of the Cu coatings attained by application of the electrodeposition technique determine the advantage of this technique over the other methods of synthesis, such as physical vapor deposition (PVD) [4], chemical vapor deposition (CVD) [5], and magnetron sputtering [6]. Electrodeposition is a low-equipment- and -product-cost, environmentally friendly, time-saving and facile technique [7]. This method also offers a possibility of the easy control of the thickness of coatings, as well as the obtaining of coatings of desired features by a suitable selection of parameters and regimes of electrodeposition [8]. The parameters of electrodeposition determine the morphological and structural characteristics, and thus, the mechanical characteristics of metal coatings are the type and composition of electrolyte, the addition of additives in the electrolyte, the type of substrate (cathode), mixing of electrolyte, temperature, etc.

In the constant galvanostatic (DC) regime, the compact adherent coatings of copper with fine-grained structure are mainly obtained in the presence of various additives. The acid sulfate electrolytes consisting of copper sulfate and sulfuric acid are the most often-used electrolytes for Cu electrodeposition. Thiourea is a traditionally used additive for obtaining fine-grained deposits of Cu [9,10]. In the last two decades, the combination of additives based on chlorides and PEG poly(ethylene glycol) with an addition of bis-3-sulfopropyl-disulfide (SPS) [11–13] or 3-mercapto-2-propanesulphonic acid (MPSA) [14–16] also found wide application. The concentration of the leveling and brightening addition agents is incomparable with the concentration of the basic components of electrolyte, which causes their fast consumption and the need for the permanent control and correction of the composition of the electrolyte. In order to avoid the use of additives, various periodically changing regimes of electrodeposition, such as pulsating and reversing current regimes, were proposed for obtaining compact uniform coatings [8,17–20]. The lower-porosity and fine-grained structure of deposits are achieved by the simple regulation of parameters constructing these regimes. The mixing of electrolytes also contributes to an improvement of the quality of coatings produced by various electrodeposition techniques [18,21–23].

All the above-mentioned parameters and regimes of the electrodeposition affecting the quality, i.e., morphological and structural characteristics, of the coatings simultaneously determine their mechanical characteristics. The hardness of the coatings is one of the most important mechanical characteristics, and it can be determined by directly using low indentation loads, or indirectly by application of the composite hardness models. Both these ways have advantages and disadvantages, and a balance between them is necessary. The direct approach is suitable for thick coatings excluding any contribution of a substrate to measure the hardness value, but the limit of this approach is the insufficient precision of a diagonal size measurement at the low indentation load. The indirect approach takes into account the contribution of substrate hardness to measure the hardness value [24,25]. The main disadvantage of the application of these models is the absence of their universal character, with numerous limitations to the calculation of true (or absolute) hardness from the measured composite hardness.

Various composite hardness models, such as Burnett-Rickerby (B-R) [26,27], Chicot-Lesage (C-L) [28–31], Chen-Gao (C-G) [32–35] and Korsunsky (K-model) [36–39], are proposed for the determination of the true hardness of metal coatings. The choice of composite hardness the model depends on the coating/substrate hardness ratios, and some of them are applied for a "hard coating-soft substrate" system, such as the Korsunsky model, while some other models are suitable for the analysis of "soft film-hard substrate" systems, such as the Chen-Gao [32–35] and the Chicot-Lesage [28–31] composite hardness models. For the same metal coating, the choice of the composite hardness model is determined by the type of used substrate. The following substrates are the often applied in the Cu electroplating processes: silicon [20,25,35,38], nickel coatings [14,21], copper [9,10,19,38], polyimide [5], graphite [23] or brass [21].

The other very important mechanical characteristic of coatings is their creep resistance. The creep resistance of the coatings gives very valuable information related to the time-dependent flow of materials [40], i.e., to an evaluation of their reliability. It is necessary to stress that aside from the data obtained for the Cu coatings electrodeposited on the Si(111) substrate by the PC regime [25], other data dealing with the analysis of this mechanical characteristic are not found in the literature for the copper coatings.

Regarding the role of the substrate in relation to the mechanical characteristics of metal coatings, the aim of this study is to examine the contribution of brass as the type of cathode on the hardness and creep resistance of the Cu coatings electrodeposited by the PC regime. In order to better perceive the role of the substrate, the obtained results will be discussed and compared with those recently observed for the Cu coatings electrodeposited under the same conditions on the Si(111) substrate belonging to the group of very hard substrates. Special attention will be devoted to a determination of the precise boundary that separates the area with absolute coating hardness from the area where the contribution of substrate hardness must be taken into account. In spite of numerous investigations related to the hardness analysis of electrolytically deposited coatings, the results dealing with a definition of the limiting value separating these two areas are not reported in the literature, and for that reason, it will be achieved in this study by application of the C-L model for the first time.

2. Materials and Methods

2.1. Preparation of Samples by Electrodeposition Process for Mechanical Characterization

The electrodeposition of copper was performed from 240 g/L $CuSO_4 \cdot 5 H_2O$ in 60 g/L H_2SO_4 at room temperature in an open square-shaped electrochemical cell. For the electrodeposition process, the regime of pulsating current (PC) with the following parameters was applied: $j_A = 100$ mA cm^{-2}, $t_c = 5$ ms and $t_p = 5$ ms. In the PC regime, the electrodeposition process occurs at the average current density (j_{av}, in mA cm^{-2}) defined by Equation (1) [8]:

$$j_{av} = \frac{j_A \cdot t_c}{t_c + t_p} \tag{1}$$

where j_A (in mA cm^{-2}) is the current density amplitude, t_c (in ms) is the deposition pulse, and t_p (in ms) is the pause duration. With these parameters of the PC regime, j_{av} was 50 mA cm^{-2}. The thicknesses of the Cu coatings were 10, 20, 40 and 60 μm. Brass ($260_{1/2}$ hard, ASTM B36, K&S Engineering) with a 1.0 × 1.0 cm^2 surface area was used as a cathode, and copper plate with a 8.0 × 5.0 cm^2 surface area was used as an anode. The cathode was situated in the middle of the cell between two parallel Cu plates. The distance between anode and cathode was 2.0 cm. Preparation of the brass electrodes for electrodeposition was performed as follows: The brass cathode was ground by # 800, # 1000 and # 1200 SiC sandpapers and rinsed in water. Then, it was degreased at a temperature of 70 °C, followed by acid etching (20% H_2SO_4) at 50 °C. After each phase, the cathodes were rinsed with distilled water. For a preparation of the electrolyte, doubly distilled water and analytical-grade reagents were used.

2.2. Characterization of the Produced Cu Coatings

The following techniques were used for characterization:

(a) Scanning electron microscope (SEM), model JEOL JSM-6610LV (JEOL Ltd., Tokyo, Japan)—morphological analysis;

(b) Atomic force microscope (AFM), model Auto Probe CP Research. TM Microscopes, Veeco Instruments, Santa Barbara, CA, USA—topographical analysis of the coatings. The values of the arithmetic average of the absolute (R_a) roughness parameters were measured from the mean image data plane, using software SPLab (SPMLab NT Ver. 6.0.2., Veeco Instruments, Santa Barbara, CA, USA);

(c) Optical microscope (OM), model Olympus CX41 connected to the computer—analysis of the internal structure (cross section analysis). The Cu coatings were immersed in self-curing acrylate (Veracril® New Stetic S. A., Antioquia, Colombia) using a mold. Three parts of self-cure polymer Veracril® and one part of self-cure monomer Veracril® were used for the mixture. The self-polymerization time at room temperature was 20 min. After polymerization, the samples were removed from the Teflon mold and mechanically polished by SiC sandpapers # 2000 and with Al_2O_3 powder emulsion with different grain sizes (1 and 0.3 µm). After rinsing in water and drying in nitrogen flow, the cross section was observed on an optical microscope and the coating thicknesses were measured.

2.3. Examination of the Mechanical Characteristics of the Cu Coatings

The mechanical characteristics of the Cu coatings were examined by use of a Vickers microhardness tester "Leitz Kleinert Prufer DURIMET I" (Leitz, Oberkochen, Germany). The number of applied loads and the dwell time depended on the type of analyzed mechanical characteristic. For the analysis of the hardness of the coatings, applied loads (P) in the 0.049–2.94 N range and a constant dwell time of 25 s were applied. The indentation creep characteristics of the Cu coatings were analyzed, varying the dwell time in the 15–65 s range with applied loads of 0.49 and 1.96 N.

3. Results

3.1. Characterization of the Copper Coatings Obtained by the PC Regime

The fine-grained copper coatings were formed by a square-wave pulsating current (PC) regime at a j_{av} of 50 mA cm^{-2} (ν = 100 Hz), which was attained by application of the following parameters of this regime: t_c = 5 ms, t_p = 5 ms and j_A = 100 mA cm^{-2} (Figure 1). With an overpotential amplitude response in the 290–350 mV range, the formation of this structure corresponds to the very beginning of the mixed activation–diffusion control, which represents the optimum for the formation of compact and uniform coatings [8,20].

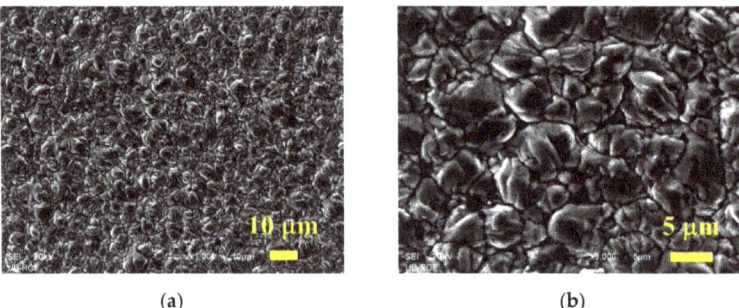

(a) (b)

Figure 1. Morphology of the copper coating electrodeposited on the brass by the PC regime at a j_{av} of 50 mA cm^{-2}. The thickness of the coating: 40 µm. The parameters of the PC regime: t_c = 5 ms, t_p = 5 ms and j_A = 100 mA cm^{-2} ((a) ×1000, and (b) ×3000).

Figure 2 shows 70 × 70 µm² surface areas and the corresponding line section analyses of the Cu coatings with thicknesses of 10, 20, 40 and 60 µm obtained by applying the above-mentioned PC regime. The values of the arithmetic average of the absolute (R_a) roughness determined by the accompanied software are given in Table 1. The data were presented as the mean ± standard deviation for 12 measuring points. The increase in roughness of the coatings with increasing the thickness is clearly visible from both Figure 2 and Table 1, and this increase in roughness was about seven times.

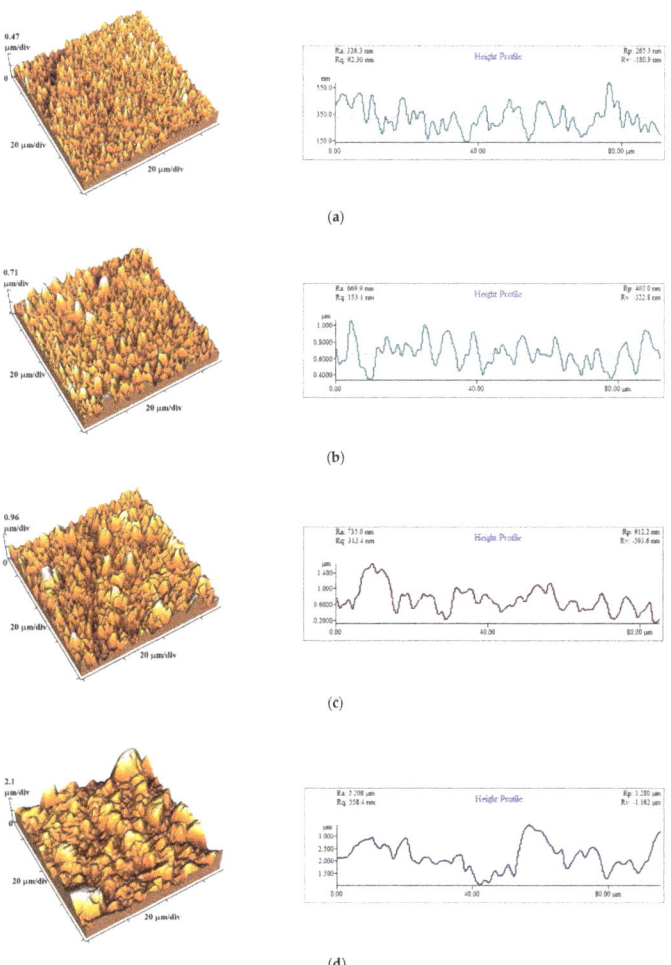

Figure 2. The surface topography and the line section analysis of the Cu coatings of thicknesses from: (**a**) 10 µm, (**b**) 20 µm, (**c**) 40 µm, and (**d**) 60 µm. Scan size: (70 × 70) µm².

Table 1. The values of the arithmetic average of the absolute (R_a) roughness with a standard deviation of the Cu coatings of various thicknesses. Scan size: (70 × 70) µm².

δ/µm	10	20	40
R_a/nm	75.05 ± 8.1	146.0 ± 7.83	215.6 ± 8.62

A cross section analysis of the same Cu coatings is presented in Figure 3, from which the uniform and compact structure of the coatings of the projected thickness can be seen.

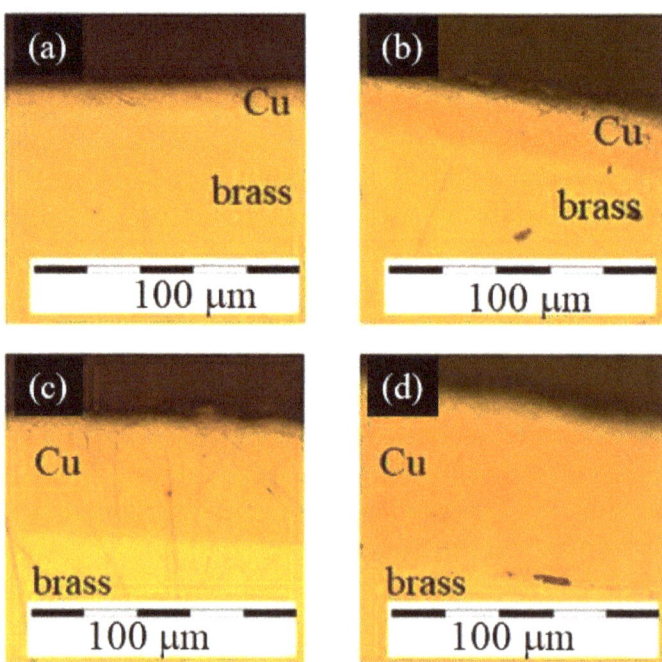

Figure 3. Cross section analysis of the Cu coatings electrodeposited by the PC regime on the brass of thicknesses from: (**a**) 10 µm, (**b**) 20 µm, (**c**) 40 µm, and (**d**) 60 µm.

3.2. Analysis of the Mechanical Characteristics of the Cu Coatings

3.2.1. Determination of Absolute Hardness of Substrate (Brass)

In the application of various composite hardness models, the first step is a determination of the absolute (or true) hardness of a substrate. The composite hardness (H_c, in Pa) depends on the applied load (P, in N) and the measured diagonal size (d, in m) according to Equation (2) [20]:

$$H_c = \frac{1.8544 \cdot P}{d^2} \qquad (2)$$

The Vicker's test is normalized by ASTM E384 and ISO 6507 standards [41,42], where P is measured in kgf and d in mm. If the applied load is expressed in N, then Equation (2) should be divided by 9.8065.

The model named PSR (proportional specimen resistance) is widely used for the determination of the absolute hardness of the substrate [43]. According to this model, the applied load and the measured diagonal size are related by Equation (3):

$$P = a_1 \cdot d + \frac{P_c}{d_0^2} \cdot d^2 \tag{3}$$

where P_c (in N) is the critical applied load above which microhardness becomes load-independent and d_0 (in m) is the corresponding diagonal length of the indents. Figure 4 shows the dependence of P/d (in N·µm^{-1}) on d (in µm), from which slope an absolute hardness of the brass B36 substrate (H_s) of 1.41 GPa was calculated.

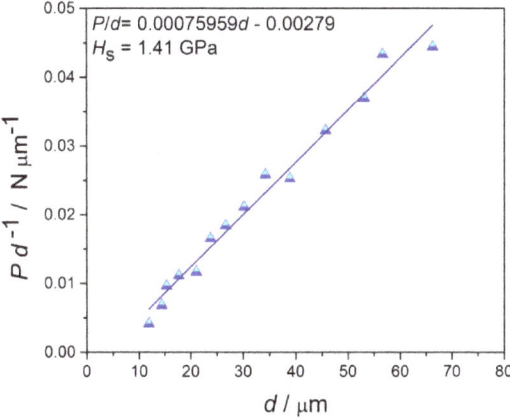

Figure 4. Determination of the absolute (true) hardness of the brass B36 substrate.

3.2.2. Hardness Analysis of Copper Coatings Electrodeposited on the Brass

The dependencies of the composite hardness, H_c, on the relative indentation depth (RID) for the Cu coatings thicknesses of 10, 20, 40 and 60 µm are shown in Figure 5a. The RID is defined as the ratio between indentation depth, h, and the thickness of the coating, δ (RID = h/δ), and RID values between 0.01 and 0.1 indicate the dominant effect of the hardness of the coating on the composite hardness [20,21,25,36–38,44]. For RID values between 0.1 and 1, both the substrate and the coating contribute to the composite value, and finally, RID values larger than 1 indicate the dominant effect of the substrate hardness on the composite hardness. The indentation depth is related with a measured diagonal size as $h = d/7$ [20,44].

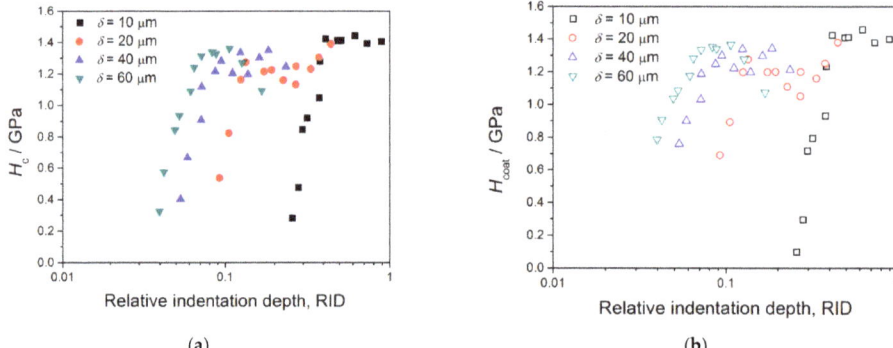

Figure 5. The dependencies of (**a**) the composite hardness, and (**b**) the coating hardness, on the RID (relative indentation depth) for the Cu coatings of thicknesses of 10, 20, 40 and 60 µm, obtained by the PC regime at a j_{av} of 50 mA cm^{-2}. The coating hardness was calculated by application of the Chicot–Lesage (C–L) model.

For the coatings of 10 and 20 µm thicknesses, the RID values were between 0.1 and 1, indicating the contribution of both the brass and the electrodeposited Cu to the composite hardness. With increasing the coating thickness, the contribution of the coating hardness to the composite hardness (RID < 0.1) was increased, which can be seen from Figure 5a. Simultaneously, the highest value of composite hardness was shown by the 10 µm thick Cu coating.

Figure 5b shows the dependencies of the coating hardness (H_{coat}) on the RID calculated according to the Chicot–Lesage (C-L) model. A similar shape of the dependencies to those obtained for the composite hardness on the RID was observed. A detailed presentation of the C-L model has been already given in Ref. [20].

An additional analysis was made with the aim of establishing a precise boundary of applicability of the C-L model, i.e., to establish a boundary whereat begins a strong contribution of the substrate to the composite hardness. For that purpose, the dependencies of the $(\delta/d)^m$ on the RID for the Cu coatings of thicknesses of 10, 20, 40 and 60 µm were measured and are shown in Figure 6. The exponent *m* represents the composite Meyer's index for a composite system, and it is calculated by the linear regression performed on all experimental points for the examined coating–substrate system [30,38,45,46]. The values of exponent *m* with R-squared values on ln(P)-ln(d) charts, obtained for the coatings of various thicknesses and 12 applied load points, are given in Table 2.

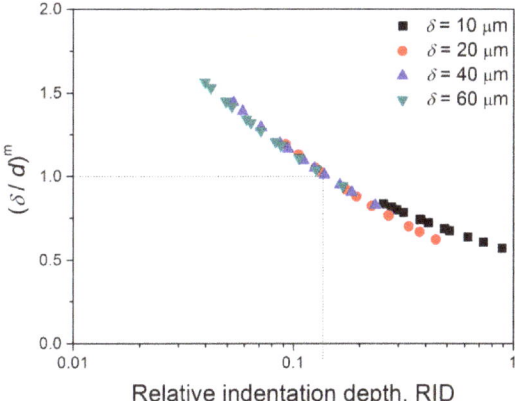

Figure 6. The dependencies of the $(\delta/d)^m$ on the RID obtained for the 10, 20, 40 and 60 μm thick Cu coatings.

Table 2. The values of exponent m and regression coefficient R^2, for the Cu coatings of various thicknesses.

δ/μm	10	20	40	
m	0.3082	0.4141	0.3744	0
R^2	0.9185	0.9767	0.9589	0

Taking a value of δ/d of 1 as a limit up to which the C-L model is applicable [28–31], the RID value of 0.14 was obtained (Figure 6). For the RID > 0.14, it is necessary to apply the composite hardness model in order to evaluate the coating hardness from the measured composite hardness, because the substrate hardness strongly affects the composite hardness value. For an RID < 0.14, the composite hardness corresponds to the coating hardness. The limitation of the application of the C-L model for the RID < 0.14 is confirmed by the fact that for the RID values smaller than this value the coating hardness becomes larger than the composite hardness, and this difference increases with a decreasing RID value (Figure 5).

To determine the true coating hardness values, another composite hardness model was applied. The shapes of the dependencies of the H_c on the RID shown in Figure 5a, as well as the value of the brass hardness of 1.41 GPa, indicate that the coatings of copper on brass belong to the "soft film on hard substrate" type of composite hardness system. For this reason, the Cheng-Gao (C-G) model [32–35] was used for a determination of the true hardness of the coating from the measured composite hardness. The C-G model has been developed for this type of composite system [33,35], and it is successfully implemented in a determination of the hardness of the copper coatings obtained by electrodeposition by the PC regime on a very hard Si(111) substrate [25]. According to the C-G model, a correlation between the composite hardness and the indentation depth is given by Equation (4) [33]:

$$H_c = A + B \cdot \frac{1}{h} + C \cdot \frac{1}{h^{n+1}} \quad (4)$$

where, as already mentioned, H_c (in Pa) is the composite hardness, h (in m) is indentation depth, A, B and C are fitting parameters used for the calculation of the absolute or true coating hardness, and n is the power index. For the "soft film on hard substrate" composite hardness system, the value for n is 1.8 [33,35].

Then, the absolute coating hardness, H_{coat}, is calculated by applying Equation (5) [33]:

$$H_{coat} = A \pm \sqrt{\frac{[n \cdot |B|/(n+1)]^{n+1}}{n \cdot |C|}} \quad (5)$$

The parameters A, B and C, and the calculated values of the coating hardness (H_{coat}), are given in Table 3. In Equation (5), the sign "−" is used for the "soft film−hard substrate" system (Cu/brass) [25,32–35].

Table 3. The values of fitting parameters (A, B and C), error fitting (*RMSE*—root mean square error) and the coating hardness (H_{coat}) obtained by application of the Cheng-Gao (C-G) model for the 10, 20, 40 and 60 μm thick Cu coatings obtained on the brass substrate by the PC regime at a j_{av} of 50 mA cm^{-2}.

δ/μm	A	B	C	RMSE	H_{coat}/GPa
10	1.148	20.24	−6647	0.08768	1.1399
20	1.14	7.138	−1457	0.1018	1.1295
40	1.34	−2.575	−109	0.05591	1.1180
60	0.9513	20.01	−4906	0.0721	0.9418

From Table 3, a decrease in the coating hardness with an increase in the thickness of the coating can be seen.

3.2.3. Creep Resistance Analysis of the Cu Coatings

Figure 7 gives the dependencies of the composite hardness on the dwell time for the Cu coatings of various thicknesses obtained with applied loads of 0.49 N (Figure 7a) and 1.96 N (Figure 7b). The decrease in the composite hardness when increasing the dwell time is clearly observed for all thicknesses of the coatings and for both the applied loads.

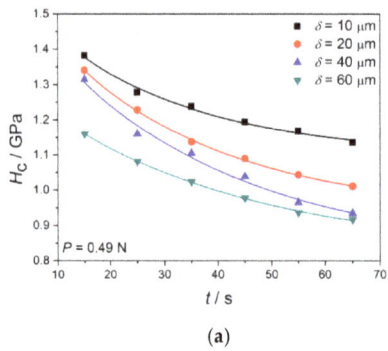
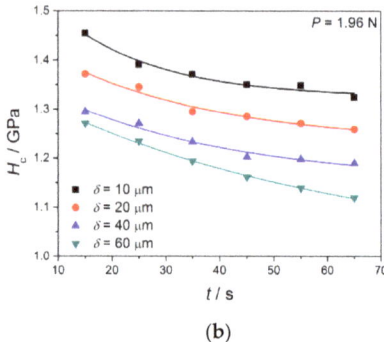

Figure 7. The dependencies of the composite hardness (H_c) on the dwell time (t) for the 10, 20, 40 and 60 μm thick Cu coatings obtained by the PC regime with an applied load of (**a**) 0.49 N and (**b**) 1.96 N.

The main parameter defining the creep resistance characteristics of the coatings is a stress exponent, μ, and this parameter can be determined by the application of the Sargent-Ashby model [47,48]. According to this model, the dependence of H_c on t can be presented by Equation (6):

$$H_c = \frac{\sigma_0}{(c \cdot \mu \cdot \varepsilon_0 \cdot t)^{\frac{1}{\mu}}} \tag{6}$$

In Equation (6), ε_0 is the strain rate at reference stress σ_0, c is constant, t (in s) is dwell time and μ is the stress exponent. The values of the stress exponent around one indicate that the dominant mechanism affecting deformation is diffusion creep; for a value close to 2 it is a grain boundary sliding, and for μ values between 3 and 10 the dominant mechanism is dislocation creep and dislocation climb [48].

The stress exponent can be determined from the linear dependence of $\ln(H_c)$ on $\ln(t)$ (Figure 8), where the slope of a straight line corresponds to a negative inverse stress exponent $(-1/\mu)$. The obtained values of stress exponents are summarized in Table 4.

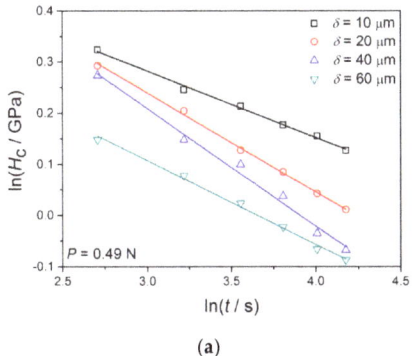
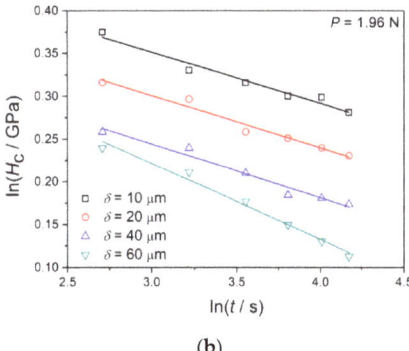

Figure 8. The dependencies of $\ln(H_c)$ on $\ln(t)$ for the 10, 20, 40 and 60 μm thick Cu coatings obtained by the PC regime with an applied load of (**a**) 0.49 N and (**b**) 1.96 N.

Table 4. The values of the stress exponent (μ) obtained for the Cu coatings thicknesses of 10, 20, 40 and 60 μm with applied loads of 0.49 and 1.96 N.

The Thickness of Coatings/μm	The Stress Exponent (μ) Obtained with Applied Load (P in N) of:	
	0.49	1.96
10	7.69	16.78
20	5.14	16.39
40	4.35	15.90
60	6.01	11.28

For an applied load of 0.49 N, the stress exponent decreases with an increase in the coating thickness from 10 to 40 μm. After the minimum was attained with the coating thickness of 40 μm, an increase in the value of this exponent was observed. On the other hand, with the high applied load (1.96 N), the stress exponents for the coatings of thicknesses of 10, 20 and 40 μm were close to each other, and were considerably higher than those obtained at a load of 0.49 N. The value obtained with the 60 μm thick Cu coating was lower, but was still significantly higher than those obtained at the low load.

As such, at the low applied load of 0.49 N, the dominant mechanism is the dislocation creep and the dislocation climb. At the high applied load, the values of the stress exponent that are high and close to each other indicate the existence of some other phenomena, which will be discussed later.

4. Discussion

The Cu coatings' hardness values in the (0.9418–1.1399) GPa range obtained by application of the C-G model were smaller than the hardness of brass (H_s = 1.41 GPa), confirming the assumption that the coatings of Cu on the brass belong to the "soft film on hard substrate" composite hardness system. The values of the coating hardness were also smaller than those obtained for the Cu coatings on Si(111) produced under the same electrodeposition conditions [25]. For the Cu coatings of thicknesses of 10, 20, 40 and 60 μm, the H_{coat} values on Si(111) obtained by application of the C-G model were 2.119, 1.914, 1.5079 and 1.164 GPa, respectively. Comparing the coating hardness values of these two substrates, it is clear that the difference between them decreases with increasing the thickness of the coating.

The largest difference was obtained for the coatings of thickness of 10 μm (about 46%), while the smallest difference was obtained for those the thickness of which was 60 μm (about 19%). This clearly indicates that the difference in the coatings' hardness values can be attributed to the type of used substrate, i.e., the various contributions of the hardness of the substrate to the determined hardness of the coating. Namely, although both substrates, the Si(111) and the brass B36, belong to the "hard" type of substrates relative to the Cu coatings, the hardness of Si(111) was about five times larger than that of brass B36 (7.42 GPa [20] vs. 1.41 GPa, respectively). Additionally, the measured composite hardness values of the Si(111) substrate were up to 0.70 GPa larger than those of the brass B36, with a tendency for this difference to decrease with the increasing thickness of coating. For the Cu coating, with the thickness of 60 μm, this difference was only about 0.050 GPa. The decrease in the difference in the composite hardness values with increasing the coating thickness is another proof of the strong influence of the substrate hardness, i.e., the type of substrate, on the measured value of the hardness of coatings.

The obtained values for the coating hardness were in line with those found in the literature for electrolytically deposited Cu coatings. The usual values for the hardness of the Cu coatings produced by galvanostatic regimes of electrodeposition were between 0.70 and 1.65 GPa [22,49,50]. The application of electrodeposition at a periodically changing rate led to the formation of Cu coatings with slightly greater hardness than those obtained by the constant galvanostatic regimes. For example, the composite hardness values of the coatings prepared by application of periodically changing regimes, such as the pulsating current (PC) and the reversing current (RC) regimes, were between 1.10 and 2.0 GPa [18]. Ultrasonic-assisted copper electrodeposition in TSV (through silicon via) gave Cu deposits the hardness values of which were between 1.58 and 1.99 GPa [51]. The high hardness value of 2.37 GPa was obtained for jet electrodeposited copper in the PS regime [52]. Aside from regimes of electrodeposition, the parameters affecting the quality, and thus the, hardness of the coatings are composition and type of electrolytes, temperature, the presence of additives in the electrolyte, time, mixing of electrolyte, the type of substrates, etc. [8,53]. Certainly, the substrate type is one of the most important parameters affecting coating hardness, and the approach taken in this investigation enabled us to determine precisely the limiting RID value separating the area where the substrate strongly effects the measured composite hardness from the area in which the measured composite hardness can be considered as the true (absolute) hardness of the coating.

On the other hand, the values of the stress exponent for the Cu coatings electrodeposited on the brass substrate were larger than the corresponding values obtained on the Si(111) substrate. These exponents were between 4.35 and 7.69 for the brass substrate with the load of 0.49 N, while those on the Si(111) substrate were between 2.79 and 5.29 [25] for the same applied load. In the case of the Si(111) substrate, the creep mechanism changed from grain boundary sliding to both dislocation climb and dislocation creep when increasing the thickness of the coatings. For the brass substrate, the dominant mechanism was the dislocation creep and the dislocation climb. The common characteristic for both the substrates was the minimal value of this exponent for the coating of 40 μm thickness.

Then, larger values of the stress exponent were obtained on the brass than on the Si(111) substrate, which caused the change in creep mechanism from the grain boundary sliding to the dislocation climb and the dislocation creep for the Cu coatings electrodeposited on the Si(111) to only the dislocation climb and the dislocation creep for the Cu coatings electrodeposited on the brass, can be discussed as follows: on both the substrates, "soft" copper films of the same characteristics were formed. Simultaneously, the brass was about five times softer than the Si(111) substrate. This means that during the indentation process at the low load, a softer brass substrate provides less resistance to indentation force than the harder Si(111) substrate. This causes the depth of penetration to be greater in the Cu coating formed on the softer brass than on the harder Si(111) substrate. As a final result of this process, the size of the diagonals of the indents was smaller on the surface area of the Cu coatings electrodeposited on the Si(111) than on the brass substrate. For example, at an applied load of 0.49 N, for the 10 μm thick Cu coating, the diagonal is 26.67 μm (H_c = 1.278 GPa) on the brass and 23.36 μm (H_c = 1.665 GPa) on the Si(111) substrate.

Aside from the substrate hardness, the roughness of the coatings is also an important parameter affecting the coating hardness and the stress exponent values. For the 10, 20, 40 and 60 μm thick Cu coatings electrodeposited on the Si(111) substrate, the R_a values of roughness were between 52.42 and 286.3 nm [20]. For the Cu coatings of the same thicknesses electrodeposited on the brass, the R_a values were between 75.05 and 512.03 nm (Table 1), indicating an increase in the roughness between 50 and 100% relative to the Si(111) substrate. This differences can be attributed to the different roughness values of the brass and the Si(111) substrates. Namely, every surface area which represents a cathode, i.e., a substrate for the electrodeposition process, possesses a certain roughness [8]. In our case, the roughness of the brass substrate was considerably greater than that for the Si(111) substrate. As is already known, owing to the production method, the Si(111) substrate represents one of the smoothest substrates.

Now we can additionally explain the various creep mechanisms of the Cu coatings formed on these two substrates. From a macro-morphological point of view, there is no a difference between the Cu coatings electrodeposited on the Si(111) and the brass substrates by the PC regime at the average current density of 50 mA cm^{-2}. Both the deposits are fine-grained, formed in the mixed activation–diffusion control, with a dominant presence of grains of about 5 μm in size. The only difference between these two deposits is the larger number of grains of a size of about 5 μm in the Cu coating electrodeposited on the brass than on the Si(111) substrate. Anyway, the Cu coatings electrodeposited on the Si(111) and the brass substrates represent the typical deposits that have approximately the same coarseness, although are formed on the substrates of various roughness values. The difference in the number of the grains can be attributed to the fact that the initial stage of the electrodeposition processes is determined by the initial state of the electrode surface, i.e., its roughness, while the final morphology of deposits is determined by parameters and the regime of electrodeposition. The substrate of greater roughness contains a larger number of irregularities, with sharp peaks representing preferential sites for the nucleation process and the initial stage of the electrodeposition process. During the deposition process, due to the current density distribution effect, electrodeposition primarily occurs on these sites, i.e., the current density is larger on the peaks than on the other parts of the electrode surface [8]. This process will lead to the formation of deposits with larger number of grains of approximately the same size (about 5.0 μm in our case) on a substrate of greater roughness (the brass) than on that with smaller roughness (the Si(111)). The increase in number of larger grains simultaneously means a decrease in overall number of formed grains, and thus, a decrease in the number of grain boundaries at the surface area of the deposit.

The decrease in the number of grain boundaries is sufficient reason for the change of the creep mechanism from grain boundary sliding to dislocation climb and dislocation creep, characterizing the Cu coatings electrodeposited on the Si(111) substrate, to the dominant dislocation climb and dislocation creep mechanism for those electrodeposited on the brass with an applied load of 0.49 N.

The very high and very close together creep exponents obtained at the Cu coatings of thicknesses of 10, 20 and 40 μm with a high applied load of 1.96 N (16.3 ± 0.50) clearly indicate that these values are determined by the features of the brass as a substrate, but not by the morphological features of Cu deposits. A similar conclusion is also valid for the 60 μm thick Cu coating. In this case, there is some contribution of the Cu deposit to the creep features, but the value of the stress exponent is still high (≈11.3), so we cannot claim to have a relevant value of this exponent for this applied load. As such, the contribution of the substrate to the creep characteristics was predominant for the Cu coating of this thickness.

5. Conclusions

Composite hardness models, such as the Chicot-Lesage (C-L) and the Cheng-Gao (C-G), were used for the determination of the hardness of the Cu coatings from the measured composite hardness. The mechanism of creep resistance was also considered. For the production of Cu coatings of various thicknesses (10, 20, 40 and 60 μm), the PC regime with the following parameters was applied: j_{av} = 50 mA cm^{-2}, j_A = 100 mA cm^{-2}, t_c = 5 ms, and t_p = 5 ms. On the basis of the obtained results, the following conclusions were derived:

- Applying the C-L model, the limiting value of RID of 0.14 was determined for the applied load range. For RID > 0.14, it is necessary to apply the composite hardness model for a determination of the absolute or true coating hardness. For RID < 0.14, the composite hardness corresponds to the coating hardness;
- The quantification of the values of the coating hardness was done by application of the C-G model. The obtained values between 0.9418 and 1.1399 GPa confirmed the assumption that the coatings of Cu on the brass belongs to the "soft film on hard substrate" composite hardness system;
- The stress exponents between 4.35 and 7.69 obtained with an applied load of 0.49 N indicated that the dominant creep mechanism is dislocation creep and dislocation climb;
- By comparison of the obtained morphological and mechanical characteristics of the Cu coatings with those obtained on the Si(111) substrate under the same electrodeposition conditions, the effect of the characteristics of the substrate on the coating hardness and the creep resistance behavior of the coatings has been additionally explained and discussed.

Author Contributions: I.O.M. performed experiments and contributed to the analysis of microhardness; J.S.L. performed analysis of microhardness; D.V.-R. performed the AFM analysis; V.R. contributed in discussion of microhardness measurement; and N.D.N. conceived and wrote the paper. All authors have read and agreed to the published version of the manuscript.

Funding: This work was financially supported by the Ministry of Education, Science and Technological Development of the Republic of Serbia (Grants No. 451-03-68/2020-14/200026 and 451-03-68/2020-14/200135).

Acknowledgments: This work was funded by Ministry of Education, Science and Technological Development of Republic of Serbia.

Conflicts of Interest: The authors declare no conflict of interest.

References

1. Copper Plating for Engineering. Available online: https://www.bendplating.com/copper-plating-for-engineering-applications/ (accessed on 1 December 2020).
2. Wei, H.L.; Huang, H.; Woo, C.H.; Zheng, R.K.; Wen, G.H.; Zhang, X.X. Development of ⟨110⟩ texture in copper thin films. *Appl. Phys. Lett.* **2002**, *80*, 2290–2292. [CrossRef]

3. Miura, S.; Honma, H. Advanced copper electroplating for application of electronics. *Surf. Coat. Technol.* **2003**, *169–170*, 91–95. [CrossRef]
4. Elrefaey, A.; Wojarski, L.; Tillmann, W. Preliminary investigation on brazing performance of Ti/Ti and Ti/steel joints using copper film deposited by PVD technique. *J. Mater. Eng. Perform.* **2012**, *21*, 696–700. [CrossRef]
5. Jeon, N.L.; Nuzzo, R.G. Physical and spectroscopic studies of the nucleation and growth of copper thin films on polyimide surfaces by chemical vapor deposition. *Langmuir* **1995**, *11*, 341–355. [CrossRef]
6. Zheng, B.C.; Meng, D.; Che, H.L.; Lei, M.K. On the pressure effect in energetic deposition of Cu thin films by modulated pulsed power magnetron sputtering: A global plasma model and experiments. *J. Appl. Phys.* **2015**, *117*, 203302. [CrossRef]
7. Wei, C.; Wu, G.; Yang, S.; Liu, Q. Electrochemical deposition of layered copper thin films based on the diffusion limited aggregation. *Sci. Rep.* **2016**, *6*, 34779. [CrossRef] [PubMed]
8. Popov, K.I.; Djokić, S.S.; Nikolić, N.D.; Jović, V.D. *Morphology of Electrochemically and Chemically Deposited Metals*; Springer International Publishing: New York, NY, USA, 2016. [CrossRef]
9. Nikolić, N.; Stojilković, E.; Djurović, D.; Pavlović, M.; Knežević, V. The preferred orientation of bright copper deposits. *Mater. Sci. Forum* **2000**, *352*, 73–78. [CrossRef]
10. Tantavichet, N.; Damronglerd, S.; Chailapakul, O. Influence of the interaction between chloride and thiourea on copper electrodeposition. *Electrochim. Acta* **2009**, *55*, 240–249. [CrossRef]
11. Song, S.J.; Choi, S.R.; Kim, J.G.; Kim, H.G. Effect of molecular weight of polyethylene glycol on copper electrodeposition in the presence of bis-3-sulfopropyl-disulfide. *Int. J. Electrochem. Sci.* **2016**, *151*, 10067–10079. [CrossRef]
12. Moffat, T.P.; Wheeler, D.; Josell, D. Electrodeposition of copper in the SPS-PEG-Cl additive system I. Kinetic measurements: Influence of SPS. *J. Electrochem. Soc.* **2004**, *151*, C262–C271. [CrossRef]
13. Bozzini, B.; D'Urzo, L.; Romanello, V.; Mele, C. Electrodeposition of Cu from acidic sulfate solutions in the presence of bis-(3-sulfopropyl)-disulfide (SPS) and chloride ions. *J. Electrochem. Soc.* **2006**, *153*, C254–C257. [CrossRef]
14. Nikolić, N.; Rakočević, Z.; Popov, K. Structural characteristics of bright copper surfaces. *J. Electroanal. Chem.* **2001**, *514*, 56–66. [CrossRef]
15. Nikolic, N.D.; Rakočević, Z.; Popov, K.I. Reflection and structural analyses of mirror-bright metal coatings. *J. Solid State Electrochem.* **2004**, *8*, 526–531. [CrossRef]
16. Pasquale, M.; Gassa, L.; Arvia, A. Copper electrodeposition from an acidic plating bath containing accelerating and inhibiting organic additives. *Electrochim. Acta* **2008**, *53*, 5891–5904. [CrossRef]
17. Marro, J.B.; Darroudi, T.; Okoro, C.A.; Obeng, Y.S.; Richardson, K. The influence of pulse plating frequency and duty cycle on the microstructure and stress state of electroplated copper films. *Thin Solid Films* **2017**, *621*, 91–97. [CrossRef]
18. Kristof, P.; Pritzker, M. Improved copper plating through the use of current pulsing & ultrasonic agitation. *Plat. Surf. Finish.* **1998**, *85*, 237–240. Available online: http://www.nmfrc.org/pdf/9811237.pdf (accessed on 5 December 2020).
19. Tantavichet, N.; Pritzker, M.D. Effect of plating mode, thiourea and chloride on the morphology of copper deposits produced in acidic sulphate solutions. *Electrochim. Acta* **2005**, *50*, 1849–1861. [CrossRef]
20. Mladenović, I.; Lamovec, J.S.; Radović, D.V.; Vasilić, R.; Radojevic, V.; Nikolić, N.D. Morphology, structure and mechanical properties of copper coatings electrodeposited by pulsating current (PC) regime on Si(111). *Metals* **2020**, *10*, 488. [CrossRef]
21. Mladenović, I.O.; Lamovec, J.S.; Jović, V.B.; Obradov, M.; Radović, D.V.; Nikolić, N.D.; Radojević, V.J. Mechanical characterization of copper coatings electrodeposited onto different substrates with and without ultrasound assistance. *J. Serbian Chem. Soc.* **2019**, *84*, 729–741. [CrossRef]
22. Martins, L.; Martins, J.; Romeira, A.; Costa, M.E.V.; Costa, J.S.; Bazzaoui, M. Morphology of copper coatings electroplated in an ultrasonic field. *Mater. Sci. Forum* **2004**, *455–456*, 844–848. [CrossRef]
23. Mallik, A.; Ray, B.C. Morphological study of electrodeposited copper under the influence of ultrasound and low temperature. *Thin Solid Films* **2009**, *517*, 6612–6616. [CrossRef]

24. Bull, S.J. Microstructure and indentation response of TiN coatings: The effect of measurement method. *Thin Solid Films* **2019**, *688*, 137452. [CrossRef]
25. Mladenović, I.O.; Lamovec, J.S.; Vasiljević-Radović, D.G.; Radojević, V.J.; Nikolić, N.D. Mechanical features of copper coatings electrodeposited by the pulsating current (PC) regime on Si(111) substrate. *Int. J. Electrochem. Sci.* **2020**, *148*, 12173–12191. [CrossRef]
26. Burnett, P.; Rickerby, D. The mechanical properties of wear-resistant coatings. II: Experimental studies and interpretation of hardness. *Thin Solid Films* **1987**, *148*, 51–65. [CrossRef]
27. Bull, S.; Rickerby, D. New developments in the modelling of the hardness and scratch adhesion of thin films. *Surf. Coat. Technol.* **1990**, *42*, 149–164. [CrossRef]
28. Lesage, J.; Pertuz, A.; Chicot, D. A new method to determine the hardness of thin films. *Matéria* **2004**, *9*, 13–22. Available online: http://www.materia.coppe.ufrj.br/sarra/artigos/artigo10294 (accessed on 7 December 2020).
29. Lesage, J.; Pertuz, A.; Puchi-Cabrera, E.; Chicot, D. A model to determine the surface hardness of thin films from standard micro-indentation tests. *Thin Solid Films* **2006**, *497*, 232–238. [CrossRef]
30. Lesage, J.; Chicot, D.; Pertuz, A.; Jouan, P.Y.; Horny, N.; Soom, A. A model for hardness determination of thin coatings from standard micro-indentation tests. *Surf. Coat. Technol.* **2005**, *200*, 886–889. [CrossRef]
31. Chicot, D.; Lesage, J. Absolute hardness of films and coatings. *Thin Solid Films* **1995**, *254*, 123–130. [CrossRef]
32. Chen, M.; Gao, J. The adhesion of copper films coated on silicon and glass substrates. *Mod. Phys. Lett. B* **2000**, *14*, 103–108. [CrossRef]
33. He, J.L.; Li, W.Z.; Li, H.D. Hardness measurement of thin films: Separation from composite hardness. *Appl. Phys. Lett.* **1996**, *69*, 1402–1404. [CrossRef]
34. Hou, Q.; Gao, J.; Li, S. Adhesion and its influence on micro-hardness of DLC and SiC films. *Eur. Phys. J. B* **1999**, *8*, 493–496. [CrossRef]
35. Magagnin, L.; Maboudian, R.; Carraro, C. Adhesion evaluation of immersion plating copper films on silicon by microindentation measurements. *Thin Solid Films* **2003**, *434*, 100–105. [CrossRef]
36. Korsunsky, A.M.; McGurk, M.R.; Bull, S.J.; Page, T.F. On the hardness of coated systems. *Surf. Coat. Technol.* **1998**, *99*, 171–183. [CrossRef]
37. Tuck, J.R.; Korsunsky, A.M.; Bull, S.J.; Davidson, R.I. On the application of the work-of-indentation approach to depth-sensing indentation experiments in coated systems. *Surf. Coat. Technol.* **2001**, *137*, 217–224. [CrossRef]
38. Lamovec, J.; Jović, V.; Randjelović, D.; Aleksić, R.; Radojevic, V. Analysis of the composite and film hardness of electrodeposited nickel coatings on different substrates. *Thin Solid Films* **2008**, *516*, 8646–8654. [CrossRef]
39. Ma, Z.; Zhou, Y.; Long, S.; Lu, C. On the intrinsic hardness of a metallic film/substrate system: Indentation size and substrate effects. *Int. J. Plast.* **2012**, *34*, 1–11. [CrossRef]
40. Chudoba, T.; Richter, F. Investigation of creep behaviour under load during indentation experiments and its influence on hardness and modulus results. *Surf. Coat. Technol.* **2001**, *148*, 191–198. [CrossRef]
41. *ASTM E384—16: Standard Test Method for Microindentation Hardness of Materials*; ASTM International: West Conshohocken, PA, USA, 2016; Available online: https://www.astm.org/DATABASE.CART/HISTORICAL/E384-16.htm (accessed on 1 December 2020).
42. *ISO 6507-1-2005: Metallic Materials—Vickers Hardness Test—Part 1: Test Method*; International Organization for Standardization: Geneva, Switzerland, 2005; Available online: https://www.iso.org/standard/37746.html (accessed on 1 December 2020).
43. Li, H.; Bradt, R.C. Knoop microhardness anisotropy of single-crystal LaB6. *Mater. Sci. Eng. A* **1991**, *142*, 51–61. [CrossRef]
44. Buckle, H. *The Science of Hardness Testing and Its Research Applications*; Westbrook, J.W., Conrad, H., Eds.; American Society for Metals: Metals Park, OH, USA, 1973; p. 453.
45. Lamovec, J.; Jovic, V.; Aleksic, R.; Radojevic, V. Micromechanical and structural properties of nickel coatings electrodeposited on two different substrates. *J. Serbian Chem. Soc.* **2009**, *74*, 817–831. [CrossRef]
46. Petrík, J.; Blaško, P.; Vasilňaková, A.; Demeter, P.; Futáš, P. Indentation size effect of heat treated aluminum alloy. *Acta Met. Slovaca* **2019**, *25*, 166–173. [CrossRef]
47. Sargent, P.M.; Ashby, M.F. Indentation creep. *Mater. Sci. Technol.* **1992**, *8*, 594–601. [CrossRef]

48. Farhat, S.; Rekaby, M.; Awad, R. Vickers microhardness and indentation creep studies for erbium-doped ZnO nanoparticles. *SN Appl. Sci.* **2019**, *1*, 546. [CrossRef]
49. Kasach, A.A.; Kurilo, I.; Kharitonov, D.S.; Radchenko, S.L.; Zharskii, I.M. Sonochemical electrodeposition of copper coatings. *Russ. J. Appl. Chem.* **2018**, *91*, 207–213. [CrossRef]
50. Tao, S.; Li, D.Y. Tribological, mechanical and electrochemical properties of nanocrystalline copper deposits produced by pulse electrodeposition. *Nanotechnology* **2006**, *17*, 65–78. [CrossRef]
51. Wang, F.; Yang, Y.; Wang, Y.; Ren, X.; Li, X. Study on physical properties of ultrasonic-assisted copper electrodeposition in through silicon via. *J. Electrochem. Soc.* **2020**, *167*, 022507. [CrossRef]
52. Fan, H.; Zhao, Y.; Jiang, J.; Wang, S.; Shan, W.; Li, Z. Effect of the pulse duty cycle on the microstructure and properties of a jet electrodeposited nanocrystalline copper coating. *Mater. Trans.* **2020**, *61*, 795–800. [CrossRef]
53. Pena, P.E.M.D.; Roy, S. Electrodeposited copper using direct and pulse currents from electrolytes containing low concentration of additives. *Surf. Coat. Technol.* **2018**, *339*, 101–110. [CrossRef]

© 2021 by the authors. Licensee MDPI, Basel, Switzerland. This article is an open access article distributed under the terms and conditions of the Creative Commons Attribution (CC BY) license (http://creativecommons.org/licenses/by/4.0/).

Article

A Modified Phase-Field Damage Model for Metal Plasticity at Finite Strains: Numerical Development and Experimental Validation

Jelena Živković [1], Vladimir Dunić [1], Vladimir Milovanović [1,*], Ana Pavlović [2] and Miroslav Živković [1]

[1] University of Kragujevac, Faculty of Engineering, Sestre Janjić 6, 34000 Kragujevac, Serbia; jelena.zivkovic@kg.ac.rs (J.Ž.); dunic@kg.ac.rs (V.D.); miroslav.zivkovic@kg.ac.rs (M.Ž.)
[2] Department of Industrial Engineering, University of Bologna, Viale Risorgimento 2, 40136 Bologna, Italy; ana.pavlovic@unibo.it
* Correspondence: vladicka@kg.ac.rs; Tel.: +381-34-300-790

Received: 28 November 2020; Accepted: 23 December 2020; Published: 28 December 2020

Abstract: Steel structures are designed to operate in an elastic domain, but sometimes plastic strains induce damage and fracture. Besides experimental investigation, a phase-field damage model (PFDM) emerged as a cutting-edge simulation technique for predicting damage evolution. In this paper, a von Mises metal plasticity model is modified and a coupling with PFDM is improved to simulate ductile behavior of metallic materials with or without constant stress plateau after yielding occurs. The proposed improvements are: (1) new coupling variable activated after the critical equivalent plastic strain is reached; (2) two-stage yield function consisting of perfect plasticity and extended Simo-type hardening functions. The uniaxial tension tests are conducted for verification purposes and identifying the material parameters. The staggered iterative scheme, multiplicative decomposition of the deformation gradient, and logarithmic natural strain measure are employed for the implementation into finite element method (FEM) software. The coupling is verified by the 'one element' example. The excellent qualitative and quantitative overlapping of the force-displacement response of experimental and simulation results is recorded. The practical significances of the proposed PFDM are a better insight into the simulation of damage evolution in steel structures, and an easy extension of existing the von Mises plasticity model coupled to damage phase-field.

Keywords: phase-field modeling; modified damage model; large-strain plasticity; S355J2+N steel; ductile fracture; two-stage yield function

---

1. Introduction

Engineering steel structures are designed to satisfy the demands of structural safety [1,2]. During their use, structures are intended to be exposed to predicted loading conditions depending on their purpose [3]. However, in some cases, due to unpredicted loading conditions (static, dynamic, or cyclic loading [4]), environments (corrosive [5,6] or high temperature [7]), or deviations in the design process, a non-permissible deformation and strain in the structure can be noticed [8]. Such behavior can lead to the initiation and evolution of damage, which often terminates in the structure's failure [9].

A cracking during the fracture phenomena in steel structures is delicate for numerical simulation [10–12]. Based on the Griffith theory, crack growth is related to a balance between bulk elastic energy and surface energy for brittle materials, which can be extended by plastic deformation energy and plastic dissipative energy for ductile materials [13,14]. A phase-field damage model (PFDM) can be used to overcome the limitations of the Griffith theory (a preexisting crack and a well-defined crack path) [15] and to show that a diffusive crack modeling is suitable for the numerical modeling of fracture [16–18]. Most PFDMs are based on variational principles [15]. Miehe et al. in [16,17] presented such a framework for diffusive fracture based on the phase-field approach and considered its numerical implementation by operator split scheme. A local history field is defined to govern the evolution of the crack field. Ambati et al. [19] discussed the problem of phase-field modeling of brittle fracture based on the Griffith theory. They proposed a hybrid formulation of a 'staggered' scheme. Molnar et al. [20] also proposed a phase-field model for brittle fracture based on the rate-independent variational principle of diffuse fracture. The displacement field and phase-field are solved separately by the 'staggered' approach. The phase-field variable separates the damaged, and virgin material states smoothly [21]. It is a scalar value, zero for virgin, and one for damaged material [20]. The displacement and phase field can be coupled in the 'monolithic' or 'staggered' approach [22]. The 'staggered' algorithm needs a well-defined stopping criterion [19].

Various authors developed and implemented PFDM in commercial or in-house finite element method (FEM) software. Azinpour et al. [23] considered the analogy between temperature and failure models and proposed unified and straightforward implementation of PFDM into Abaqus FEM software. Liu et al. [22] used commercial FEM software Abaqus to explore the monolithic and staggered coupling approaches for the phase-field model. They used user material and user element subroutines. The coupling between plasticity and PFDM is essential for the efficient modeling of cracking processes [13]. Plasticity is related to the development of inelastic strains, while damage is associated with reducing the material's stiffness [24]. A contribution to the energy dissipation is related to plastic strains [13]. Ambati et al. in [18] presented a phase-field model for ductile fracture capable of capturing the behavior of J2-plasticity material and crack initiation, propagation, and failure. Ambati et al. in [25] extended the phase-field model to the finite strain regime. Fracture is controlled by a critical scalar value of plastic strain. This model allowed simulation of various phenomena such as necking and fracture of flat specimens. They used the S355 type of steel for validation purposes. Badnava et al. [26] proposed a phase-field and rate-dependent plasticity coupling by energy function. The influence of plastic strain energy on crack propagation was defined by a threshold variable. Miehe et al. [27] presented a variational formulation for the phase-field modeling of ductile fracture for large strains. They introduced independent length scales for the plastic zone and cracks to guarantee mesh objectivity in post-critical ranges. Paneda et al. in [28] presented the possibility of the phase-field approach for fracture as a suitable solution for hydrogen-assisted cracking to predict catastrophic failures in corrosive environments. A fine mesh is necessary for the smooth phase-field distribution what can be computationally expensive [21]. Zhang et al. [29] investigated the accuracy of crack path simulation by the small length scale, leading to an unrealistic force–displacement relationship. Seles et al. [12,21] investigated and presented a stopping criterion based on the residual norm's control within a fracture analysis staggered scheme. Ribeiroa et al. [30] presented possibilities of the Abaqus damage plasticity model for simulation of S355NR+J2 experimentally investigated specimens. They obtained the specific behavior of this type of steel in the yielding zone [31].

The main research findings presented in this article are enhanced simulation of steel structure behavior by the coupled PFDM and von Mises plasticity model for ductile fracture presented by Ambati et al. [18,25], and Miehe et al. [16,17,32]. The improvements of coupled multifield three-dimensional finite element and Simo's hardening function for plasticity are implemented into the in-house FEM software PAK developed at the Faculty of Engineering, University of Kragujevac, Serbia. In Section 2.1, the overview of governing equations evolution for the PFDM and the von Mises plasticity with Simo hardening function is presented. The main improvements are:

- the modified coupling variable presented in Section 2.2, which considers the damage influence induced by plastic strains after the saturation hardening stress is achieved, and
- the two-stage yield function presented in Section 2.3 for the simulation of metallic materials behavior, which exhibits the stress plateau after yielding occurs. It consists of (1) perfect plasticity and (2) extended Simo-type yield hardening function. The evolution of the extended Simo-type yield function from its basic form for simple implementation into the standard von Mises plasticity constitutive model is given.

In Section 2.4, the large-strain theory based on the multiplicative decomposition of the deformation gradient, and logarithmic natural strain measure is given in the algorithmic form. The theory is adopted to fit the staggered iterative scheme for the displacement and damage phase-field solution. In Section 2.5, the FEM implementation details including, the staggered coupling scheme, are presented.

Section 3 is focused on the verification of staggered iterative scheme by benchmark one element example available in the literature and the experimental investigation of S355J2+N steel flat dog-bone specimens and validation of PFDM and von Mises plasticity model. The material parameters are identified by the calibration of the force-displacement diagram comparing the experimental and numerical results. By comparing the force-displacement response of the experimental investigation and simulation results presented an excellent qualitative and quantitative prediction, while the equivalent plastic strain field compared to the deformed configuration of the specimen gives good qualitative comparison results. At the end of the article, in Section 4, the main conclusions are presented.

2. Phase-Field Damage Model for Ductile Fracture

2.1. Overview of Governing Equations Evolution

In this subsection, the governing equations for coupled damage phase-field and plasticity are derived according to the literature [15–17,20,28,32–35] to clarify the proposed modifications. According to Griffith's theory, a fracture is defined by a criterion based on the equilibrium of the surface energy and the elastic energy stored in the material. It is possible to predict a crack initiation for existing cracks, but the nucleation and the crack propagation is not possible. There are two primary methods for modeling crack propagation in structures: (a) discrete and (b) diffuse [20]. The diffuse method considers crack as smeared damage. Francfort and Marigo [15] proposed a variational fracture model based on minimizing an energy functional for the displacement field and discontinuous crack set. The model of Bourdin et al. [33] is regularized with a smeared crack by the introduction of a phase-field to describe fully broken and intact material phases.

To introduce the phase-field model as a type of diffuse method for crack modeling, let us consider a bar given in Figure 1 with a constant cross-section. A damage phase-field variable d along the coordinate x of the bar can be formulated as local discontinuity for sharp crack topology as follows [16,17,20,28]

$$d(x) = \begin{cases} 1 & \text{if } x = 0 \\ 0 & \text{if } x \neq 0 \end{cases}, \quad (1)$$

however, for the diffusive crack topology, the damage can be given as an exponential function of the bar length in the form

$$d(x) = e^{-\frac{|x|}{l_c}}, \quad (2)$$

where l_c is defined as a characteristic length-scale parameter. This formulation converges to Equation (1), when $l_c \to 0$.

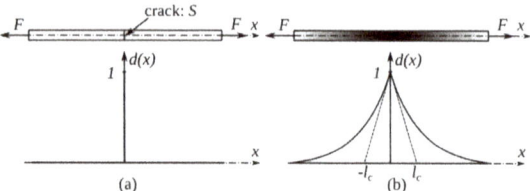

Figure 1. A bar loaded by forces F on both sides with a crack surface S at the middle: damage phase-field (a) for sharp crack and (b) for diffusive crack topology [17,20].

By following Miehe et al. [17] formulation for cracks in one-dimensional solids, and the extension of the regularized crack functional to multi-dimensional problems as

$$S(d) = \int_V \gamma(d, \nabla d) dV, \quad (3)$$

a crack surface density function γ per unit volume is defined as [16,17,20,28,32]

$$\gamma(d, \nabla d) = \frac{d^2}{2l_c} + \frac{l_c}{2} |\nabla d|^2, \quad (4)$$

where ∇ is the gradient operator. For the application of the phase-field model to the ductile behavior of the materials, the internal potential energy density ψ for the ductile fracture is considered as the sum of elastic $\psi^E(\varepsilon_E, d)$ and plastic $\psi^P(\bar{\varepsilon}_P, d)$ energy density, fracture surface energy density $\varphi^S(d)$ and plastic dissipated energy density $\varphi^P(\bar{\varepsilon}_P, d)$ as in [27]

$$\psi = \psi^E(\varepsilon_E, d) + \psi^P(\bar{\varepsilon}_P, d) + \varphi^S(d) + \varphi^P(\bar{\varepsilon}_P, d), \quad (5)$$

where ε_E is the elastic strain tensor, and $\bar{\varepsilon}_P$ is the equivalent plastic strain. Let us define each term in Equation (5). The elastic energy density of virgin material ψ_0^E is multiplied by degradation function $g(d)$ to define the elastic energy density ψ^E as [20,27,28]

$$\psi^E(\varepsilon_E, d) = g(d) \psi_0^E(\varepsilon_E) = g(d) \frac{1}{2} \varepsilon_E^T : \mathbf{C}_0 : \varepsilon_E = g(d) \frac{1}{2} \varepsilon_E^T : \sigma_0, \quad (6)$$

where $\mathbf{C}_0$ is the fourth-order elastic constitutive matrix, and σ_0 is the Cauchy stress tensor of an undamaged solid. Similarly, the "damaged" Cauchy stress σ is given in the following form [17]

$$\sigma = g(d) \sigma_0 = g(d) \mathbf{C}_0 : \varepsilon_E. \quad (7)$$

By using Equation (3), the fracture surface energy Φ^S at the crack surface S is defined as [16,17]

$$\Phi^S = \int_S G_c dS \approx \int_V G_c \gamma(d, \nabla d) dV = \int_V \varphi^S(d) dV, \quad (8)$$

where the fracture surface energy density dissipated by the formation of the crack is defined as

$$\varphi^S(d) = G_c \gamma(d, \nabla d). \quad (9)$$

In Equations (8) and (9), the Griffith-type critical fracture energy release rate G_c is used, also known as the fracture toughness of the material described as the amount of energy required to produce a unit area of fracture surface [33,34]. The plastic energy density for Simo hardening is [17]

$$\psi^P(\bar{\varepsilon}_P, d) = g(d)\left(\sigma_{y0,\infty} - \sigma_{yv}\right)\left(\bar{\varepsilon}_P + \frac{1}{n}e^{-n\bar{\varepsilon}_P}\right) + g(d)\frac{1}{2}H\bar{\varepsilon}_P^2, \qquad (10)$$

where σ_{yv} is the initial yield stress, $\sigma_{y0,\infty}$ is the saturation hardening stress, n is the hardening exponent, and H is the hardening modulus [36]. The plastic dissipated energy density is [35]

$$\varphi^P(\bar{\varepsilon}_P, d) = g(d)\sigma_{yv}\bar{\varepsilon}_P. \qquad (11)$$

By using the Equations (5), (6) and (9)–(11) the total internal potential energy Ψ functional is defined as [35]

$$\Psi = \int_V \psi dV = \int_V \left\{ g(d)\frac{1}{2}\varepsilon_E^T : \sigma_0 + \psi^P(\bar{\varepsilon}_P, d) + G_V\left[\frac{d^2}{2} + \frac{l_c^2}{2}|\nabla d|^2\right] + g(d)\sigma_{yv}\bar{\varepsilon}_P \right\} dV, \qquad (12)$$

where the authors introduced a critical fracture energy release rate per unit volume as $G_V = G_c/l_c$. The variation of the internal potential energy in Equation (12) over the elastic strain, damage and equivalent plastic strain is given as [35]

$$\delta\Psi = \int_V \left(\frac{\partial \psi}{\partial \varepsilon_E} : \delta\varepsilon_E + \frac{\partial \psi}{\partial d} : \delta d + \frac{\partial \psi}{\partial \bar{\varepsilon}_P} : \delta\bar{\varepsilon}_P \right) dV \qquad (13)$$

and for the Simo hardening function [17], one can obtain [27,35]

$$\delta\Psi = \int_V \left\{ \sigma : \delta\varepsilon_E + \frac{1}{2}g'(d)\varepsilon_E^T : \sigma_0\delta d + g'(d)\left(\sigma_{y0,\infty} - \sigma_{yv}\right)\left(\bar{\varepsilon}_P + \frac{1}{n}e^{-n\bar{\varepsilon}_P}\right)\delta d + g'(d)\frac{1}{2}H\bar{\varepsilon}_P^2\delta d + \right.$$
$$+ g'(d)\sigma_{yv}\bar{\varepsilon}_P\delta d + G_V\left[d\delta d + l_c^2\nabla d\nabla\delta d\right] +$$
$$\left. + \left(-g(d)\sigma_0 : \frac{\partial \varepsilon_P}{\partial \bar{\varepsilon}_P} + g(d)\left(\sigma_{y0,\infty} - \sigma_{yv}\right)\left(1 - e^{-n\bar{\varepsilon}_P}\right) + g(d)H\bar{\varepsilon}_P + g(d)\sigma_{yv}\right)\delta\bar{\varepsilon}_P \right\} dV \qquad (14)$$

where ε_P is the plastic strain tensor, and $g'(d)$ is the derivative of the degradation function $g(d)$ over d. A variation of the external potential energy W_{ext} is known as [37]

$$\delta W_{ext} = \int_V \mathbf{b} \cdot \delta\mathbf{u} dV + \int_A \mathbf{h} \cdot \delta\mathbf{u} dA, \qquad (15)$$

where $\mathbf{b}$ is a body force field per unit volume, $\mathbf{h}$ is a boundary traction per unit area, and $\mathbf{u}$ is the displacements vector. The equilibrium of the internal (14) and external (15) potential energy for the Simo hardening function gives [35]

$$\int_V \left\{ \sigma : \delta\varepsilon_E + \frac{1}{2}g'(d)\varepsilon_E^T : \sigma_0\delta d + g'(d)\left(\sigma_{y0,\infty} - \sigma_{yv}\right)\left(\bar{\varepsilon}_P + \frac{1}{n}e^{-n\bar{\varepsilon}_P}\right)\delta d + \right.$$
$$+ g'(d)\frac{1}{2}H\bar{\varepsilon}_P^2\delta d + g'(d)\sigma_{yv}\bar{\varepsilon}_P\delta d + G_V\left[d\delta d + l_c^2\nabla d \cdot \nabla\delta d\right] +$$
$$\left. + \left(-g(d)\sigma_0 : \frac{\partial \varepsilon_P}{\partial \bar{\varepsilon}_P} + g(d)\left(\sigma_{y0,\infty} - \sigma_{yv}\right)\left(1 - e^{-n\bar{\varepsilon}_P}\right) + g(d)H\bar{\varepsilon}_P + g(d)\sigma_{yv}\right)\delta\bar{\varepsilon}_P \right\} dV \qquad (16)$$
$$= \int_V \mathbf{b} \cdot \delta\mathbf{u} dV + \int_A \mathbf{h} \cdot \delta\mathbf{u} dA$$

By the application of total derivatives of the following terms

$$\nabla \cdot [\boldsymbol{\sigma} \cdot \delta \mathbf{u}] = \nabla \cdot [\boldsymbol{\sigma}] \cdot \delta \mathbf{u} + \boldsymbol{\sigma} : \nabla \cdot [\delta \mathbf{u}] = Div[\boldsymbol{\sigma}] \cdot \delta \mathbf{u} + \boldsymbol{\sigma} : \delta \boldsymbol{\varepsilon}_E \quad (17)$$

$$\boldsymbol{\sigma} : \delta \boldsymbol{\varepsilon}_E = \nabla \cdot [\boldsymbol{\sigma} \cdot \delta \mathbf{u}] - Div[\boldsymbol{\sigma}] \cdot \delta \mathbf{u} \quad (18)$$

$$\nabla \cdot [\nabla d \delta d] = \nabla \cdot [\nabla d] \, \delta d + \nabla d \cdot \nabla [\delta d] \quad (19)$$

$$\nabla d \cdot \nabla [\delta d] = \nabla \cdot [\nabla d \delta d] - \nabla \cdot [\nabla d] \, \delta d = \nabla \cdot [\nabla d \delta d] - \nabla^2 d \delta d \quad (20)$$

and by using the Gauss theorem, it can be obtained [27,35]

$$\int_V \left\{ -\left[g'(d) \psi + G_V \left[d - l_c^2 \nabla^2 d \right] \right] \delta d - [Div[\boldsymbol{\sigma}] + \mathbf{b}] \cdot \delta \mathbf{u} + \right.$$
$$\left. + \left(-g(d) \boldsymbol{\sigma}_0 : \frac{\partial \varepsilon_P}{\partial \bar{\varepsilon}_P} + g(d) \left(\sigma_{y0,\infty} - \sigma_{yv} \right) \left(1 - e^{-n \bar{\varepsilon}_P} \right) + g(d) H \bar{\varepsilon}_P + g(d) \sigma_{yv} \right) \delta \bar{\varepsilon}_P \right\} dV \quad (21)$$
$$+ \int_A \left\{ [\boldsymbol{\sigma} \cdot \mathbf{n} - \mathbf{h}] \cdot \delta \mathbf{u} \right\} dA + \int_A \left\{ \left[G_V l_c^2 \nabla d \cdot \mathbf{n} \right] \delta d \right\} dA = 0$$

where $\mathbf{n}$ is the unit outer normal to the surface A, and the internal potential energy density given in Equation (5) is

$$\psi = \frac{1}{2} \boldsymbol{\varepsilon}_E : \boldsymbol{\sigma}_0 + \left(\sigma_{y0,\infty} - \sigma_{yv} \right) \left(\bar{\varepsilon}_P + \frac{1}{n} e^{-n \bar{\varepsilon}_P} \right) + \frac{1}{2} H \bar{\varepsilon}_P^2 + \sigma_{yv} \bar{\varepsilon}_P. \quad (22)$$

The Neumann-type boundary conditions are [20]

$$\boldsymbol{\sigma} \cdot \mathbf{n} - \mathbf{h} = 0, \quad (23)$$

$$\nabla d \cdot \mathbf{n} = 0, \quad (24)$$

what leads to the governing balance equations of the coupled PFDM-von Mises plasticity problem for Simo's hardening function [35]

$$\nabla d \cdot \mathbf{n} = 0, \quad (25)$$

$$G_V \left[d - l_c^2 \nabla^2 d \right] + g'(d) \psi = 0, \quad (26)$$

$$\bar{\sigma}_{eq} - \sigma_{yv} - \left(\sigma_{y0,\infty} - \sigma_{yv} \right) \left(1 - e^{-n \bar{\varepsilon}_P} \right) - H \bar{\varepsilon}_P = 0, \quad (27)$$

where the equivalent stress is defined as $\bar{\sigma}_{eq} = \boldsymbol{\sigma}_0 : \frac{\partial \varepsilon_P}{\partial \bar{\varepsilon}_P}$.

2.2. Modifications of Coupling Variable

The degradation function and its derivative over d are proposed by Ambati et al. [18] for the phase-field damage modeling of ductile fracture as

$$g(d) = (1-d)^{2p}, \quad (28)$$

$$g'(d) = -2p(1-d)^{2p-1}. \quad (29)$$

where p is the coupling variable. The same degradation function in Equation (28) can be used for the brittle fracture by setting $p = 1$ what will be used for the one element example in Section 3. The coupling variable can be defined to depend on the critical value of equivalent plastic strain $\bar{\varepsilon}_P^{crit}$ [18]. In this paper, the authors propose the modification of the Ambati et al. [18] because the material is considered to be intact (undamaged) until the equivalent plastic strain achieves the critical value $\bar{\varepsilon}_P = \bar{\varepsilon}_P^{crit}$. The critical value of the equivalent plastic strain can be registered in the experimental stress–strain diagram, when the stress starts to decrease (Figure 2).

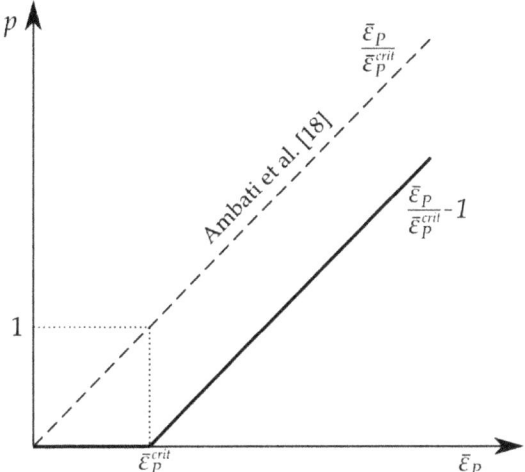

Figure 2. New coupling variable p (continuous line) in relation to the equivalent plastic strain $\bar{\varepsilon}_p$ compared to Ambati et al. [18].

At that critical point C (Figure 3b), the material can be considered damaged due to the plastic strains and the damage-plasticity coupling variable is activated. Therefore, the coupling variable p can be defined for equivalent plastic strain as it is given in Figure 3 by the function

$$p = \begin{cases} 0 & ; \frac{\bar{\varepsilon}_p}{\bar{\varepsilon}_p^{crit}} < 1 \\ \frac{\bar{\varepsilon}_p}{\bar{\varepsilon}_p^{crit}} - 1 & ; \frac{\bar{\varepsilon}_p}{\bar{\varepsilon}_p^{crit}} \geq 1 \end{cases}. \quad (30)$$

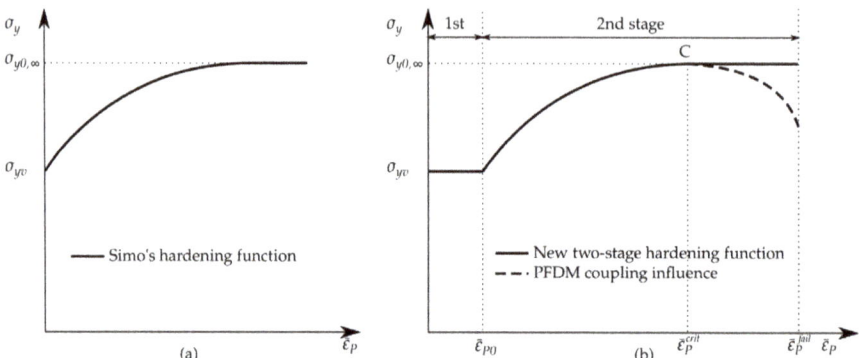

Figure 3. Yield stress of current yield surface σ_y vs. equivalent plastic strain $\bar{\varepsilon}_p$ diagram for metallic materials: (**a**) the Simo's hardening function Equations (27) and (36), (**b**) the proposed new two-stage hardening function for simulation of metallic materials with or without plateau after yielding occurs (σ_{yv}—the initial yield stress, $\sigma_{y0,\infty}$—the saturation hardening stress, $\bar{\varepsilon}_{p0}$—the maximal equivalent plastic strain for perfect plasticity stage, $\bar{\varepsilon}_p^{crit}$—the critical equivalent plastic strain, $\bar{\varepsilon}_p^{fail}$—the failure equivalent plastic strain).

The stored internal potential energy density ψ is considered as the elastic energy density ψ_0^E because the influence of the plastic part is taken into account by coupling variable p, so [25]

$$\psi = \begin{cases} \psi_0^E; & \psi_0^E > {}^t\psi \\ {}^t\psi; & \text{otherwise} \end{cases}, \tag{31}$$

where ${}^t\psi$ is the previously stored internal potential energy density.

2.3. Two-Stage Yield Hardening Function

For the simulation of the metallic material's behavior, which exhibits stress plateau after yielding occurs, such as S355J2+N steel, the extended two-stages yielding function is necessary to describe the idealized response given in Figure 3 (continuous line). In the first stage, the yielding occurs, the plastic strain increases, while the stress is constant. In the second stage, the stress increases nonlinearly until the saturation hardening stress $\sigma_{y0,\infty}$. After the stress achieves the maximal value at the end of the second stage, the stress decreases due to the phase-field damage model influence on the material (dashed line). To simulate the described behavior, 'perfect plasticity' is employed for the first stage of loading ($\bar{\varepsilon}_p < \bar{\varepsilon}_{p0}$) until the plastic strain $\bar{\varepsilon}_{p0}$ is achieved. In the first stage, the yield condition is given in the following form [37]

$$f_y = \bar{\sigma}_{eq} - \sigma_{yv} \leq 0, \tag{32}$$

while in the second period, the yield condition is defined based on Equation (28) as

$$f_y = \bar{\sigma}_{eq} - \sigma_{yv} - (\sigma_{y0,\infty} - \sigma_{yv})\left(1 - e^{-n(\bar{\varepsilon}_p - \bar{\varepsilon}_{p0})}\right) - H(\bar{\varepsilon}_p - \bar{\varepsilon}_{p0}) \leq 0$$
$$f_y = \bar{\sigma}_{eq} - \sigma_{y1} \leq 0 \tag{33}$$

where the yield stress equal to equivalent stress in the extended Simo-type yield function is given as

$$\sigma_{y1} = \sigma_{yv} + (\sigma_{y0,\infty} - \sigma_{yv})\left(1 - e^{-n(\bar{\varepsilon}_P - \bar{\varepsilon}_{P0})}\right) + H(\bar{\varepsilon}_P - \bar{\varepsilon}_{P0}). \tag{34}$$

The complete two-stage yield function shown in Figure 3 by continuous line can be defined by the equation

$$f_y = \begin{cases} \bar{\sigma}_{eq} - \sigma_{yv} & ; \bar{\varepsilon}_P < \bar{\varepsilon}_{P0} \\ \bar{\sigma}_{eq} - \left[\sigma_{yv} + (\sigma_{y0,\infty} - \sigma_{yv})\left(1 - e^{-n(\bar{\varepsilon}_P - \bar{\varepsilon}_{P0})}\right) + H(\bar{\varepsilon}_P - \bar{\varepsilon}_{P0})\right] & ; \bar{\varepsilon}_P \geq \bar{\varepsilon}_{P0} \end{cases}. \tag{35}$$

If the yield function (35) is less than zero, the solution is elastic. If the condition is violated, the equivalent plastic strain increment $\Delta\bar{\varepsilon}_P$ of the function $f_y = 0$ must be determined in an iterative Newton–Raphson procedure given in the stress integration algorithm in the next subsection. After the correct value of equivalent plastic strain $\bar{\varepsilon}_P$ is computed, the deviatoric stress tensor, the total stress tensor, the elastic strain tensor, and the elastic-plastic constitutive matrix must be updated.

The yield stress term given by Equation (34) in the extended Simo-type yield function is derived from its basic form obtained from in Equation (27) as

$$\sigma_{yS} = \sigma_{yv} + (\sigma_{y0,\infty} - \sigma_{yv})\left(1 - e^{-n\bar{\varepsilon}_P}\right) + H\bar{\varepsilon}_P \tag{36}$$

The Equation (34) can be transformed into the form

$$\sigma_{y1} = \sigma_{yv} + (\sigma_{y0,\infty} - \sigma_{yv})\, e^{-n\bar{\varepsilon}_P}\left(e^{n\bar{\varepsilon}_P} - e^{n\bar{\varepsilon}_{P0}}\right) + H(\bar{\varepsilon}_P - \bar{\varepsilon}_{P0}). \tag{37}$$

To provide a simple implementation into the existing von Mises constitutive model for metal plasticity, the basic form of Simo's yield function σ_{yS} in Equation (36) can be transformed by subtraction of the stress increment $\Delta\sigma_{yS}$ for the plastic strain $\bar{\varepsilon}_{P0}$, when the constant stress was registered during yielding given as

$$\Delta\sigma_{yS} = (\sigma_{y0,\infty} - \sigma_{yv})\left(1 - e^{-n\bar{\varepsilon}_{P0}}\right) + H\bar{\varepsilon}_{P0}. \tag{38}$$

To obtain the stress function in Equation (37) it is necessary to calculate the difference between the basic form of Simo's hardening function in Equation (36) and the stress value given by Equation (38) as

$$\sigma_{y2} = \sigma_{yS} - \Delta\sigma_{yS} = \sigma_{yv} + (\sigma_{y0,\infty} - \sigma_{yv})\left(1 - e^{-n\bar{\varepsilon}_P}\right) + H\bar{\varepsilon}_P - (\sigma_{y0,\infty} - \sigma_{yv})\left(1 - e^{-n\bar{\varepsilon}_{P0}}\right) + H\bar{\varepsilon}_{P0}, \tag{39}$$

which finally gives

$$\sigma_{y2} = \sigma_{yv} + (\sigma_{y0,\infty} - \sigma_{yv})\, e^{-n\bar{\varepsilon}_P} e^{-n\bar{\varepsilon}_{P0}}\left(e^{n\bar{\varepsilon}_P} - e^{n\bar{\varepsilon}_{P0}}\right) + H(\bar{\varepsilon}_P - \bar{\varepsilon}_{P0}). \tag{40}$$

By comparing the yield stress in Equations (37) and (40), it can be noticed that the second term in Equation (40) is multiplied by $e^{-n\bar{\varepsilon}_{P0}}$, so to propose the same form as Equation (37), and to allow the possibility of using the Equation (39) for the implementation purpose, the coefficient in the second term in Equation (40), $(\sigma_{y0,\infty} - \sigma_{yv})$ needs to be multiplied by $e^{n\bar{\varepsilon}_{P0}}$ what will give the equivalent equation to Equation (37).

2.4. Stress Integration Algorithm for von Mises Large Strain Plasticity

In this section, an overview of the well-known stress integration algorithm for von Mises large strain plasticity is presented with the two new steps 11 and 12 which are necessary for solving the PFDM governing equation given in Equation (26). The value of elastic strain energy ψ_0^E is calculated at the integration point level as well as the coupling variable p. In the following equation, the complete algorithm is given to provide the implementation. The deviatoric strain can be obtained using the multiplicative decomposition of the deformation gradient [37]

$$\mathbf{F} = \mathbf{F}_E \mathbf{F}_P, \tag{41}$$

where $\mathbf{F}_E$ and $\mathbf{F}_P$ are the elastic and plastic deformation gradient, respectively. The elastic deformation gradient can also be decomposed into the isochoric deformation gradient $\overline{\mathbf{F}}_E$ and volumetric portion as [37]

$$\overline{\mathbf{F}}_E = (\det \mathbf{F}_E)^{-\frac{1}{3}} \mathbf{F}_E, \tag{42}$$

so, the elastic left Cauchy–Green strain tensor $\overline{\mathbf{b}}_E$ can be calculated as [37,38]

$$\overline{\mathbf{b}}_E = \overline{\mathbf{F}}_E \overline{\mathbf{F}}_E^T. \tag{43}$$

The elastic deviatoric strain $\mathbf{e}_E$ can be calculated using the Hencky strain measure $\overline{\mathbf{h}}_E$ as [37,38]

$$\mathbf{e}_E = \overline{\mathbf{h}}_E = \frac{1}{2} \ln \overline{\mathbf{b}}_E, \tag{44}$$

and the mean strain e_m, in that case, is [37,38]

$$e_m = \frac{1}{3} \det \mathbf{F}. \tag{45}$$

The total stress tensor can be decomposed on the elastic deviatoric $\mathbf{S}_E$ and the volumetric part σ_m as [37]

$$\boldsymbol{\sigma} = \mathbf{S}_E + \sigma_m \mathbf{I}, \tag{46}$$

where $\mathbf{I}$ is the unit tensor. The elastic deviatoric stress can be defined as [37]

$$\mathbf{S}_E = 2G \mathbf{e}_E, \tag{47}$$

and the mean stress is [37]

$$\sigma_m = c_m e_m, \tag{48}$$

where shear and bulk modulus are

$$G = \frac{E}{2(1+\nu)}; \quad c_m = \frac{E}{1 - 2\nu}, \tag{49}$$

while E is the Young's modulus, and ν is the Poisson's ratio.

The detailed algorithm of the von Mises plasticity for large strain problems, is given in details below [37]:

t—time at the beginning of time step; Δt—time increment

1. Input values: ${}_0^{t+\Delta t}\mathbf{F}, {}_0^{t}\mathbf{F}, {}^{t}\overline{\mathbf{b}}_E, {}^{t}\psi, {}^{t}\overline{\varepsilon}_P, E, \nu, \sigma_{y\nu}, \sigma_{y0,\infty}, n, \overline{\varepsilon}_P^{crit}, \overline{\varepsilon}_{P0}$

2. Initial conditions (save at the integration point level):

$$d = {}^t d; \; \psi = {}^t \psi; \; \bar{\varepsilon}_P = {}^t \bar{\varepsilon}_P \tag{50}$$

3. Calculate trial elastic deviatoric strain:

$$^{t+\Delta t}_{0}\mathbf{F} = {}^{t+\Delta t}_{t}\mathbf{F}{}^{t}_{0}\mathbf{F} \tag{51}$$

$$^{t+\Delta t}_{t}\bar{\mathbf{F}} = \left(\det {}^{t+\Delta t}_{t}\mathbf{F}\right)^{-\frac{1}{3}} {}^{t+\Delta t}_{t}\mathbf{F} \tag{52}$$

$$\bar{\mathbf{b}}_E^* = {}^{t+\Delta t}_{t}\bar{\mathbf{F}}\,{}^t\bar{\mathbf{b}}_{Et}^* \,{}^{t+\Delta t}_{t}\bar{\mathbf{F}}^T \tag{53}$$

$$\mathbf{e}_E^* = \frac{1}{2}\ln\bar{\mathbf{b}}_E^* \tag{54}$$

$$e_m = \frac{1}{3}\det\left({}^{t+\Delta t}_{0}\mathbf{F}\right) \tag{55}$$

4. Trial elastic deviatoric stress:

$$\mathbf{S}_E^* = 2G\mathbf{e}_E^* \text{ where } G = \frac{E}{2(1+\nu)} \tag{56}$$

5. Check for yielding:

$$^t\sigma_y = \begin{cases} \sigma_{yv} & ; \bar{\varepsilon}_P < \bar{\varepsilon}_{P0} \\ {}^t\sigma_{y1} & ; \bar{\varepsilon}_P \geq \bar{\varepsilon}_{P0} \end{cases} \tag{57}$$

$$\bar{\sigma}_{eq}^* = \sqrt{\frac{3}{2}}\|\mathbf{S}_E^*\| \tag{58}$$

$$f_y^* = \bar{\sigma}_{eq}^* - {}^t\sigma_y \leq 0 \tag{59}$$

If the condition is satisfied, the solution is $\mathbf{S}_E = \mathbf{S}_E^*$ and $\Delta\bar{\varepsilon}_P = 0$, and go to 7.

6. Find equivalent plastic strain increment $\Delta\bar{\varepsilon}_P$ of the function $f_y(\Delta\bar{\varepsilon}_P) = 0$

$$\bar{\varepsilon}_P = {}^t\bar{\varepsilon}_P + \Delta\bar{\varepsilon}_P; \tag{60}$$

$$\sigma_y = \begin{cases} \sigma_{yv} & ; \bar{\varepsilon}_P < \bar{\varepsilon}_{P0} \\ \sigma_{y1} & ; \bar{\varepsilon}_P \geq \bar{\varepsilon}_{P0} \end{cases}; \; \Delta\lambda = \frac{3}{2}\frac{\Delta\bar{\varepsilon}_P}{\sigma_y} \tag{61}$$

$$\hat{C} = \frac{2}{3}\left(ne^{n\bar{\varepsilon}_{P0}}\left(\sigma_{y0,\infty} - \sigma_{yv}\right)e^{-n\bar{\varepsilon}_P} + H\right); \; \mathbf{S}_E = \frac{\mathbf{S}_E^*}{1 + (2G + \hat{C})\Delta\lambda}; \; \bar{\sigma}_{eq} = \sqrt{\frac{3}{2}}\|\mathbf{S}_E\| \tag{62}$$

$$f_y(\Delta\bar{\varepsilon}_P) = |\bar{\sigma}_{eq} - \sigma_y| > tol \text{ go to step 6.} \tag{63}$$

7. Update of left Cauchy-Green strain tensor:

$$\bar{\mathbf{b}}_E = {}^t\bar{\mathbf{b}}_E^* e^{-2\Delta\bar{\varepsilon}_P} \tag{64}$$

8. Mean stress and total stress:

$$\sigma_m = c_m e_m; \; \boldsymbol{\sigma}_0 = \mathbf{S}_E + \sigma_m \mathbf{I}; \; c_m = \frac{E}{1-2\nu} \tag{65}$$

9. Calculate elastic deviatoric strain:
$$\mathbf{e}_E = \frac{\mathbf{S}_E}{2G} \qquad (66)$$

10. Total elastic strain is:
$$\boldsymbol{\varepsilon}_E = \mathbf{e}_E + e_m \mathbf{I} \qquad (67)$$

11. (NEW STEP) Elastic strain energy density:
$$\psi_0^E = \tfrac{1}{2}\boldsymbol{\varepsilon}_E^T : \mathbf{C}_0 : \boldsymbol{\varepsilon}_E = \tfrac{1}{2}\boldsymbol{\varepsilon}_E^T : \boldsymbol{\sigma}_0 = \tfrac{1}{2}(\mathbf{S}_E + \sigma_m \mathbf{I}) : (\mathbf{e}_E + e_m \mathbf{I}) =$$
$$= \tfrac{1}{2}(\mathbf{S}_E + \sigma_m \mathbf{I}) : \left(\tfrac{\mathbf{S}_E}{2G} + \tfrac{\sigma_m \mathbf{I}}{c_m}\right) = \tfrac{1}{2}\left(\tfrac{1}{2G}\mathbf{S}_E : \mathbf{S}_E + \tfrac{3\sigma_m^2}{c_m}\right) = \tfrac{1}{2}\left(\tfrac{\bar{\sigma}_{eq}^2}{3G} + \tfrac{3\sigma_m^2}{c_m}\right) \qquad (68)$$

If $\psi_0^E > {}^t\psi$ then $\psi = \psi_0^E$

12. (NEW STEP) Coupling variable:
$$p = \begin{cases} 0 & ; \dfrac{\bar{\varepsilon}_p}{\varepsilon_p^{crit}} < 1 \\ \dfrac{\bar{\varepsilon}_p}{\varepsilon_p^{crit}} - 1 & ; \dfrac{\bar{\varepsilon}_p}{\varepsilon_p^{crit}} \geq 1 \end{cases} \qquad (69)$$

13. Calculate elasto-plastic matrix: $\mathbf{C}_{EP}$
14. Return: $\boldsymbol{\sigma}_0, \psi, \mathbf{C}_{EP}, p$

2.5. Implementation into FEM Software

In this section, the discretization using standard Lagrange finite elements is considered and the standard Galerkin finite element method is used. The nodal displacements and phase-field damage variable are unknowns that need to be determined. These finite elements are also known as "multifield" finite elements extensively applied in multiphysics FE simulations.

2.5.1. Finite Element Discretization

The interpolation matrices for displacement $\mathbf{N}^u$ and damage $\mathbf{N}^d$ and derivatives $\mathbf{B}^d$ and $\mathbf{B}^u$ are given as follows [20]:
$$\mathbf{N}^d = [N_1 \ldots N_8], \qquad (70)$$

$$\mathbf{B}^d = \begin{bmatrix} N_{1,x} & \ldots & N_{8,x} \\ N_{1,y} & \ldots & N_{8,y} \\ N_{1,z} & \ldots & N_{8,z} \end{bmatrix}, \qquad (71)$$

$$\mathbf{N}^u = \begin{bmatrix} N_1 & 0 & 0 & \ldots & \ldots & \ldots & N_8 & 0 & 0 \\ 0 & N_1 & 0 & \ldots & \ldots & \ldots & 0 & N_8 & 0 \\ 0 & 0 & N_1 & \ldots & \ldots & \ldots & 0 & 0 & N_8 \end{bmatrix}, \qquad (72)$$

$$\mathbf{B}^u = \begin{bmatrix} N_{1,x} & 0 & 0 & \ldots & \ldots & \ldots & N_{8,x} & 0 & 0 \\ 0 & N_{1,y} & 0 & \ldots & \ldots & \ldots & 0 & N_{8,y} & 0 \\ 0 & 0 & N_{1,z} & \ldots & \ldots & \ldots & 0 & 0 & N_{8,z} \\ N_{1,y} & N_{1,x} & 0 & \ldots & \ldots & \ldots & N_{8,y} & N_{8,x} & 0 \\ 0 & N_{1,z} & N_{1,y} & \ldots & \ldots & \ldots & 0 & N_{8,z} & N_{8,y} \\ N_{1,z} & 0 & N_{1,x} & \ldots & \ldots & \ldots & N_{8,z} & 0 & N_{8,x} \end{bmatrix}. \qquad (73)$$

The damage phase field value at an integration point is described as [20]
$$d = \mathbf{N}^d \mathbf{d}, \qquad (74)$$

where **d** is the damage phase-field vector of nodal values. The local damage gradient which can also be referred to as "damage strain" is [20]

$$\varepsilon^d = \nabla d = \mathbf{B}^d \mathbf{d}. \tag{75}$$

The total strain vector ε is interpolated in terms of the nodal displacements **u** [20,37]

$$\varepsilon = \mathbf{B}^u \mathbf{u}. \tag{76}$$

The internal $\mathbf{f}_e^{int}$ and external forces $\mathbf{f}_e^{ext}$ are [20,37]

$$\mathbf{f}_e^{int} = \int_V \left[(1-d)^{2p}\right] (\mathbf{B}^u)^T \sigma_0 dV, \tag{77}$$

$$\mathbf{f}_e^{ext} = \int_V (\mathbf{N}^u)^T \mathbf{b} dV + \int_A (\mathbf{N}^u)^T \mathbf{h} dA. \tag{78}$$

The residue vector $\mathbf{r}_e^d$ for the phase-field degrees is [20,37]

$$\mathbf{r}_e^d = \int_V \left\{ \left[G_V d - 2p\left(1-d\right)^{2p-1} \psi\right] \left(\mathbf{N}^d\right)^T + G_V l_c^2 \left(\mathbf{B}^d\right)^T \varepsilon^d \right\} dV. \tag{79}$$

The tangent matrices for damage $\mathbf{K}_e^d$ and displacement $\mathbf{K}_e^u$ field are [20,37]

$$\mathbf{K}_e^d = \int_V \left\{ \left[G_V + 2p\left(2p-1\right)\left(1-d\right)^{2p-2}\psi\right] \left(\mathbf{N}^d\right)^T \left(\mathbf{N}^d\right) + G_V l_c^2 \left(\mathbf{B}^d\right)^T \left(\mathbf{B}^d\right) \right\} dV \tag{80}$$

$$\mathbf{K}_e^u = \int_V \left\{ \left[(1-d)^{2p}\right] (\mathbf{B}^u)^T \mathbf{C}_{EP} \mathbf{B}^u \right\} dV. \tag{81}$$

The linear equation system can be solved by using of Newton–Raphson algorithm [20]

$$\begin{bmatrix} \mathbf{K}^u & 0 \\ 0 & \mathbf{K}^d \end{bmatrix} \begin{bmatrix} \delta \mathbf{u} \\ \delta \mathbf{d} \end{bmatrix} = \begin{bmatrix} \mathbf{f}^{ext} \\ 0 \end{bmatrix} - \begin{bmatrix} \mathbf{f}^{int} \\ \mathbf{r}^d \end{bmatrix}. \tag{82}$$

2.5.2. Staggered Solution Strategy

By following Miehe et al. [16], the weak formations of mechanical field given in Equations (25) and (27) and phase-field given in Equation (26) have to be solved by the staggered algorithm shown in Figure 4. The Equation (27) is the Simo yield hardening function used as a part of the local stress integration algorithm for the computation of the mechanical field. At the beginning of the iterative procedure (t_0), both, displacement and damage vectors are equal to the vectors from the previous time step. Initial conditions at the structure level are $\mathbf{u}^{(0)} = {}^t\mathbf{u}$; $\mathbf{d}^{(0)} = {}^t\mathbf{d}$.

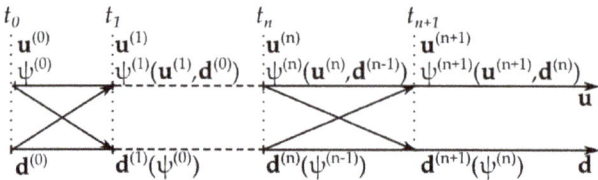

Figure 4. Staggered iterative scheme for phase-field modeling in FEM software (u—displacement vector, d—damage vector, ψ—internal potential energy density, t—time) [20].

Both fields are solved simultaneously by minimization of the two residual equations for the displacement $\mathbf{r}_e^u$ and the damage phase-field $\mathbf{r}_e^d$ to determine the vectors in the next time step [18,20]

$$\mathbf{r}_e^u = \mathbf{f}_e^{int} - \mathbf{f}_e^{ext} = \int_V \left[(1-d)^{2p}\right] (\mathbf{B}^u)^T \boldsymbol{\sigma}_0 dV - \int_V (\mathbf{N}^u)^T \mathbf{b} dV - \int_A (\mathbf{N}^u)^T \mathbf{h} dA. \tag{83}$$

$$\mathbf{r}_e^d = \int_V \left\{ \left[G_V d - 2p(1-d)^{2p-1} \psi\right] (\mathbf{N}^d)^T + G_V l_c^2 (\mathbf{B}^d)^T \varepsilon^d \right\} dV. \tag{84}$$

The staggered iterative scheme across the time steps is given in Figure 4, where it can be observed that the procedure starts with both displacement and damage vectors equal to zero. The Newton–Raphson iterative procedure is given with the convergence criterion and the implementation details below [20,37]:

Input values: E[MPa], ν[−], l_c[mm], G_V[MPa]
Loop A. Initial condition at structure level for increment Δt

$$\begin{array}{c} \text{Displacement}: \mathbf{u}^{(0)} = {}^t\mathbf{u}; \text{ damage}: \mathbf{d}^{(0)} = {}^t\mathbf{d}; \\ \text{external loads}: \mathbf{f}_e^{ext} = \int_V (\mathbf{N}^u)^T \mathbf{b} dV + \int_A (\mathbf{N}^u)^T \mathbf{h} dA \end{array} \tag{85}$$

Loop B. Iteration at the structure level
$i = 0$
$i = i + 1$
    Loop C. Integration points loop
    Strain-displacement matrix $\mathbf{B}^u$ and damage matrix $\mathbf{B}^d$
    Strain

$$\boldsymbol{\varepsilon}^{(i)} = \mathbf{B}^u \mathbf{u}^{(i)};\ d^{(i)} = \mathbf{N}^d \mathbf{d}^{(i)};\ \boldsymbol{\varepsilon}^{d(i)} = \mathbf{B}^d \mathbf{d}^{(i)} \tag{86}$$

Stress integration $\boldsymbol{\sigma}_0^{(i)}$ (plasticity model) $\rightarrow \psi^{(i)} = {}^t\psi,\ p^{(i)} = {}^tp$
Internal forces

$$\mathbf{f}_e^{int(i)} = \int_V \left[\left(1 - d^{(i)}\right)^{2p^{(i)}}\right] (\mathbf{B}^u)^T \boldsymbol{\sigma}_0^{(i)} dV,\ [\text{N}] \tag{87}$$

$$\mathbf{r}_e^{d(i)} = \int_V \left\{ \left[G_V d^{(i)} - 2p^{(i)} \left(1 - d^{(i)}\right)^{2p^{(i)}-1} \psi^{(i)}\right] (\mathbf{N}^d)^T + G_V l_c^2 (\mathbf{B}^d)^T \varepsilon^{d(i)} \right\} dV,\ [\text{N mm}] \tag{88}$$

Stiffness matrices

$$\mathbf{K}_e^{u(i)} = \int_V \left\{ \left[\left(1 - d^{(i)}\right)^{2p^{(i)}} \right] (\mathbf{B}^u)^T \mathbf{C}_{EP} \mathbf{B}^u \right\} dV, \quad \left[\frac{N}{mm} \right] \qquad (89)$$

$$\mathbf{K}_e^{d(i)} = \int_V \left\{ \left[G_V + 2p^{(i)} \left(2p^{(i)} - 1\right) \left(1 - d^{(i)}\right)^{2p^{(i)} - 2} \psi^{(i)} \right] (\mathbf{N}^d)^T (\mathbf{N}^d) + G_V l_c^2 (\mathbf{B}^d)^T (\mathbf{B}^d) \right\} dV, \text{[Nmm]} \qquad (90)$$

Displacement and damage increment, update of displacement and damage

$$\mathbf{K}^{u(i)} \delta \mathbf{u} = \mathbf{f}^{\text{ext}} - \mathbf{f}^{\text{int}(i)} \qquad \mathbf{K}^{d(i)} \delta \mathbf{d} = -\mathbf{r}^{d(i)} \qquad (91)$$

$$\mathbf{u}^{(i+1)} = \mathbf{u}^{(i)} + \delta \mathbf{u}, \text{ [mm]}; \qquad \mathbf{d}^{(i+1)} = \mathbf{d}^{(i)} + \delta \mathbf{d}, \text{[−]} \qquad (92)$$

If convergence criteria are not satisfied, go to step B.
If $\left\| \mathbf{r}^{u(i)} \right\| \leq tol$ and $\left\| \mathbf{r}^{d(i)} \right\| \leq tol$ go to next time step A.

3. Validation Examples

3.1. One Element—Brittle Fracture Benchmark Example

The simplest model that can be used to verify and understand the proposed staggered iterative scheme (Section 2.5.2) for coupling of damage phase field and displacement field is suggested by Molnar and Gravouil in [20] for the brittle fracture. The model is one three-dimensional hexahedral element of unit dimensions (1 mm × 1 mm × 1 mm).

The parameters necessary for brittle fracture simulation are the same as in [9]: the Young's modulus $E = 210$ GPa, the Poisson's ratio $v = 0.3$, the critical energy release rate $G_V = 5 \cdot 10^{-2}$ GPa and the length scale parameter $l_c = 0.1$ mm. The plasticity material parameters are large enough so the material is in the elastic regime, while the coupling variable is $p = 1$ to simulate the brittle fracture [9]. The analytical stress σ vs. strain ε, as well as damage d vs. strain ε relationships, are given as follows [9]:

$$c_{22} = \frac{E(1-v)}{(1+v)(1-2v)}, \qquad (93)$$

$$d = \frac{\varepsilon^2 c_{22}}{G_V + \varepsilon^2 c_{22}}, \qquad (94)$$

$$\sigma = (1-d)^2 c_{22} \varepsilon, \qquad (95)$$

where c_{22} is the element of the elastic constitutive matrix $\mathbf{C}_0$. The nodes on the bottom of the cube are constrained in all three directions, while the top nodes are allowed to slide vertically. The loading is realized in a displacement control regime in 1000 time steps until the total displacement of 0.1 mm. The obtained results are quantitatively compared to the analytical results [9]. The comparison given in Figures 5 and 6 confirms the functionality of the proposed iterative scheme and its correctness.

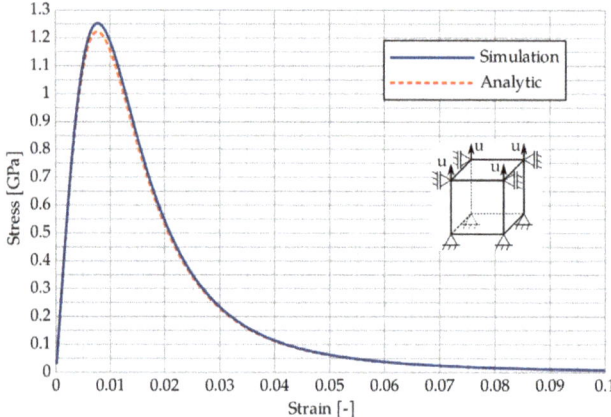

Figure 5. Axial stress vs axial strain relationship for the one element example.

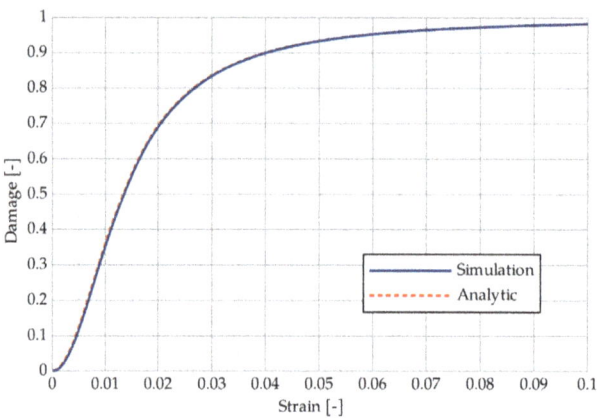

Figure 6. Damage vs. axial strain relationship for the one element example.

3.2. Experimental Investigation and FEM Simulation of S355J2+N Specimens

The S355J2+N is widely used in engineering structures due to the good weldability and machinability. It also exhibits constant stress plateau after yielding occurs, so it is chosen as a representative steel type to validate the two-stage hardening yield function. This material is also used in experimental investigation and simulation of fracture by various authors [25,30,39]. For the validation purposes of modified PFDM, three steel S355J2+N specimens are investigated by using servo-hydraulic testing machine—EHF-EV101 K3-070-0A (Shimadzu Corporation, Tokyo, Japan), with force ±100 kN and stroke ±100 mm. The specimen's chemical composition given in Table 1.

Table 1. Chemical composition of the examined S355J2+N specimens (wt %)

C	Si	Mn	P	S	Cr	Ni	Mo	Cu	N	Al
0.161	0.046	1.488	0.0224	0.0086	0.040	0.014	0.005	0.005	0.004	0.049

Uniaxial tensile tests are performed on representative flat specimens with the same thickness in all cross-sections. The tests are carried out according to standards EN ISO 6892-1 [40] and ASTM E8M-01 [41] at room temperature (23 ± 5 °C) in constant stroke control rate of 4 mm/min. Specimen's shape and dimensions are given in Figure 7. For the elongation measurement and identification of Young modulus, the extensometer MFA25 (MF Mess- & Feinwerktechnik GmbH, Velbert, Germany), with a gauge length of 50 mm is used as given in Figure 8.

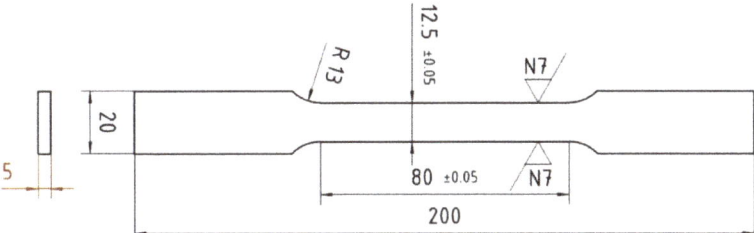

Figure 7. S355J2+N steel specimen shape and dimensions.

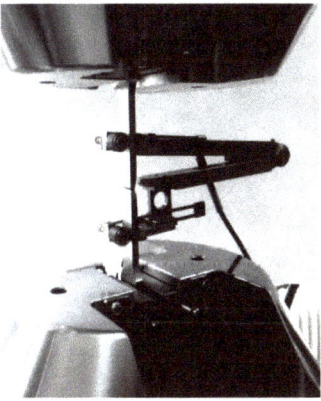

Figure 8. Mess & Feinwerktechnic GmbH MFA25 extensometer.

The investigated specimens are presented in Figure 9 after the experiment. The force-displacement responses are recorded and the comparison of the obtained results is given in Figure 10. The response of "Specimen 3" is selected as the representative for the PFDM validation purpose.

Figure 9. S355J2+N steel specimens after the experimental uniaxial tests.

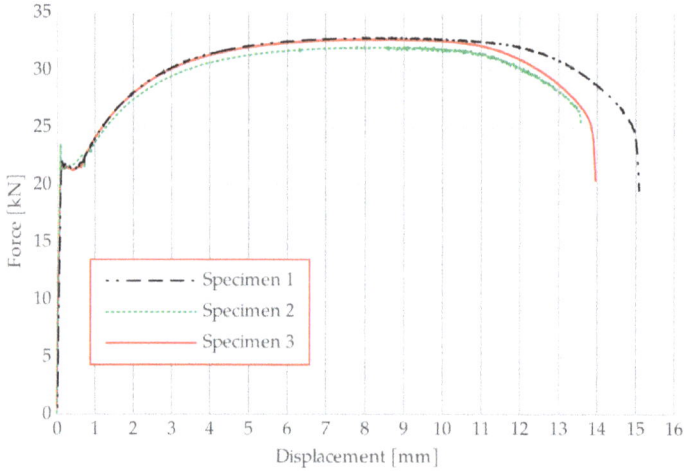

Figure 10. Force–displacement response of experimentally investigated S355J2+N steel specimens.

The FE model is prepared for the straight part of the specimen same as the gauge length (50 mm), according to the specimen's dimensions. One-eighth of the specimen is modeled due to the existence of three symmetry planes. Dimensions of the FE model are 25 mm × 6.25 mm × 2.5 mm. The geometrical imperfection necessary to trigger the plastic deformation process in a zone of 10 mm (L_2) from the middle of the specimen, where necking is expected, is prescribed as 0.01% a linear decrease of the specimen width D and thickness (Figure 11). The FE model is created using 2100 standard full integrated 8-node hexahedral finite elements with mesh refinement in the expected necking zone. Boundary conditions include the constraint of nodes in symmetry planes in a direction perpendicular to the symmetry plane they belong to. The FE model tensile loading is applied to the top surface nodes by displacement increment of 0.02 mm for 350 steps.

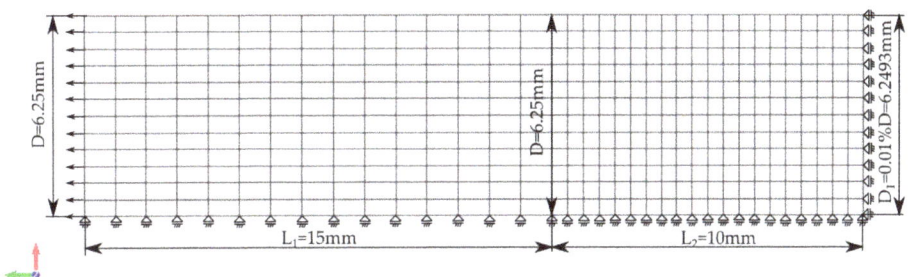

Figure 11. Finite element mesh, imperfection, loading, and boundary conditions.

The material parameters used for simulations are given in Table 2: E [MPa]—Young's modulus, ν [-]—Poisson's ratio, σ_{yv} [MPa]—initial yield stress, $\sigma_{y0,\infty}$ [MPa]—saturation hardening stress, H [MPa]—hardening modulus, n [-]—hardening exponent, G_V [MPa]—fracture energy release rate, l_c [mm]—characteristic length, $\bar{e}_p^{crit}$—critical equivalent plastic strain, $\bar{e}_{P0}$—perfect plasticity equivalent plastic strain.

Table 2. Material parameters used for phase-field damage model simulation.

E [MPa]	ν [-]	σ_{yv} [MPa]	$\sigma_{y0,\infty}$ [MPa]	H [MPa]	n [-]	G_V [MPa]	l_c [mm]	$\bar{e}_p^{crit}$	$\bar{e}_{P0}$
199,000	0.29	345	635	9	18	9.09	0.01	0.188	0.01

Figure 12 shows the deformed S355J2+N specimen after the experimental investigation along with the equivalent plastic strain field obtained from the PFDM FEM simulation. The distribution of the equivalent plastic strain field qualitatively simulates the deformed configuration of the experimentally obtained deformations. The equivalent plastic strain field obtained by phase-field plasticity coupled simulation is localized in a zone where the fracture occurs.

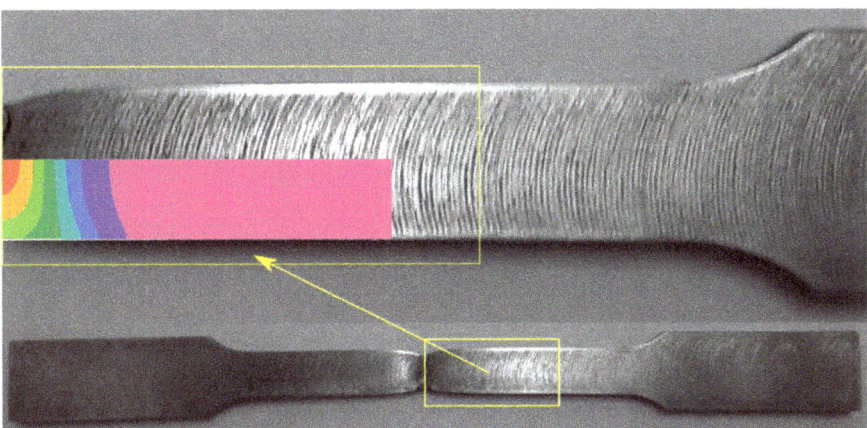

Figure 12. S355J2+N steel specimen with the equivalent plastic strain field at the critical zone after the experiment.

The strong relationship between the damage field and the equivalent plastic strain field obtained by the coupled phase-field simulation given in Figure 13 imposes that damage is a governing phenomenon that leads to the fracture of the specimen. The equivalent plastic strain field for the pure plasticity without phase-field (Figure 13a) and the PFDM simulation (Figure 13b) are presented to show the influence of the damage field on the localization of the plastic strains. In Figure 13a, the plastic strains are localized in the middle of the model, with the minimal difference between the minimal and maximal value, which does not suggest the fracture zone location. On the other side, the distribution of the damage field given in Figure 13c corresponds to the equivalent plastic strain field in Figure 13b, so it can be considered as a generator of the fracture process.

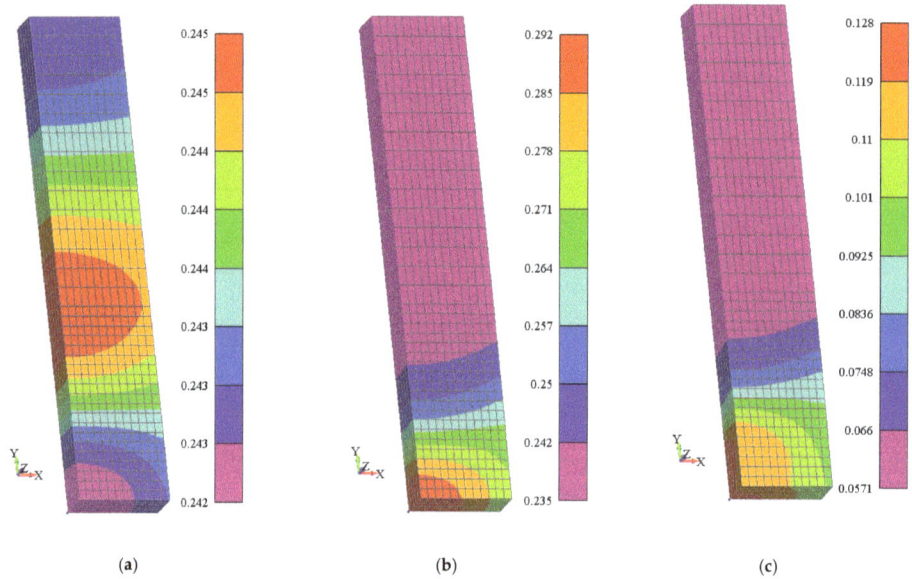

Figure 13. FEM simulation results: (a) Effective plastic strain field—plasticity; (b) Effective plastic strain field—phase-field and plasticity; and (c) damage field—phase-field and plasticity.

The comparison of the force-displacement relationship between experimental and simulation results is given in Figure 14. By comparing the diagrams, one can notice that 'pure' plasticity without damage cannot follow the behavior recorded by the experiment after the maximal force is achieved. Using the PFDM approach and the proposed modified coupling variable p, the influence of plastic strain development is activated after the loading force attains the maximum value and starts to decrease. The coupling variable p linearly increases as described in Figure 3 and simultaneously, the force starts to decrease until the specimen's fracture. In Figure 14, the value of the damage is given in the middle of the specimen in relation to the displacement of the specimen. It can be noticed that the damage is zero until the critical value of plastic strain is achieved. After the damage starts to increase, the force decreases following the slope of the damage change.

The element's dimension of the coarse FE mesh along the specimen length is 0.5 mm. This mesh has been used in previous simulations to show that the force-displacement response can be obtained even for the coarse FE meshes. However, if we want to simulate the evolution of damage phase-field with moving interface, the FE mesh must be refined further in the zone where the crack is expected. The dimension of the first two rows of elements (1.0 mm of specimen length) is reduced 4 times for the medium mesh (the element length—0.125 mm), while for the fine mesh the reduction is 10 times (the element length—0.05 mm). The force-displacement responses for the same material parameters given in Table 2 are presented in Figure 15, where it can be noticed that the finer mesh gives softer response in post-critical zone as suggested in Ambati et al. [18]. The evolution of damage field for the post-critical behavior is given in Figure 16 as well as the equivalent plastic strain field development in Figure 17. As it can be noticed, both the damage field and the equivalent plastic strain field evolves in the cracking zone of the specimen. For the visualization purpose, the further research will be criteria of "element death" which need to be satisfied to remove the elements as suggested in [19].

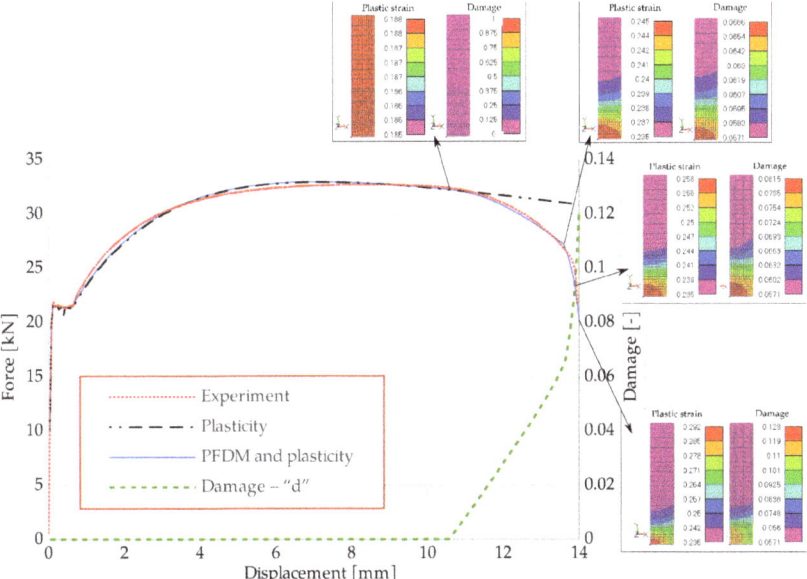

Figure 14. Force–displacement response of experiment and simulations vs. maximal damage value for S355J2+N steel specimens.

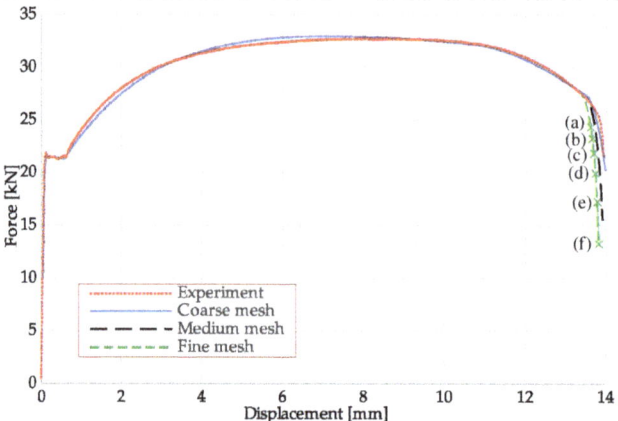

Figure 15. Effect of the mesh size on the force–displacement response: the specimen's top surface displacement for the fine mesh (**a**) 13.64 mm, (**b**) 13.68 mm, (**c**) 13.72 mm, (**d**) 13.76 mm, (**e**) 13.80 mm, (**f**) 13.84 mm.

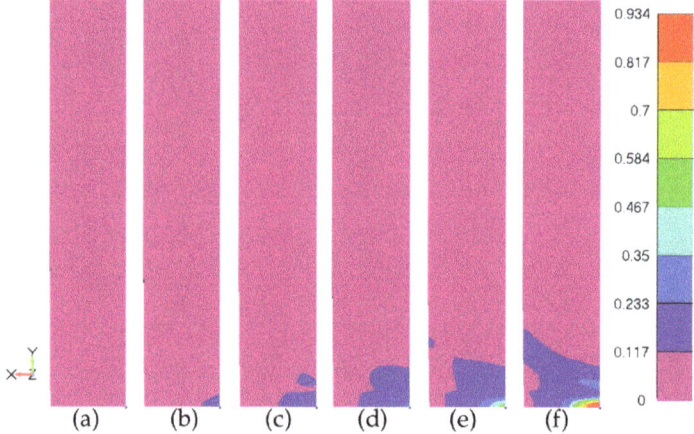

Figure 16. Damage phase-field d development in post-critical zone with moving interface for the specimen's top surface displacement of (**a**) 13.64 mm, (**b**) 13.68 mm, (**c**) 13.72 mm, (**d**) 13.76 mm, (**e**) 13.80 mm, (**f**) 13.84 mm.

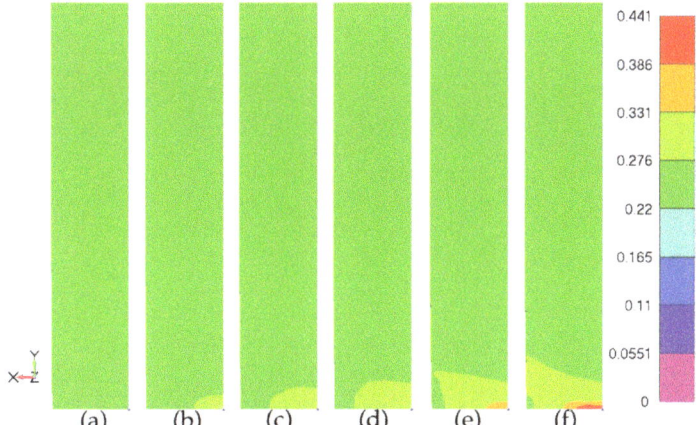

Figure 17. Equivalent plastic strain field $\bar{\varepsilon}_P$ development in post-critical zone for the specimen's top surface displacement of (**a**) 13.64 mm, (**b**) 13.68 mm, (**c**) 13.72 mm, (**d**) 13.76 mm, (**e**) 13.80 mm, (**f**) 13.84 mm.

Finally, the von Mises model with the same hardening function, but without coupled damage field (PFDM), cannot capture the post-critical behavior resulting from the material deterioration. The S355J2+N type of steel exhibits the specific plateau after the yielding occurs and it is captured by the proposed two-stage hardening function. The standard Simo's hardening function cannot be used for the simulation of such materials, as shown in Figure 18. The standard Simo's hardening function cannot follow the experimental force-displacement response at the same quantitatively and qualitatively level as the proposed two-stage hardening yield function.

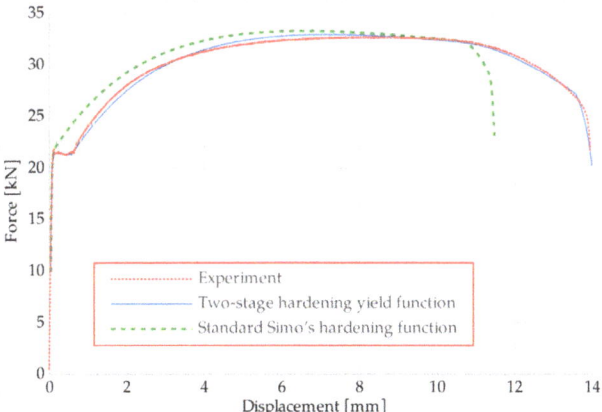

Figure 18. Comparison of the standard and the proposed two-stage hardening function for the same material parameters given in Table 1.

4. Conclusions

(1) The purpose of this article is to offer a better insight into the damage of materials and fracture of structures, propose necessary modifications of the existing PFDM, and allow better control over the damage simulation process.

(2) The PFDM for ductile fracture is one of the hottest topics in computational mechanics of solid structures. Well-established algorithms are already developed and implemented into the commercial and in-house FEM software documented in the cited literature.

(3) The authors proposed the two-stage yield function (perfect-plasticity and extended Simo-type hardening) to offer more realistic simulations of metallic materials behavior, which exhibit constant stress plateau after yielding.

(4) To control the start of the damage field development, the modification of the coupling between the plastic strain and the damage field is determined by the coupling variable, which starts to increase linearly after the equivalent plastic strain achieves the critical value.

(5) The main differences and advantages of the proposed method are possibilities to control: (a) the onset of hardening after the initial constant stress plateau is ended, (b) the initiation of damage phase-field development due to the plastic strain development, (c) the distribution of critical fracture energy release rate in damaged zone.

(6) The successful implementation of the staggered coupling scheme has been verified by one element benchmark example from literature. The same results have been obtained for both stress–strain response, but also the damage–strain relationship.

(7) The main application of the implementation is shown by simulation of S355J2+N test specimen behavior investigated by the universal testing machine. The experimentally captured force-displacement response is compared to the FEM simulation by proposed modified PFDM and excellent results are achieved. Also, the evolutions of the equivalent plastic strain field and the damage phase field are presented and the zone of maximal plastic strain qualitatively corresponds to the main deformed zone of the experimentally investigated specimens.

(8) These modifications will allow better control over the simulation of damage initiation and development and possibility to simulate various types of metallic materials in engineering practice.

Author Contributions: Conceptualization, V.D. and J.Ž.; Methodology, V.D. and M.Ž.; Software, M.Ž. and V.D.; Validation, J.Ž. and V.M.; Formal analysis, J.Ž. and A.P.; Funding, M.Ž.; Wxperimental investigation, J.Ž. and V.M.; Supervision, V.D.; Writing—original draft preparation, V.D. and J.Ž.; Writing—review and editing, V.D. and A.P.; Validation, M.Ž. and A.P.; Visualization, J.Ž. and A.P. All authors have read and agreed to the published version of the manuscript.

Funding: This research was funded by project TR32036 of Ministry of Educations, Science and Technological Development, Republic of Serbia.

Acknowledgments: The authors would like to thank IMW Institute doo Lužnice, Serbia for support in the investigation of chemical composition of S355J2+N specimens.

Conflicts of Interest: The authors declare no conflict of interest. The funders had no role in the design of the study; in the collection, analyses, or interpretation of data; in the writing of the manuscript, or in the decision to publish the results.

Nomenclature

FEM	Finite Element Method	ψ	internal potential energy density
PFDM	Phase-Field Damage Model	ψ^E	elastic energy density
S	crack surface	ψ_0^E	elastic energy density of virgin material
d	damage phase-field variable	ψ^P	plastic energy density,
x	coordinate along the bar	φ^S	fracture surface energy density
l_c	characteristic length-scale parameter	φ^P	plastic dissipated energy density
γ	crack surface density function	W_{ext}	external potential energy
∇	gradient operator	Ψ	total internal potential energy
Δ	increment	ε	total strain
δ	variation of variable	$\mathbf{F}_P$	plastic deformation gradient
V	volume	$\bar{\mathbf{F}}_E$	isochoric elastic deformation gradient
A	surface	$\bar{\mathbf{b}}_E$	elastic left Cauchy-Green strain
ε_E	elastic strain	$\bar{\mathbf{h}}_E$	Hencky strain
ε_P	plastic strain	e_m	mean strain
$\bar{\varepsilon}_P$	equivalent plastic strain	σ_m	mean stress
g	degradation function	G	shear modulus
$\mathbf{C}_0$	elastic constitutive matrix	c_m	bulk modulus
$\mathbf{C}_{EP}$	elastic-plastic constitutive matrix	E	Young's modulus
$\boldsymbol{\sigma}$	"damaged" Cauchy stress	υ	Poisson's ratio
$\boldsymbol{\sigma}_0$	"undamaged" Cauchy stress	t	time
Φ^S	fracture surface energy	$\mathbf{I}$	unit tensor
G_c	Griffith-type critical fracture energy release rate per unit area	$\mathbf{e}_E$	elastic deviatoric strain
G_V	critical fracture energy release rate per unit volume	$\mathbf{N}^u$	interpolation matrix for displacements
σ_{yv}	initial yield stress	$\mathbf{N}^d$	interpolation matrix for damage phase-field
$\sigma_{y0,\infty}$	saturation hardening stress	$\mathbf{B}^u$	matrix of interpolation functions derivatives for displacements
n	hardening exponent	$\mathbf{B}^d$	matrix of interpolation functions derivatives for damage phase-field
H	hardening modulus	$\mathbf{d}$	damage phase-field vector of nodal values
$\mathbf{b}$	body force field per unit volume	ε^d	damage strain
$\mathbf{h}$	boundary traction per unit area	$\mathbf{f}^{int}$	internal forces vector
$\mathbf{n}$	unit outer normal to the surface A	$\mathbf{f}^{ext}$	external forces vector

$\bar{\sigma}_{eq}$	equivalent stress	$\mathbf{r}^d$	residue vector for the damage phase-field
p	coupling variable	$\mathbf{r}^u$	residue vector for the displacement field
$\bar{\varepsilon}_p^{crit}$	critical equivalent plastic strain	$\mathbf{K}^d$	tangent stiffness matrix for damage phase-field
$\bar{\varepsilon}_{p0}$	maximal equivalent plastic strain for perfect plasticity stage	$\mathbf{K}^u$	tangent stiffness matrix for displacement field
$\bar{\varepsilon}_p^{fail}$	failure equivalent plastic strain	$\mathbf{u}$	nodal displacements vector
σ_y	yield stress of current yield surface	c_{22}	element of the elastic matrix $\mathbf{C}_0$
σ_{yS}	basic Simo's yield stress of current yield surface	D	width
σ_{y1}	second stage extended Simo's yield stress of current yield surface	L	length
σ_{y2}	internal variable related σ_{y1}	σ	stress
$\mathbf{S}_E$	elastic deviatoric stress	ε	strain
f_y	yield function		
F	force		
$\mathbf{F}$	total deformation gradient		
$\mathbf{F}_E$	elastic deformation gradient		

References

1. Trahair, N.S.; Bradford, M.A.; Nethercot, D.; Gardner, L. *The Behaviour and Design of Steel Structures to EC3*, 4th ed.; Taylor & Francis: London, UK, 2008.
2. Gaylord, E.H.; Gaylord, C.N.; Stallmeyer, J.E. *Design of Steel Structures*, 3rd ed.; McGraw-Hill: New York, NY, USA, 1992.
3. Subramanian, N. *Steel Structures-Design and Practice*; Oxford University Press: Oxford, UK, 2011.
4. Fragassa, C.; Minak, G.; Pavlovic, A. Measuring deformations in the telescopic boom under static and dynamic load conditions. *Facta Univ. Ser. Mech. Eng.* **2020**, *18*, 315–328. [CrossRef]
5. Vukelic, G.; Brnic, J. Predicted Fracture Behavior of Shaft Steels with Improved Corrosion Resistance. *Metals* **2016**, *6*, 40. [CrossRef]
6. Lesiuk, G.; Smolnicki, M.; Mech, R.; Ziety, A.; Fragassa, C. Analysis of fatigue crack growth under mixed mode (I + II) loading conditions in rail steel using CTS specimen. *Eng. Fail. Anal.* **2020**, *109*, 104354. [CrossRef]
7. Soliman, M.; El Rayes, M.; Abbas, A.; Pimenov, D.; Erdakov, I.; Junaedi, H. Effect of tensile strain rate on high-temperature deformation and fracture of rolled Al-15 vol% B4C composite. *Mater. Sci. Eng. A* **2019**, *749*, 129–136. [CrossRef]
8. Englekirk, R.E. *Steel Structures: Controlling Behavior through Design*; John Wiley & Sons: Hoboken, NJ, USA, 1994.
9. Busby, J.S. Characterizing failures in design activity. *Proc. Inst. Mech. Eng. Part B J. Eng. Manuf.* **2001**, *10*, 1417–1424. [CrossRef]
10. Dimaki, A.; Shilko, E.; Psakhie, S.; Popov, V. Simulation of fracture using a mesh-dependent fracture criterion in the discrete element method. *Facta Univ. Ser. Mech. Eng.* **2018**, *16*, 41–50. [CrossRef]
11. Srnec Novak, J.; De Bona, F.; Benasciutti, D. Benchmarks for Accelerated Cyclic Plasticity Models with Finite Elements. *Metals* **2020**, *10*, 781. [CrossRef]
12. Seleš, K.; Jurčević, A.; Tonković, Z.; Sorić, J. Crack propagation prediction in heterogeneous microstructure using an efficient phase-field algorithm. *Theor. Appl. Fract. Mech.* **2019**, *100*, 289–297. [CrossRef]
13. Alessi, R.; Ambati, M.; Gerasimov, T.; Vidoli, S.; De Lorenzis, L. Comparison of Phase-Field Models of Fracture Coupled with Plasticity. In *Advances in Computational Plasticity: A Book in Honour of D. Roger J. Owen*; Oñate, E., Peric, D., de Souza Neto, E., Chiumenti, M., Eds.; Springer International Publishing: London, UK, 2018; pp. 1–21. [CrossRef]
14. Alessi, R.; Marigo, J.; Maurini, C.; Vidoli, S. Coupling damage and plasticity for a phase-field regularisation of brittle, cohesive and ductile fracture: One-dimensional examples. *Int. J. Mech. Sci.* **2018**, *149*, 559–576. [CrossRef]
15. Francfort, G.; Marigo, J.-J. Revisiting brittle fracture as an energy minimization problem. *J. Mech. Phys. Solids* **1998**, *46*, 1319–1342. [CrossRef]

16. Miehe, C.; Hofacker, M.; Welschinger, F. A phase field model for rate-independent crack propagation: Robust algorithmic. *Comput. Methods Appl. Mech. Eng.* **2010**, *199*, 2765–2778. [CrossRef]
17. Miehe, C.; Welschinger, F.; Hofacker, M. Thermodynamically consistent phase-field models of fracture: Variational principles and multi-field FE implementations. *Int. J. Numer. Methods Eng.* **2010**, *83*, 1273–1311. [CrossRef]
18. Ambati, M.; Gerasimov, T.; De Lorenzis, L. Phase-field modeling of ductile fracture. *Comput. Mech.* **2015**, *55*, 1017–1040. [CrossRef]
19. Ambati, M.; Gerasimov, T.; De Lorenzis, L. A review on phase-field models of brittle fracture and a new fast hybrid formulation. *Comput. Mech.* **2015**, *55*, 383–405. [CrossRef]
20. Molnár, G.; Gravouil, A. 2D and 3D Abaqus implementation of a robust staggered phase-field solution for modeling brittle fracture. *Finite Elem. Anal. Des.* **2017**, *130*, 27–38. [CrossRef]
21. Seleš, K.; Lesičar, T.; Tonković, Z.; Sorić, J. A residual control staggered solution scheme for the phase-field modeling of brittle fracture. *Eng. Fract. Mech.* **2019**, *205*, 370–386. [CrossRef]
22. Liu, G.; Li, Q.; Msekh, M.; Zuo, Z. Abaqus implementation of monolithic and staggered schemes for quasi-static and dynamic fracture phase-field model. *Comput. Mater. Sci.* **2016**, *121*, 35–47. [CrossRef]
23. Azinpour, E.; Ferreira, J.P.S.; Parente, M.P.L.; Cesar de Sa, J. A simple and unified implementation of phase field and gradient damage models. *Adv. Model. Simul. Eng. Sci.* **2018**, *5*, 15. [CrossRef]
24. Alessi, R.; Marigo, J.-J.; Vidoli, S. Gradient Damage Models Coupled with Plasticity and Nucleation of Cohesive Cracks. *Arch. Ration. Mech. Anal.* **2014**, *214*, 575–615. [CrossRef]
25. Ambati, M.; Kruse, R.; De Lorenzis, L. A phase-field model for ductile fracture at finite strains and its experimental verification. *Comput. Mech.* **2016**, *57*, 149–167. [CrossRef]
26. Badnava, H.; Etemadi, E.; Msekh, M.A. A Phase Field Model for Rate-Dependent Ductile Fracture. *Metals* **2017**, *7*, 180. [CrossRef]
27. Miehe, C.; Aldakheel, F.; Raina, A. Phase field modeling of ductile fracture at finite strains: A variational gradient-extended plasticity-damage theory. *Int. J. Plast.* **2016**, *84*, 1–32. [CrossRef]
28. Pañeda, E.M.; Golahmar, A.; Niordson, C.F. A phase field formulation for hydrogen assisted cracking. *Comput. Methods Appl. Mech. Eng.* **2018**, *342*, 742–761. [CrossRef]
29. Zhang, X.; Vignes, C.; Sloan, S.W.; Sheng, D. Numerical evaluation of the phase-field model for brittle fracture with emphasis on the length scale. *Comput. Mech.* **2017**, *59*, 737–752. [CrossRef]
30. Ribeiroa, J.; Santiagoa, A.; Riguei, C. Damage model calibration and application for S355 steel. *Procedia Struct. Integr.* **2016**, *2*, 656–663. [CrossRef]
31. Balokhonov, R.; Romanova, V. On the problem of strain localization and fracture site prediction in materials with irregular geometry of interfaces. *Facta Univ. Ser. Mech. Eng.* **2019**, *17*, 169–180. [CrossRef]
32. Miehe, C.; Schänzel, L.-M.; Ulmer, H. Phase field modeling of fracture in multi-physics problems. Part I. Balance of crack surface and failure criteria for brittle crack propagation in thermo-elastic solids. *Comput. Methods Appl. Mech. Eng.* **2015**, *294*, 449–485. [CrossRef]
33. Bourdin, B.; Francfort, G.A.; Marigo, J.-J. Numerical experiments in revisited brittle fracture. *J. Mech. Phys. Solids* **2000**, *48*, 797–826. [CrossRef]
34. Wu, J.-Y.; Nguyen, V.P.; Nguyen, C.T.; Sutula, D.; Sinaie, S.; Bordas, S. Chapter One—Phase-field modeling of fracture. In *Advances in Applied Mechanics*; Bordas, S., Balint, D., Eds.; Elsevier: Amsterdam, The Netherlands, 2020; Volume 53, pp. 1–183. [CrossRef]
35. Fang, J.; Wu, C.; Li, J.; Liu, Q.; Wu, C.; Sun, G.; Li, Q. Phase field fracture in elasto-plastic solids: Variational formulation for multi-surface plasticity and effects of plastic yield surfaces and hardening. *Int. J. Mech. Sci.* **2019**, *156*, 382–396. [CrossRef]
36. Simo, J.C.; Miehe, C. Associative coupled thermoplasticity at finite strains: Formulation, numerical analysis and implementation. *Comput. Methods Appl. Mech. Eng.* **1992**, *98*, 41–104. [CrossRef]
37. Kojić, M.; Bathe, K.J. *Inelastic Analysis of Solids and Structures*; Springer: Berlin/Heidelberg, Germany, 2005.
38. Dunić, V.; Busarac, N.; Slavković, V.; Rosić, B.; Niekamp, R.; Matthies, H.; Slavković, R.; Živković, M. A thermo-mechanically coupled finite strain model considering inelastic heat generation. *Continuum. Mech. Thermodyn.* **2016**, *28*, 993–1007. [CrossRef]

39. Dzioba, I.; Lipiec, S. Fracture Mechanisms of S355 Steel—Experimental Research, FEM Simulation and SEM Observation. *Materials* **2019**, *12*, 3959. [CrossRef] [PubMed]
40. EN ISO 6892-1. *Metalic Materials—Tensile Testing—Part. 1: Method of Test at Room Temperature (ISO 6892-1:2009)*; International Organization for Standardization: Geneva, Switzerland, 2009.
41. ASTM: E8M-01. *Standard Test. Method for Tension Testing of Metalic Material*; ASTM International: West Conshohocken, PA, USA, 2002.

 © 2020 by the authors. Licensee MDPI, Basel, Switzerland. This article is an open access article distributed under the terms and conditions of the Creative Commons Attribution (CC BY) license (http://creativecommons.org/licenses/by/4.0/).

Article

Non-Destructive Micromagnetic Determination of Hardness and Case Hardening Depth Using Linear Regression Analysis and Artificial Neural Networks

Rahel Jedamski [1,*] and Jérémy Epp [1,2]

1 Leibniz-Institute for Materials Engineering—IWT, 28359 Bremen, Germany
2 MAPEX Center for Materials and Processes, University of Bremen, 28359 Bremen, Germany; epp@iwt-bremen.de
* Correspondence: jedamski@iwt-bremen.de; Tel.: +49-421-218-51326

Received: 20 November 2020; Accepted: 19 December 2020; Published: 24 December 2020

Abstract: Non-destructive determination of workpiece properties after heat treatment is of great interest in the context of quality control in production but also for prevention of damage in subsequent grinding process. Micromagnetic methods offer good possibilities, but must first be calibrated with reference analyses on known states. This work compares the accuracy and reliability of different calibration methods for non-destructive evaluation of carburizing depth and surface hardness of carburized steel. Linear regression analysis is used in comparison with new methods based on artificial neural networks. The comparison shows a slight advantage of neural network method and potential for further optimization of both approaches. The quality of the results can be influenced, among others, by the number of teaching steps for the neural network, whereas more teaching steps does not always lead to an improvement of accuracy for conditions not included in the initial calibration.

Keywords: artificial neural network; linear regression; micromagnetic testing; hardness; case hardening depth

1. Introduction

The heat treatment state of a case-hardened steel workpiece especially surface hardness and case hardening depth, which also often correlates with the surface oxidation depth, are important properties for the final service properties of high-performance parts, which have to be continuously controlled in industrial production. Further, these surface properties influence the grindability of the components as well as the micromagnetic detectability of grinding damages [1,2]. Therefore, the knowledge of these properties can allow determining optimal grinding parameters as a function of the component's initial state and can enable a reliable in-process micromagnetic monitoring of the grinding processes [3]. Besides the known destructive testing methods which are precise but time-consuming, the heat treatment condition can also be assessed non-destructively using micromagnetic methods such as Barkhausen noise analysis or the 3MA-technique [4].

Using micromagnetic methods, hardness is in most cases determined by calibration using linear regression [4,5]. To determine the case hardening depth by means of Barkhausen noise analysis, several additional approaches based on the different properties of soft core and hard surface layer exist. For example, Send et al. observed additional peaks and asymmetries of the Barkhausen peak generated by the hardened case that could be described with different parameters and correlated with the case hardening depth [6]. In [7] Santa-aho et al. employed sweeps of the magnetization voltage to determine the maximum slope of Barkhausen noise for two magnetizing frequencies (and therefore different penetration depth) and correlated the ratio with the case hardening depth. The 3MA-technique combines four micromagnetic methods with different penetration depth. Besides the Barkhausen noise, it includes, among others, the method of harmonic analysis of the tangential magnetic field. Here, the case hardening could be correlated to the frequency limit at which the distortion factor is near to an asymptote [8]. Further influences of case depth on the frequency characteristics of various micromagnetic signals were observed in [9], but could not be fully explained.

In [10] and [11], among others, artificial neural networks were used to determine hardness from hysteresis loop and eddy current measurement respectively Barkhausen noise and tangential magnetic field analysis. Besides hardness, Liu et al. determined residual stresses and case depth from a combination of Barkhausen noise, tangential magnetic field, and hysteresis measurement by use of artificial neural networks [12]. Sorsa et al. compared the suitability of linear regression, artificial neural networks, and fuzzy models for determination of residual stresses from Barkhausen noise measurements. They identified better performance indices RMSE (root mean square error) and R^2 (coefficient of determination) for artificial neural networks than for linear regression but also pointed out an advantage of linear regression. Indeed, this method has a low risk of overfitting and is therefore well suited for small calibration data sets and extrapolation. To achieve good regression results, a preselection of significant features was performed, as a too high number of inputs led to overfitting of the network [13]. Furthermore, in [10] only a single digit number of measured variables was used for hardness determination. Criterion for the selection was the ratio of standard deviation and mean value. An improvement of the performance could be achieved by additional measured variables. The influence of the volume of the data set is described in [11]. An extension of the calibration data set leads to decrease of the standard error. For the given example of 17 measurement points over a Jominy sample with three hardness zones, the standard error starting with nine data points in calibration was below 5% of the maximum hardness.

Further actual studies handle with the correlation of magnetic properties with hardness, residual stresses, and other properties. In general, decreasing hardness [14] and increasing (tensile) residual stresses [15,16] led to an increase in the Barkhausen noise level. While increase of Barkhausen noise with residual stresses in some cases is not linear due to poor magnetization, the averaged permeability derived from hysteresis loop shows a much better linear correlation [16]. Other works describe linear increase with rising tensile stresses for Barkhausen noise as well as permeability [17,18]. Besides the maximum and average values of Barkhausen noise and permeability, various other measured variables are determined from the raw signal depending on the measuring device. The coercivity develops roughly opposite to the Barkhausen noise or permeability level and correlates well with the hardness as well as residual stresses [17,19]. Peak width decreases with increasing tensile stresses. Remanence decreases with increasing hardness [19].

The aim of the present work is the non-destructive evaluation of the heat treatment state of case-hardened steel workpieces by means of micromagnetic 3MA-measurements with variation of magnetization and analysis frequencies. Instead of the first mentioned often very complex approaches for the evaluation of frequency sweeps, different calibration strategies were developed. In order to use the possibilities of frequency-dependent analyses for the determination of gradient properties, different from previous works, a large number of measured variables were included. After calibration using samples with known properties, these can simply be applied to the following measurements. For this purpose a comparison of classical regression analysis and calibration using artificial neural networks was carried out. A sample set with two-stage variation of three variables and two samples per state was prepared for calibration and analysis. Special attention was therefore paid to the challenges of—compared to number of measured variables and influences on heat treatment state—small calibration data sets. Standard errors of calibration data and unknown test data were compared. Possible influences of different parameters on the quality of the result were evaluated and compared.

1.1. Micromagnetic Measurements and Data Evaluation

The 3MA-II-technology (Micromagnetic Multiparametric Microstructure- and Stress-Analysis) developed by Fraunhofer Institute for Nondestructive Testing (IZFP) combines the four micromagnetic methods Barkhausen Noise (BN), Harmonic Analysis of the tangential magnetic field strength (HA), Incremental permeability (IP), and multi-frequency Eddy Current analysis (EC). Together these methods provide 41 measured variables with different sensitivity to microstructure and stress state. Due to the frequency ranges the methods have different analyzing depths. The combination of measuring parameters allows a separation of different influences such as residual stresses and hardness as well as the compensation of disturbances [20]. An additional software tool, the so-called sweep module, enables the automatized and fast variation of measurement parameters such as frequency and magnetization amplitude [4,9].

BN results from the stepwise shift of Bloch walls, separating magnetic domains, during magnetization of ferromagnetic materials. If no further Bloch wall shifts are possible, the domains are aligned in direction of the field by rotation processes as the magnetization continues to increase [21]. In this region the magnetic hysteresis loop, $B(H)$ becomes flatter until it becomes horizontal in the saturation state [22]. A schematical hysteresis curve is shown in Figure 1. After amplifying, filtering, and rectifying the recorded BN signal, its envelope or profile curve is displayed above the magnetic field strength H. The shape of the hysteresis curve, the profile curve, and thus the characteristic parameters are influenced by the microstructure, the mechanical properties, and residual stress state [9,20]. Mechanically hard materials are magnetically hard with high coercivity H_{CM} and low remanence M_R and maximum Barkhausen noise amplitude M_{MAX} because Bloch wall shifts interact analogous to dislocation movements. Other measured variables are the averaged amplitude over one magnetization cycle M_{MEAN} and the curve width at 25%, 50%, and 75% of M_{MAX} called DH_{25M}, DH_{50M}, and DH_{75M}, respectively. Compressive stresses cause a high coercivity and low amplitude, tensile stresses a low coercivity and high amplitude [4,23–25].

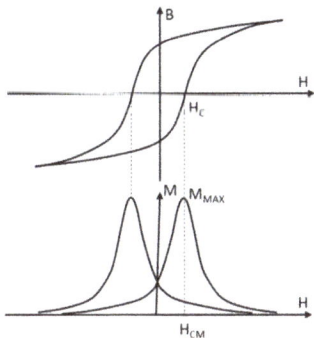

Figure 1. Schematical view of a hysteresis curve (**top**) and the envelope of the Barkhausen noise signal (**bottom**).

Depending on the frequency, different depth ranges can be investigated. According to (1), the exponential damping function results in the penetration depth at which the amplitude of the magnetic field is still $1/e$ of the intensity at the surface. Thereby, the relative permeability μ_r and the electrical conductivity σ are material and heat treatment-dependent parameters [26]. The high-pass frequency of the Barkhausen noise signal can be stepwise varied between no limitation and 1 MHz, and low-pass between 500 kHz and no limitation. Due to the base magnetization frequency between 10 Hz and 1000 Hz, harmonic analysis has the largest possible analyzing depth of the four testing methods with up to several mm probing depth [4]. The analyzing depth of the Barkhausen noise, depending on the analyzing frequency range, is in a range from about 10 µm to a few 100 µm [9]. In this particular case of hardened surface layers, an analyzing depth of up to 50 µm can be assumed [27]. The analyzing depth of the incremental permeability is in a similar range [9]. Depending on the set frequencies the eddy current analysis can reach large range of depth from about 10 µm to few hundred microns.

$$\delta = \frac{1}{\sqrt{\pi f \sigma \mu_0 \mu_r}} \tag{1}$$

In the harmonic analysis of the tangential magnetic field strength, the sample volume to be examined similar to Barkhausen noise analysis is magnetized by a sinusoidal alternating field. The development of the tangential field strength during the hysteresis loop is recorded by a Hall probe and processed by Fourier analysis. In this way, the odd higher harmonics can be determined. Measured variables generated from this are the amplitudes $A_{3\ldots 7}$ and phases $P_{3\ldots 7}$ of the harmonics, the distortion factor K(2), the sum of all upper harmonics UHS, the coercive field strength H_{CO} of the harmonic analysis (increases with hardness), the harmonic content of the magnetic field strength at zero crossing H_{r0}, and the final stage voltage of the electromagnet V_{mag} [4]. Soft magnetic materials generally lead to a high distortion factor and vice versa [4].

$$K = \sqrt{\frac{A_3^2 + A_5^2 \ldots + A_n^2}{A_1^2}} \tag{2}$$

To determine the incremental permeability $\mu_\Delta = \frac{dB}{dH}$, the hysteresis loop is superimposed by a further higher-frequency hysteresis loop. The signal picked up in this process corresponds to the incremental permeability, which is determined by alternating field strength changes at various points in the hysteresis loop, which would require an enormous amount of time in practice. The incremental permeability μ plotted against the magnetic field strength results in a similar curve shape as the Barkhausen noise M and the determination of the measured variables is carried out in the same way. While the Barkhausen noise is based on irreversible Bloch wall jumps, the incremental permeability depicts reversible processes [8].

In eddy current analysis, a high-frequency alternating weak magnetic field is generated. This primary field generates electric eddy currents, which are accompanied by a secondary field in the opposite direction to the primary field. A receiver coil measures the magnetic field as induced voltage. The eddy currents and thus the induced voltage are influenced by both the conductivity and permeability of the sample. The 3MA-II technology allows the simultaneous use of four eddy current frequencies. Real parts $Re_{1\ldots 4}$ and imaginary parts $Im_{1\ldots 4}$ as well as magnitude $Mag_{1\ldots 4}$ and phase angle $Ph_{1\ldots 4}$ of the voltage are output as measured variables [4].

1.2. Regression Analysis

For quantitative determination of a target quantity (e.g., hardness) it is necessary to find a calibration function which describes the relationship between target quantity and variables. Because the theoretical calculation of this relationship with physical models is possible at least for simple materials, 3MA is usually calibrated based on empirical data [4]. For this purpose, the target values are determined with a reference method (e.g., hardness testing) and stored in a database together with the micromagnetic measurements to analyze correlations by use of regression analysis or pattern recognition. Possible terms of the regression are the measurement parameters of 3MA as measured, their squares and square roots. The coefficients are determined by method of least squared errors [4,8,28]. Measures for the goodness of a regression are the coefficient of determination R^2 and the root mean square error RMSE [28].

An adjustment of the regression algorithm makes it possible to minimize the effect of drifts of measured variables on the regression result. This is of special interest for low-frequency and systematic stochastic errors, for example due to ageing or wear of the sensor, that cannot be reduced by a larger number of measurements in the calculation of averages. The range of values W gives the minimum and maximum value of a partial expression in the calibration data set.

$$M = 100\% * \Delta W / W_{max} \quad (3)$$

$$F_1 = W_{max}/100 = W/M \quad (4)$$

(3) gives the modulation M and (4) the 1% error effect F_1. If the measured value changes by 1% of W_{max} the regression result changes by the value F_1. The larger the error effect, the greater the reproducibility of the measured variables. The calibration wizard of the 3MA-MMS software gives the opportunity to choose the highest tolerated F_1. This limitation reduces the possible expressions to those that fulfill the condition $F_1 < F_{1,max}$. A too strong limitation of the 1% error effect leads also to a decreasing coefficient of determination R^2 and a rising standard error RMSE, as the remaining terms have not only a small error effect but only a small effect at all [8].

1.3. Artificial Neural Networks

Artificial neural network (ANNs) consist of several interconnected processing units called neurons, which can be divided in input, hidden, and output units and are arranged in layers (Figure 2a). The input units of the first layer distribute the components of the net input vector to the units of the second layer. The second till second last layer are hidden layers [29,30]. Every neuron consists of data collection (neuron input), processing, and sending results (neuron output), as shown in Figure 2b. To understand the process it is important to differentiate between the net-input or -output (Figure 2a) and the neuron-input or -output (Figure 2b). The weights of the links control the effect of the inputs on a neuron. They are changed during training of the ANN to optimize the relation of input and output and carry the information of the ANN. The weighted neuron inputs are summed and the transfer function determines the neuron output. There are several possible linear and nonlinear transfer functions to choose when constructing the network [31,32].

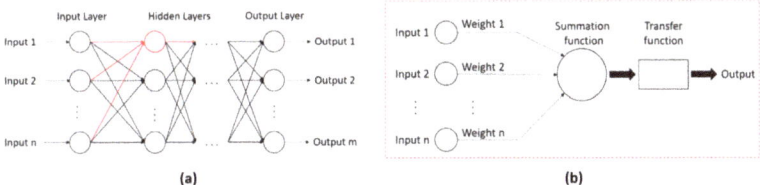

Figure 2. Schematical view of a feedforward neural network (**a**) and detail view of a neuron (**b**).

For training of an ANN (i.e., optimization of the weights) there are three general learning strategies, the supervised, unsupervised, and reinforcement learning. For supervised learning, the training set consists of (net) inputs and the related outputs. In unsupervised learning, only (net) inputs are given and the network learns without intervention of the trainer. Reinforcement learning means that the trainer only indicates whether the output of the network is correct or not. To evaluate the quality of the trained network in a test phase with unknown input, the pattern set is split up into teaching set and test set before teaching phase [29]. Furthermore it is necessary to normalize the net input and output values into the range 0–1 before teaching to get similar ranges for all variables [31,32].

A commonly used training algorithm for nets with input layer, output layer, and hidden layers is the backpropagation algorithm. Every teaching step consists of a forward pass and a backward pass. The forward pass starts with the input layer, the output of every unit is transmitted to the next layer till the output layer is reached. The estimated net outputs y_k are compared with the correct outputs y_k. Through the backward pass the differences are stepwise given back from the output layer to the first hidden layer and the weights are dated up to minimize the error of the next teaching step [31,32].

2. Materials and Methods

The sample set consists of 54 discs made from one batch of steel AISI 4820 (DIN 18CrNiMo7-6) with a diameter of 68 mm and a thickness of 20 mm. The chemical composition of the steel batch was analyzed with an optical emission spectrometer ARL 3460 (Thermo Fisher Scientific, Waltham, MA, USA) and is given in Table 1.

Table 1. Measured chemical composition of the AISI 4820 in wt. %.

C	Si	Mn	P	S	Cr	Mo	Ni
0.17	0.39	0.51	0.01	<0.002	1.56	0.26	1.43

The samples were gas carburized and oil quenched in 27 variations given in Table 2. The heat treatment was performed in a chamber furnace (Aichelin Holding GmbH, Mödling, Austria). Details for case hardening are shown in Figure 3 and Table 3. After oil quenching, the samples were tempered for two hours at the temperatures given in Table 2.

Table 2. Heat treatment variations (target values).

Surface Carbon Content/wt. %	Case Hardening Depth/mm	Tempering Temperature/°C
0.6/0.7/0.8	0.5/1/2	150/180/210

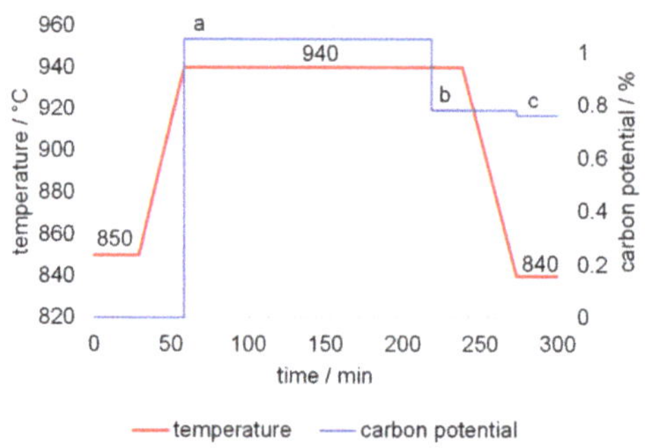

Figure 3. Exemplary process stages for the gas carburizing; details about the varying parameters are given in Table 3.

Table 3. Holding times at the different temperatures and carbon potential in the atmosphere for the different heat treatment variations.

Variation	850 °C	a	940 °C	b	840 °C	c
0.5 mm, 0.6 wt. %	30 min	0.9 vol. %	20 min	0.57 vol. %	30 min	0.57 vol. %
0.5 mm, 0.7 wt. %	30 min	0.9 vol. %	80 min	0.68 vol. %	30 min	0.68 vol. %
0.5 mm, 0.8 wt. %	30 min	0.9 vol. %	80 min	0.79 vol. %	30 min	0.79 vol. %
1 mm, 0.6 wt. %	30 min	1.0 vol. %	200 min	0.55 vol. %	30 min	0.55 vol. %
1 mm, 0.7 wt. %	30 min	1.05 vol. %	180 min	0.68 vol. %	30 min	0.66 vol. %
1 mm, 0.8 wt. %	30 min	1.05 vol. %	180 min	0.78 vol. %	30 min	0.76 vol. %
2 mm, 0.6 wt. %	120 min	1.05 vol. %	740 min	0.55 vol. %	60 min	0.55 vol. %
2 mm, 0.7 wt. %	120 min	1.1 vol. %	720 min	0.65 vol. %	60 min	0.64 vol. %
2 mm, 0.8 wt. %	120 min	1.1 vol. %	720 min	0.75 vol. %	60 min	0.75 vol. %

Surface hardness and carburization depth, as an approximation of the case hardening depth (CHD), were determined at smaller coupon samples treated in the same heat treatment batches as the samples for the investigations. Surface hardness was measured according to Vickers with a LV-700AT (LECO Instrumente GmbH, Mönchengladbach, Germany) with a test load of 9.807 N (HV1). To determine the carburization depth, carbon depth profiles were recorded using spark optical emission spectroscopy (ARL 3460, Thermo Fisher Scientific, Waltham, MA, USA). The carburization depth is defined as the depth with a carbon content of 0.3 wt. %.

Micromagnetic measurements were performed with a 3MA-II device from Fraunhofer IZFP, Saarbrücken, Germany and a standard sensor with convex pole shoes and spring-mounted transducer unit. The measurement settings are given in Table 4. The magnetization frequency, high-pass frequency, and eddy current frequency of the incremental permeability IP were varied with the sweep module to record a total of 425 measurement quantities with different analyzing depths. All samples were measured ten times at one position of the circumference with magnetization in tangential direction.

Table 4. Configuration of the 3MA-measurements.

	Standard Configuration	
Method	Magnetization Frequency	120 Hz
Harmonic analysis (HA)	Magnetization amplitude	75 A/cm
Barkhausen noise (BN)	Magnetization amplitude Highpass frequency Lowpass frequency	75 A/cm 100 kHz No limitation
Incremental permeability (IP)	Magnetization amplitude Eddy current frequency	75 A/cm 250 kHz
Eddy current (EC)	Magnetization amplitude Frequencies	65 A/cm 3,5 kHz, 1 MHz, 2 MHz, 5 MHz
	Sweeps	
HA, BN, IP	Magnetization frequency	20 Hz, 30 Hz, 40 Hz, 50 Hz, 60 Hz, 80 Hz, 100 Hz, 150 Hz, 200 Hz, 250 Hz, 300 Hz, 350 Hz
BN	High pass frequency	no limitation, 500 kHz, 1 MHz
IP	Eddy current frequency	10 kHz, 20 kHz, 60 kHz, 80 kHz, 100 kHz, 150 kHz, 200 kHz, 300 kHz, 350 kHz

After a first check of the data set and removal of samples with obvious outliers, calibration was carried out using the calibration module of the 3MA software and also by artificial neural networks. For calibration with linear regression analysis, the data obtained with magnetization frequency of 20 Hz were removed from the data set. For this frequency, instability of the magnetization meant that several measurement parameters of the IP could not be determined consistently. As this would reduce the information content of the data set, as only complete measurements with all parameters are used for regression, the magnetization frequency was removed from all data sets.

Linear regression with the "calibration wizard" of 3MA-software (Fraunhofer IZFP, Saarbrücken, Germany) [4,8] was carried out with carburization depth and hardness as target values. The maximum number of terms of the polynomial was set to 10. As mentioned in chapter 1, the limitation of the maximum error effect leads to an improvement especially for low-frequency stochastic errors (e.g., due to wear/ageing of the sensor). Nevertheless the maximum error effect was varied here to identify possible effects, e.g., in context of deviations due to contact between sensor and sample. Maximum error effect was reduced stepwise to find a setting with reduced error effect without too strong worsening of the regression quality. For both, the unlimited and the optimized error effect, additionally to these of the calibration data set R^2 and RMSE of test samples, which were not included in the calibration/training, were evaluated.

When dividing the data set into test and calibration data set, a too small calibration data set would lead to a bad regression result. In contrast, a large calibration data set reduces the test data set and thus the reliability of the control. Therefore, for the maximization of calibration and test data set the autorecognition test described in [8] was used. Each of the k samples (10 measurements) is taken out of the calibration data set and used for validation one time. After k calibrations with $k-1$ samples, every sample is used for validation one time.

For the artificial neural network analysis, 425 input neurons were employed as well as the same number of hidden neurons and output neurons with carburization depth and hardness as net output. The net was constructed in the software MemBrain [33,34] (free version for non-commercial and educational use, Thomas Jetter, Mainz, Germany). The activation function (transfer function) is a logistic function. The net was taught/trained with backpropagation method over 30 and 60 repetitions of the teaching lesson. Results were evaluated in the same way as described for regression analysis.

To get an idea about advantages and disadvantages with the use of sweeps, additional calibrations were performed with only the measured variables of the standard configuration. Linear regression was performed without limitation of the maximum error effect and network training with 30 repetitions.

3. Results and Discussion

As a result of the heat treatment variations described in Table 2, material states with carburization depth of approximately 0.55 mm, 0.9 mm, and 1.9 mm and surface hardness between 640 HV1 and 760 HV1 were generated. All measured data are presented in Table A1. Figures 4–6 show carbon profiles (a), microhardness profiles (b), and cross sections (c) of samples with target case hardening depth of 0.5 mm, 1 mm, and 2 mm. Comparison of the marked Case Hardening Depth (CHD 550), the carburization depth (0.3 wt. % carbon) and the transition between case hardened layer and bulk material in the cross section shows that carburization depth can be used as reliable closely related values to evaluate the CHD.

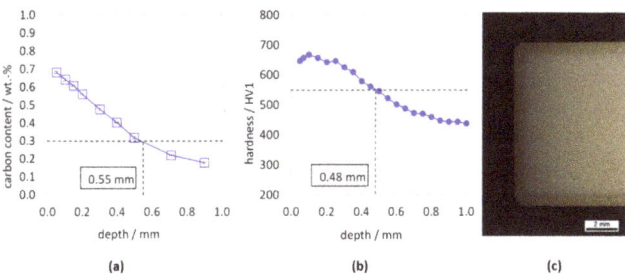

Figure 4. Carbon profile (**a**), microhardness profile (**b**), and micrograph of cross section (**c**) of a sample with target case hardening depth of 0.5 mm.

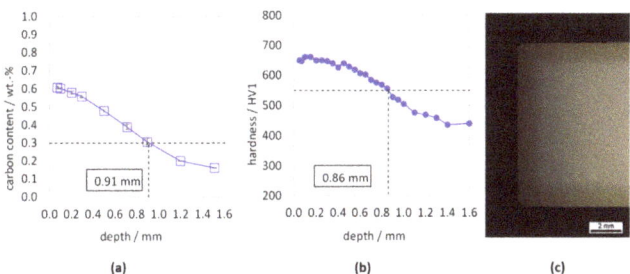

Figure 5. Carbon profile (**a**), microhardness profile (**b**), and micrograph of cross section (**c**) of a sample with target case hardening depth of 1 mm.

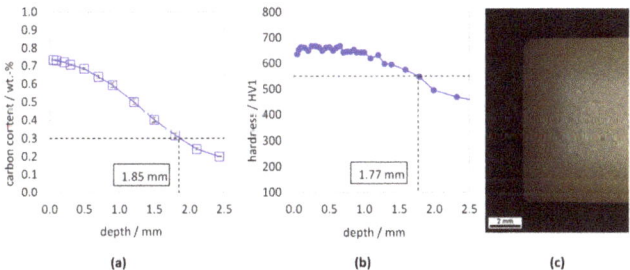

Figure 6. Carbon profile (**a**), microhardness profile (**b**), and micrograph of cross section (**c**) of a sample with target case hardening depth of 2 mm.

3.1. Calibration by Use of Regression Analysis

Linear regression of the hardness without limitation of the maximum error effect leads to an error effect of $F_1 = 9.274$ HV1. The determination coefficient of this regression is $R^2 = 0.8709$ and the standard error RMSE = 12.919. Table 5 shows the change of determination coefficient and standard error due to limitation of the error effect. A limitation to $F_1 = 3$ HV1 was chosen as optimum since with a stronger limitation there would be a too strong decrease of the determination coefficient and increase of the standard error.

Table 5. Coefficient of determination and standard error in dependence of the maximum error effect for regression analysis of hardness.

$F_{1,max}/HV1$	$F_1/HV1$	R^2	RMSE/HV1
∞	9.274	0.8709	12.919
9	5.582	0.8772	12.599
5	4.887	0.8662	13.151
4	3.489	0.8669	13.116
3	2.819	0.8571	13.59
2	1.899	0.837	14.514
1	0.961	0.723	18.923

Micromagnetic determined hardness of the calibration data set plotted over the measured hardness is shown in Figure 7. There is no pronounced difference due to the limitation of the error effect. The plotted data points all represent an averaged value for the 10 micromagnetic measurements per sample and the associated standard deviation. This average is the reason for the difference between the standard errors given in Table 5 and Figure 7.

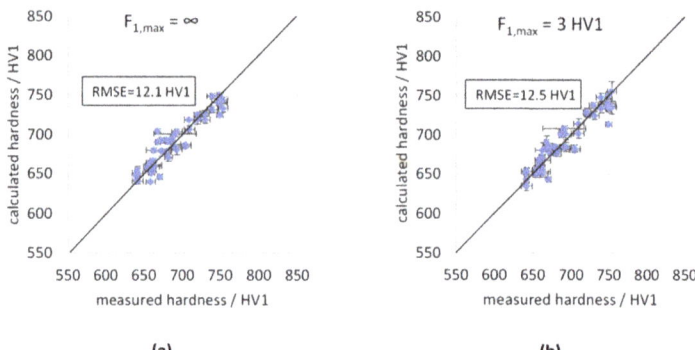

Figure 7. Correlation of micromagnetic determined and actual measured hardness of the calibration data set without limitation of the maximum error effect (a) and with limitation to 3 HV1 (b).

Figure 8 shows the measured and determined hardness for unknown samples, not included in the calibration (autorecognition test). The standard error is about 50% higher than for the calibration data set shown in Figure 7 and reaches values of 15% of the range of target values. The influence from limitation of the error effect on the result is very low. For later use of the calibration function, it would be more indicated to use the calibration function with reduced F1 values in order to reduce possible effects of sensor ageing or of a change of the sensor.

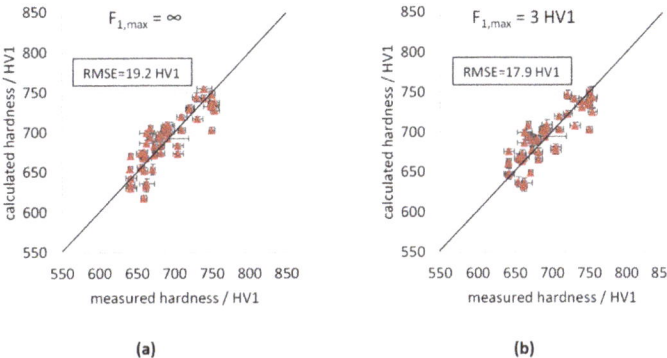

Figure 8. Correlation of micromagnetic determined and actual measured hardness of the test data set without limitation of the maximum error effect (**a**) and with limitation to 3 HV1 (**b**).

The terms of the regression are summarized in Table 6. No measured parameters from Eddy Current analysis is part of the calibration function but values from Harmonic Analysis (A_3–A_7, P_3–P_7, V_{mag}, H_{co}), Barkhausen noise (M_{MAX}, H_{CM}, M_r), and Incremental Permeability ($DH25_\mu$, μ_r) with different frequencies. As the values for the regression analysis are not normalized, the regression coefficients do not allow any conclusion about relevance of the single terms. It is obvious that some measured variables appear with a positive and negative prefix. For example, M_{MAX} is part of the regression result as negative term, what corresponds to the state of the art, but later M_{MAX}^2 appears as positive term. H_{CM}^2 is part of the result as negative term whereas it typically increases with hardness.

Table 6. Overview of the regression results for hardness (calibration data set).

$F_{1,max} = \infty$		$F_{1,max} = 3\ HV1$	
2882.68	1	1640.43	1
−44.3	A3 (120 Hz)	−697.51	A7 (100 Hz)
−3397.44	Mmax (350 Hz)	−19.5	P52 (30 Hz)
274.2	$\sqrt{P3}$ (50 Hz)	−0.36	Hcm2 (80 Hz, HP 100 kHz)
−0.12	Hco2 (50 Hz)	411.53	$\sqrt{A7}$ (100 Hz)
−142.44	$\sqrt{DH25\mu}$ (120 Hz, 60 kHz)	−5.59	P72 (120 Hz)
−772.37	$\sqrt{\mu r}$ (120 Hz, 100 kHz)	−240.26	$\sqrt{\mu r}$ (120 Hz, 250 kHz)
56.74	P32 (150 Hz)	−1582.39	$\sqrt{\mu r}$ (150 Hz)
−33.69	Vmag2 (350 Hz)	1974.5	Mr2 (350 Hz)
7305.07	Mmax2 (350 Hz)	−926.19	$\sqrt{Mr}$ (350 Hz)

Plotting the hardness over the various measured variables shows the best correlation with the remanence of the incremental permeability measurement (Figure 9). However, no obvious relationship between hardness and the measured variables from Harmonic Analysis could be observed. The different suitability of the measurement methods for determining hardness can be explained by the different penetration depths depending on the frequency. An overview of correlation of different measured variables with hardness and carburization depth is given in Table A2. The comparison with Table 6 and Table 9 shows that not all terms of the regression result are visibly correlated with hardness respectively carburization depth. On the one hand, this indicates that the calibration would be possible also with a lower number of terms. On the other hand, due to variation of surface carbon content, case hardening depth, and tempering temperature, signals are affected by multiple influences. Measured variables with coefficient of determination of more than 0.25 are remanence of Barkhausen noise and incremental permeability which correlate negatively with the hardness and peak width at 25% of the maximum of the incremental permeability. This is consistent with the known general correlations.

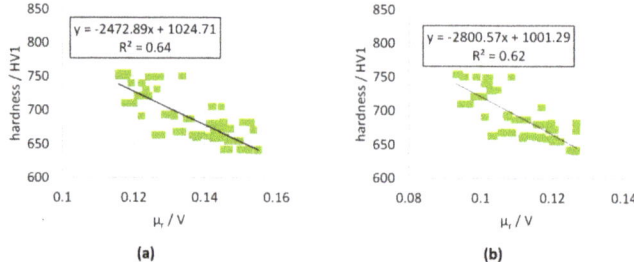

Figure 9. Correlation between hardness and amplitude at the remanence point of the incremental permeability μ_r measured with a base magnetization frequency of 120 Hz and an eddy current loop frequency of 100 kHz (**a**) and a base magnetization frequency of 150 Hz and an eddy current loop frequency of 250 kHz (**b**).

To make sure that only relevant measurement parameters are included in calibration the number of possible terms was limited to less than 10 (Table 7). A stronger limitation of the number of regression terms leads to a decrease of the coefficient of determination and an increasing standard error. But at the same time there is a clear decrease of the error effect. Regression with a high number of terms fits very well to the teaching data set but does not necessarily lead to an improvement on the result for unknown test data (exemplary shown for 20, 10, and 6 terms). A limitation of the number of terms thus has a similar effect as the limitation of the maximum error effect. The most important measured value, which is used to calculate hardness with only two terms is the remanence of the incremental permeability μ_r.

Table 7. Maximum error effect, coefficient of determination, and standard error in dependence of the maximum number of regression terms for regression analysis of hardness.

Maximum Number of Regression Terms	F_1/HV1	R^2	RMSE/HV1	RMSE$_{test}$/HV1
20	581.163	0.9457	8.375	13.5
10	9.274	0.8709	12.919	19.2
8	6.738	0.8548	13.7	-
6	7.174	0.8198	15.260	18.7
4	3.916	0.7674	17.341	-
2	3.496	0.6183	22.210	-

Additionally, a comparison of Figure 9a,b shows no difference in the qualitative evolution for different frequencies. A limitation to fewer frequency variants or a single frequency should therefore be possible without negative effects on the calibration but would reduce the measurement and calibration effort. Furthermore, this is in agreement with the results of previous studies, where calibration was carried out with only few preselected variables.

Similar to the hardness, the measurements were calibrated for carburization depth. Table 8 again shows the regression quality in dependence of the maximum error effect. A maximum error effect of 0.06 mm was chosen as optimum.

Table 8. Coefficient of determination and standard error in dependence of the maximum error effect for regression analysis of carburization depth.

$F_{1,max}$/mm	F_1/mm	R^2	RMSE/mm
∞	0.267	0.9539	0.117
0.2	0.191	0.9495	0.123
0.1	0.084	0.949	0.124
0.08	0.076	0.927	0.148
0.07	0.069	0.9246	0.150
0.06	0.059	0.9446	0.129
0.05	0.038	0.9333	0.141
0.03	0.029	0.9423	0.131

Figure 10 shows the carburization depth calculated without (a) limitation of the error effect plotted over the measured carburization depth. For this calibration data set there is a good agreement of measured and calculated values with a standard error of 0.074 mm, what is ~6% of the range of target values. The limitation of the error effect (b) only slightly affects the goodness of the prediction since small increase of RMSE is resulting.

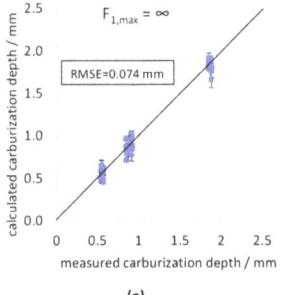
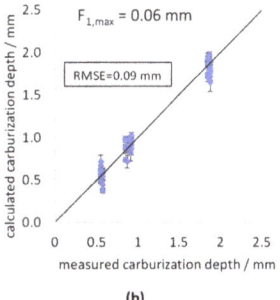

(a)   (b)

Figure 10. Correlation of micromagnetic determined and actual measured carburization depth of the calibration data set without limitation of the maximum error effect (**a**) and with limitation to 0.06 mm (**b**).

The same calibration types with and without limitation of the error effect in Figure 11 are illustrated for the test data set. The standard errors again are obviously higher than those of the calibration data set but with these are still within an acceptable range of reliability with errors around 10% of the target values. Without taking into account the outlier at 1.87 mm, the RMS is 0.105 mm for only 8% of the range of target values. As for the calibration data set, no pronounced effect due to the limitation of the maximum error effect can be observed, and the standard error increases slightly.

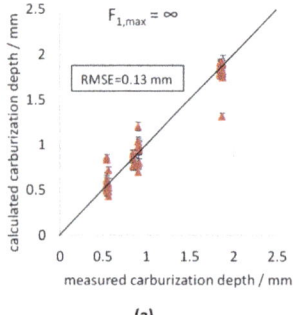
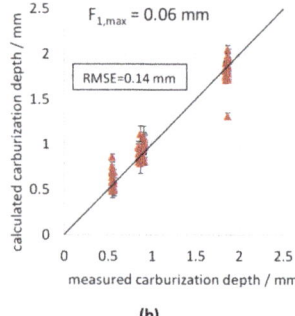

Figure 11. Correlation of micromagnetic determined and actual measured carburization depth of the test data set without limitation of the maximum error effect (**a**) and with limitation to 0.06 mm (**b**).

The regression terms for the determination of carburization depth are shown in Table 9. Again, the function contains measured variables from different methods and frequencies. Compared to the hardness determination the importance of Harmonic analysis and low frequencies (larger penetration depth) increases. Measured variables with best coefficient of determination (see Table A2) are K, H_{cm}, and H_{co} which all correlate positive with carburization depth. This corresponds to the general relationship that these measured variables increase with hardness.

Table 9. Overview of the regression results for carburization depth (calibration data set).

F1,Max = ∞		F1,Max = 0.06 mm	
−25.23	1	14.39	1
46.78	$\sqrt{V_{mag}}$ (30 Hz)	0.72	K (300 Hz)
−1.82	$\sqrt{P_3}$ (30 Hz)	−6.16	$\sqrt{P_3}$ (40 Hz)
9.25	$\sqrt{\mu_{max}}$ (30 Hz)	−0.08	$P_7{}^2$ (50 Hz)
−0.07	$P_7{}^2$ (50 Hz)	0.7	$\sqrt{H_{co}}$ (50 Hz)
0.001	$H_{cm}{}^2$ (80 Hz)	−0.0005	$DH50_m{}^2$ (80 Hz)
−4.7	$V_{mag}{}^2$ (150 Hz)	−7.87	$\sqrt{Ph_3}$ (120 Hz)
0.66	$\sqrt{K}$ (200 Hz)	0.005	$H_{cm}{}^2$ (250 Hz)
0.44	$\sqrt{P_3}$ (250 Hz)	−0.0003	$DH25_\mu{}^2$ (250 Hz)
0.36	$\sqrt{H_{co}}$ (350 Hz)	−0.31	$V_{mag}{}^2$ (350 Hz)

Again, a stronger limitation of the number of regression terms leads to lower coefficients of determination and lower error effects (Table 10). To calculate the carburization depth of unknown samples, the function with 10 terms gives good results but a limitation on six or eight regression terms could also be used, as a good compromise between the standard error and the maximum error effect. Similar to the observations of Table 9, regression results with six or less terms are based on measured variables from harmonic analysis and the coercivity H_{cm} (80 Hz).

Table 10. Maximum error effect, coefficient of determination, and standard error in dependence of the maximum number of regression terms for regression analysis of carburization depth.

Maximum Number of Regression Terms	F_1/mm	R^2	RMSE/mm	$RMSE_{test}$/mm
20	180.252	0.984	0.069	0.315
10	0.267	0.9539	0.117	0.130
8	0.178	0.9421	0.132	-
6	0.115	0.9198	0.155	0.156
4	0.06	0.842	0.217	-
2	0.047	0.5361	0.372	-

3.2. Calibration by Use of Artificial Neural Networks

The output of the artificial neural network plotted over the measured hardness is shown in Figure 12 after 30 (a) and 60 (b) iterations of the teaching lesson. The standard error is much lower than after calibration with linear regression and decreases significantly with duplication of the number of teaching lessons. RMSE = 7.1 HV1 after 30 teaching lessons are in the same range as the standard deviation of Vickers hardness measurements so as the RMSE = 3.7 HV1 after 60 lessons.

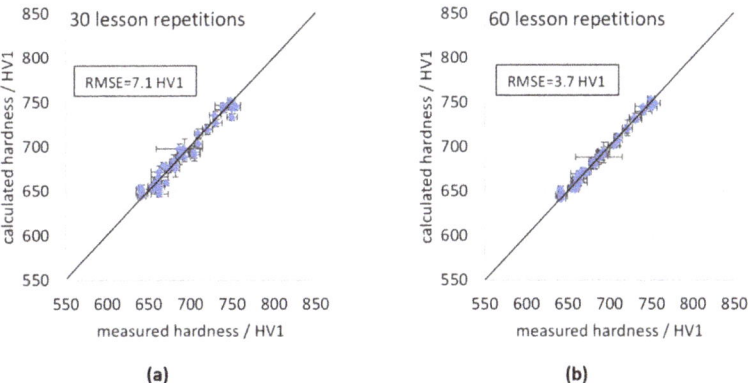

Figure 12. Correlation of micromagnetic determined and actual measured hardness of the training data set after 30 (a) and 60 (b) repetitions of the training lesson using artificial neural networks (ANNs).

For the unknown test data (autorecognition test), the standard error between measured hardness and net output is much higher than for the training data (see Figure 13). It is slightly below than after regression analysis but the difference between teaching and test data set is significantly higher. The duplication of the number of training lessons does not lead to pronounced change of the accuracy with a slight increase of the standard error. This illustrates the risk of overfitting. If the network adapts too much to the training data, this is at the expense of the quality of the prediction for unknown test data. For optimization, the number of teaching lessons should therefore be increased stepwise. As the standard error of the training data set continues to decrease or gets towards an asymptote, the standard error of the test data set will rise if the net overfits the training data. With the size of the network, the ability to fit complex solutions increases as well as the risk of overfitting [35]. Therefore, besides the number of training steps, the number of input variables also offers optimization potential.

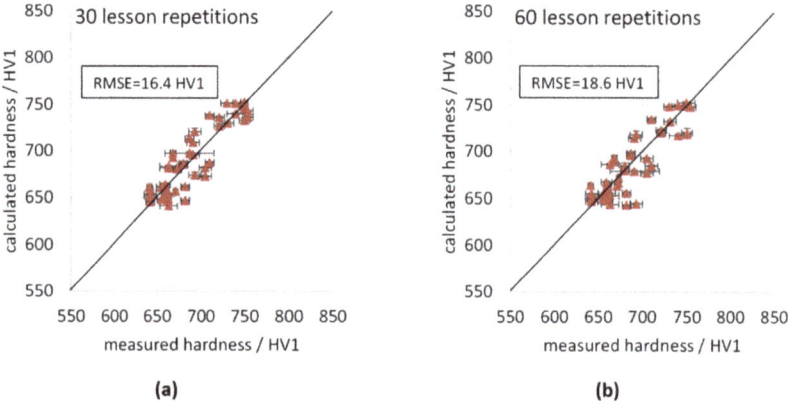

Figure 13. Correlation of micromagnetic determined and actual measured hardness of the test data set after 30 (**a**) and 60 (**b**) repetitions of the training lesson using ANN.

After doubling the training steps for calculation of hardness no improvement of the results was resulting. Figure 14 shows therefore only the network output for carburization depth after 30 repetitions of the training lesson. Nevertheless for the mentioned optimization the influence of the number of training steps on both target values should be examined more in detail. Despite some outliers at a carburizing depth of 0.9 mm, this is the lowest standard error. At the same time, the standard error increases more than three times from training to calibration data set. Possible reasons for this and potential improvements have already been mentioned for hardness. In addition, an extension of the sample set by further carburizing depths between 1 mm and 2 mm or more than 2 mm could be useful, to extend the range of the training data.

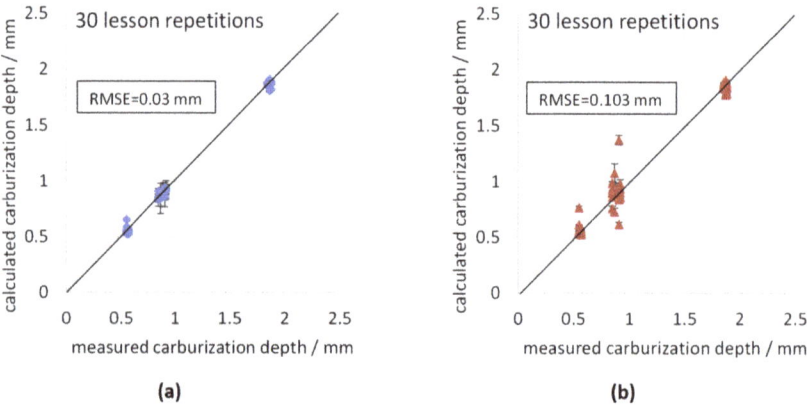

Figure 14. Correlation of micromagnetic determined and actual measured carburization depth after 30 repetitions of the training lesson for training (**a**) and test data set (**b**).

3.3. Calibration with Standard Configuration and Variation of Measurement Parameters

For comparison between calibration results for measurements with standard configuration (41 measured variables) and with use of sweeps (425 measured variables) the validation strategy was changed. In order to reduce the validation effort, autorecognition test was performed only for 14 samples (one sample of each of the states marked bold in Table A1). The deviation between the standard errors in Figure 15b, Figure 7a, and Figure 8a illustrates the problem of small test data sets. Some samples that led to outliers before are not part of the reduced test data set. Therefore, it is important that only standard errors based on the same test data set (Figure 15; Table 11) are compared with each other. With use of the frequency sweeps the standard error for hardness determined by linear regression decreases for calibration and test data set.

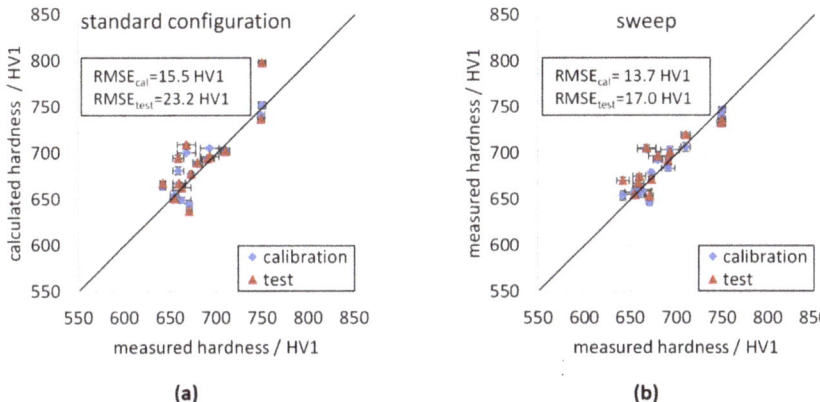

Figure 15. Correlation of micromagnetic determined (linear regression) and actual measured hardness of calibration and test data set with standard configuration (**a**) and use of sweeps (**b**).

Table 11. Standard error based on a test data set of 14 samples for calibration with standard configuration and use of sweeps.

Calibration Method	Hardness		Carburization Depth	
	$RMSE_{cal}$/HV1	$RMSE_{test}$/HV1	$RMSE_{cal}$/mm	$RMSE_{test}$/mm
linear regression (standard configuration)	15.5	23.2	0.12	0.22
linear regression (sweep)	13.7	17.0	0.07	0.13
ANN (standard configuration)	12.7	15.0	0.06	0.13
ANN (sweep)	7.2	16.5	0.02	0.04

Table 11 summarizes standard error of hardness and carburization depth for calibration with linear regression and ANN with standard configuration and use of sweeps. In general, the variation of the measurement parameters (sweep) leads to a significant reduction of the standard error. This reduction is particularly pronounced for the determination of the carburization depth by use of ANN. For determination of the hardness by use of ANN, only the standard error of the calibration data set decreases while that of the test data set increases. This is attributed to the already mentioned problem of overfitting.

Table 12 gives an overview of the different types of calibration (with use of sweeps) for prediction of the hardness and the case hardening depth. While there are large differences in the standard error of calibration data set, the quality of calibration for the test data set is slightly better for the ANN method, than for linear regression. It is noticeable that compared to the range of target values the accuracy of the hardness determination is generally lower than that of carburization depth, which means that unaccounted interfering variations of material properties may have influence on the results of the calculations.

Table 12. Overview of the standard error for the different types of calibration.

Calibration Method	Hardness		Carburization Depth	
-	$RMSE_{cal}$/HV1	$RMSE_{test}$/HV1	$RMSE_{cal}$/mm	$RMSE_{test}$/mm
linear regression ($F_{1,max}$ unlim.)	12.1	19.1	0.074	0.130
linear regression ($F_{1,max}$ lim.)	12.5	17.9	0.090	0.140
ANN (30 repetitions)	7.1	16.4	0.030	0.103
ANN (60 repetitions)	3.7	18.6	-	-

In contrast to the linear regression analysis, the ANN does not allow acquiring information about how the measured variables are included in the result, which means that it is a black box. For the mentioned reduction of the input variables, the knowledge from the regression analysis can provide an approach. For the analysis of the hardness, it could be observed that parameters with low penetration depths, as in Incremental permeability, are of great importance. To determine the carburization depth, methods and parameters with greater penetration depths such as harmonic analysis with low magnetization frequency are required. Additionally the comparison in Chapter 3.3 has shown an improvement of calibration results thanks to the use of frequency sweeps. Therefore, it seems to be useful to use more than one magnetization frequency for determination of carburization or case hardening depth, as data related to the property gradients are recorded and analyzed. For determination of surface hardness this brings no significant benefit. To avoid overfitting and reduce the measurement and calibration effort, the very large number of used frequency variants should be reduced also for carburization depth by choosing few relevant frequencies over the whole range. Apart from the selection of suitable measurement methods and frequencies depending on the target evaluation of eddy current results could be skipped for this application. Furthermore, a preselection of measured variables as described in previous work is recommended to reduce the time for calibration and the risk of overfitting of ANN. Possible criteria are the correlation of the single terms with the target value and low standard deviations—especially for similar measured variables like from Barkhausen noise and incremental permeability. Another way to reduce the complexity of linear regression would be to exclude roots and squares of the values, but it is expected that the goodness of fit will also decrease

Nevertheless, with the used strategies, hardness standard errors under 3% of the maximum target value could be achieved. Different from most of the previous studies, in the used dataset three variables (surface carbon content, case hardening depth, and tempering temperature) were used for the variation. Limitations of the maximum error effect and number of terms for linear regression and number of training steps for ANN were identified as options for optimization. For practical use, a stronger limitation seems to be useful in order to reduce the influence of small deviations, for example, due to inconsistent sensor contact. The chosen values for number of terms, maximum error effect and regression steps show opportunities but have to be optimized for each specific application.

4. Conclusions

This work has shown and compared the opportunities of calibration with linear regression and artificial neural networks for the non-destructive determination of hardness and carburization depth by micromagnetic measurements. The standard error RMSE of the calibration and test data set was used as indicators for the quality of the calibration. Even if there were larger differences in the standard errors of the calibration data sets, the standard errors of the test data sets were generally higher than in calibration set but comparable for all calibration strategies.

The best results for investigation of unknown samples were achieved with an artificial neural network and a not too high number of teaching steps, which limited the effect of overfitting in the calibration data set. Further potential for improvement lies in the optimal selection of the number of teaching steps and input variables. The last point also applies to the calibration with linear regression as well as the extension of the sample set. An increase or decrease of the number of terms in the calibration function by linear regression can improve/degrade the standard error, but also affects the maximum error effect. Therefore, optimum has to be determined based on the available data set. Great potential also lies in the targeted selection of measured variables. While here a number of 425 variables was used, a limitation can reduce the measurement and calibration effort as well as the risk of overfitting. For further studies, the sample set should also be extended to more than one steel batch in order to include the batch influence in the calibration.

Author Contributions: Conceptualization, methodology, formal analysis, investigation, writing—original draft preparation, and visualization, R.J.; writing—review and editing, supervision, and project administration, J.E. All authors have read and agreed to the published version of the manuscript.

Funding: This research was funded by German Research Foundation (DFG), grant number 401804277.

Acknowledgments: The scientific work has been supported by the German Research Foundation (DFG) within the research priority program (SPP) 2086 for project EP128/3-1. The authors thank the DFG for this funding and intensive technical support.

Conflicts of Interest: The authors declare no conflict of interest.

Appendix A

Table A1. Measured surface carbon content, carburization depth and hardness of the 27 heat treatment variations; bold printed results were used for the validation in Figure 15 and Table 11.

Target			Measured			
Surface Carbon Content/wt. %	CHD/mm	Tempering Temperature/°C	Surface Carbon Content/wt. %	Carburization Depth/mm	Hardness/HV1	
0.6	0.5	150	0.6	0.55	750 ±	4
0.6	0.5	180	0.6	0.55	687 ±	28
0.6	0.5	210	0.6	0.55	659 ±	6
0.6	1	150	0.61	0.91	721 ±	6
0.6	1	180	0.61	0.91	668 ±	10
0.6	1	210	0.61	0.91	662 ±	4
0.6	2	150	0.64	**1.87**	**710** ±	**5**
0.6	2	180	0.64	1.87	663 ±	7
0.6	2	210	0.64	**1.87**	**663** ±	**10**
0.7	0.5	150	0.68	0.55	750 ±	6
0.7	0.5	180	0.68	**0.55**	**680** ±	**5**
0.7	0.5	210	0.68	0.55	670 ±	3
0.7	1	150	0.72	**0.85**	**749** ±	**4**
0.7	1	180	0.72	0.85	705 ±	7
0.7	1	210	0.72	**0.85**	**671** ±	**3**
0.7	2	150	0.73	1.85	754 ±	6
0.7	2	180	0.73	**1.85**	**691** ±	**7**
0.7	2	210	0.73	1.85	660 ±	4
0.8	0.5	150	0.79	**0.57**	**730** ±	**7**
0.8	0.5	180	0.79	0.57	682 ±	5
0.8	0.5	210	0.79	**0.57**	**642** ±	**3**
0.8	1	150	0.77	0.87	740 ±	4
0.8	1	180	0.77	**0.87**	**673** ±	**2**
0.8	1	210	0.77	0.87	642 ±	7
0.8	2	150	0.81	**1.88**	**731** ±	**10**
0.8	2	180	0.81	1.88	693 ±	7
0.8	2	210	0.81	**1.88**	**655** ±	**5**

Table A2. Correlation of various measured variables with hardness and carburization depth.

Measured Variable x	Regression Function	R^2
Hardness		
V_{mag} (350 Hz)	$1861x + 2795$	0.03
A_3 (120 Hz)	$-10.6x + 707$	0.02
A_7 (100 Hz)	$74.3x + 688$	0.003
P_3 (50 Hz)	$67.5x + 633$	0.07
P_3 (150 Hz)	$21x + 670$	0.03
P_5 (30 Hz)	$-52.8x + 775$	0.04
P_7 (120 Hz)	$-5.7x + 696$	0.01
H_{co} (50 Hz)	$3.1x + 620$	0.08
M_{max} (350 Hz)	$-835x + 861$	0.2
M_r (350 Hz)	$-667x + 795$	0.25
H_{cm} (80 Hz, HP 100 kHz)	$-1529x + 729.92$	0.07
μ_r (120 Hz, 100 kHz)	$-2472.9x + 1025$	0.64
μ_r (120 Hz, 250 kHz)	$-916.2x + 867.7$	0.3
μ_r (150 Hz)	$-2801x + 1001$	0.62
$DH25_\mu$ (120Hz, 60 kHz)	$11.1x + 365$	0.24
Carburization Depth		
V_{mag} (30 Hz)	$7.7x + 5.5$	0.08
V_{mag} (150 Hz)	$-2.97x + 5.7$	0.16
V_{mag} (350 Hz)	$-1.98x + 6.8$	0.17
P_3 (30 Hz)	$-2.5x + 2.8$	0.18
P_3 (40 Hz)	$-2.4x + 3.03$	0.25
P_3 (250 Hz)	$-0.44x + 1.5$	0.05
P_7 (50 Hz)	$-1.15x + 4.3$	0.3
K (200 Hz)	$0.58x + 0.1$	0.3
K (300 Hz)	$1.029x - 0.83$	0.48
H_{co} (50 Hz)	$-0.1x + 3.36$	0.35
H_{co} (350 Hz)	$0.01x + 0.9$	0.01
H_{cm} (80 Hz)	$0.281x - 1.53$	0.57
H_{cm} (250 Hz)	$0.15x - 0.23$	0.31
$DH50_m$ (80 Hz)	$-0.09x + 4.26$	0.07
μ_{max} (30 Hz)	$15.9x + 6.6$	0.11
$DH25_\mu$ (250 Hz)	$-0.04x + 4.8$	0.03
Ph_3 (120 Hz)	$807x + 8.4$	0.01

References

1. Gorgels, C. *Schleifbarkeit von Einsatzstählen—Untersuchungen zur Schleifbarkeit Unterschiedlich Wärmebehandelter Einsatzstähle für die Zahnradfertigung—Abschlussbericht FVA 329 III.*; Forschungsvereinigung Antriebstechnik e.V.: Frankfurt, Germany, 2008.
2. Sackmann, D.; Epp, J. *Sichere Schädigungsdetektion von Randzonenschädigungen Antriebstechnischer Bauteile Infolge Einer Hartfeinbearbeitung Mithilfe von Zerstörungsfreien Mikromagnetischen Prüfverfahren—Abschlussbericht FVA 723 I.*; Forschungsvereinigung Antriebstechnik e.V.: Frankfurt, Germany, 2018.
3. Jedamski, R.; Heinzel, J.; Rößler, M.; Epp, J.; Eckebrecht, J.; Gentzen, J.; Putz, M.; Karpuschewski, B. Potential of Magnetic Barkhausen Noise analysis for In-Process Monitoring of Surface Layer Properties of steel components in Grinding. *TM Tech. Mess.* **2020**, *87*, 787–798. [CrossRef]
4. Wolter, B.; Gabi, Y.; Conrad, C. Nondestructive Testing with 3MA—An Overview of Principles and Applications. *Appl. Sci.* **2019**, *9*, 1068. [CrossRef]
5. Sorsa, A.; Santa-aho, S.; Aylott, C.; Shaw, B.A.; Vippola, M.; Leiviskä, K. Case Depth Prediction of Nitrided Samples with Barkhausen Noise Measurement. *Metals* **2019**, *9*, 325. [CrossRef]
6. Send, S.; Dapprich, D.; Thomas, J.; Suominen, L. Non-destructive Case Depth Determination by Means of Low-Frequency Barkhausen Noise Measurements. *J. Nondestruct. Eval.* **2018**, *37*, 82. [CrossRef]

7. Santa-Aho, S.; Vippola, M.; Sorsa, A.; Leiviskä, K.; Lindgren, M.; Lepistö, T. Utilization of Barkhausen noise magnetizing sweeps for case-depth detection from hardened steel. *NDT E Int.* **2012**, *52*, 95–102. [CrossRef]
8. Szielasko, K. Entwicklung Messtechnischer Module zur Mehrparametrischen Elektromagnetischen Werkstoffcharakterisierung und -Prüfung. Ph.D. Thesis, Universität des Saarlandes, Saarbrücken, Germany, 18 August 2009.
9. Epp, J.; Szielasko, K. *Weiterentwicklung Der Mikromagnetischen Multiparameter-Methode zur ZerstöRungsfreien Ermittlung von GefüGe- und Spannungsgradienten in RandschichtgehäRteten und Verfestigten ZustäNden—Schlussbericht IGF 18171 N.* 2017.
10. Liu, X.; Zhang, R.; Wu, B.; He, C. Quantitative Prediction of Surface hardness in 12 CrMoV Steel Plate on Magnetic Barkhausen Noise and Tangential Magnetic Field Measurements. *J. Nondestruct. Eval.* **2018**, *37*, 2.
11. Ahadi Akhlaghi, I.; Kahrobaee, S.; Akbarzadeh, A.; Kashefi, M.; Krause, T.W. Predicting hardness of steel specimens subjected to Jominy test using an artificial neural network and electromagnetic nondestructive technique. *Nondestruct. Test. Eval.* **2020**, *35*, 1–17. [CrossRef]
12. Liu, X.; Shang, W.; He, C.; Zhang, R.; Wu, B. Simultaneous quantitative prediction of tensile stress, surface hardness and case depth in medium carbon steel rods based on multifunctional magnetic testing techniques. *Measurement* **2018**, *128*, 455–463. [CrossRef]
13. Sorsa, A.; Santa-aho, S.; Vippola, M.; Leiviskä, K. Comparison of some data-driven modelling techniques applied to Barkhausen noise data sets. In Proceedings of the 11th International Conference on Barkhausen noise and Micromagnetic Testing, Aydin, Kusadasi, Turkey, 18–21 June 2015.
14. Gür, C.H. Microstructure Characterization of Heat-Treated Ferromagnetic Steels by Magnetic Barkhausen Noise Method. In Proceedings of the 5th World Congress on Mechanical, Chemical and Material Engineering, Lisbon, Portugal, 1 August 2019.
15. Hizli, H.; Gür, H.C. Applicability of the Magnetic Barkhausen Noise Method for Nondestructive Measurement of Residual Stresses in the Carburized and Tempered 19CrNi5H Steels. *Res. Nondestruct. Eval.* **2018**, *29*, 221–236. [CrossRef]
16. Srivastava, A.; Awale, A.; Vashista, M.; Yusufzai, M.Z.K. Monitoring of thermal damages upon grinding of hardened steel using Barkhausen noise analysis. *J. Mech. Sci. Technol.* **2020**, *34*, 2145–2151. [CrossRef]
17. Knyazeva, M.; Rozo Vasquez, J.; Gondecki, L.; Weibring, M.; Pohl, F.; Kipp, M.; Tenberge, P.; Theisen, W.; Walther, F.; Biermann, D. Micro-Magnetic and Microstructural Characterization of Wear Progress on Case-Hardened 16MnCr5 Gear Wheels. *Materials* **2018**, *11*, 2290. [CrossRef] [PubMed]
18. Srivastava, A.; Awale, A.; Vashista, M.; Yusufzai, M.Z.K. Characterization of Ground Steel Using Nondestructive Magnetic Barkhausen Noise Technique. *J. Mater. Eng. Perform* **2020**, *29*, 4617–4625. [CrossRef]
19. Sorsa, A.; Leiviskä, K.; Santa-aho, S.; Lepistö, T. Quantitative prediction of residual stress and hardness in case-hardened steel based on the Barkhausen noise measurement. *NDT E Int.* **2012**, *46*, 100–106. [CrossRef]
20. Altpeter, I.; Boller, C.; Kopp, M.; Wolter, B.; Fernath, R.; Hirninger, B.; Werner, S. Zerstorungsfreie Detektion von Schleifbrand. In Proceedings of the DGZfP-Jahrestagung, Bremen, Germany, 30 May–1 June 2011.
21. Kneller, E. *Ferromagnetismus—Mit einem Beitrag Quantentheorie und Elektronentheorie des Ferromagnetismus*; Springer: Berlin/Heidelberg, Germany, 1962.
22. Cullity, B.D.; Graham, C.D. *Introduction to Magnetic Materials*, 2nd ed.; John Wiley & Sons: Hoboken, NJ, USA, 2011.
23. Karpuschewski, B.; Bleicher, O.; Beutner, M. Surface integrity inspection on gears using Barkhausen noise inspection. *Procedia Eng.* **2011**, *19*, 162–171. [CrossRef]
24. Szielasko, K.; Kopp, M.; Tschuncky, K.; Lugin, S.; Altpeter, I. Barkhausenrausch- und Wirbelstrommikroskopie zur ortsaufgelösten Charakterisierung von dünnen Schichten. In Proceedings of the DGZfP-Jahrestagung 2004, Salzburg, Austria, 17–19 May 2004.
25. Altpeter, I.; Tschuncky, R.; Szielasko, K. Electromagnetic techniques for materials characterization. In *Materials Characterization Using Nondestructive Evaluation (NDE) Methods*; Woodhead Publishing: Sawston, UK, 2016.
26. Jiles, D.C. Dynamics of Domain Magnetization and the Barkhausen Effect. *Chechoslov. J. Phys.* **2000**, *50*, 893–988. [CrossRef]

27. Stupakov, A.; Perevertov, A.; Neslušan, M. Reading depth of the magnetic Barkhausen noise. II. Two-phase surface-treated steels. *J. Magn. Magn. Mater.* **2020**, *513*, 167239. [CrossRef]
28. Fahrmeir, L.; Heumann, C.; Künstler, R.; Pigeot, I.; Tutz, G. *Statistik—Der Weg zur Datenanalyse*; Springer: Berlin/Heidelberg, Germany, 2016.
29. Nalbant, M.; Gokkaya, H.; Toktas, I. Comparison of Regression and Artificial Neural Network Models for Surface Roughness Prediction with the Cutting Parameters in CNC Turning. *Model. Simul. Eng* **2007**, *2007*, 92717. [CrossRef]
30. Specht, D.F. A General Regression Neural Network. *IEEE T. Neural Networ.* **1991**, *2*, 568–576. [CrossRef]
31. Hecht-Nielsen, R. Theory of Backpropagation Neural Network. *Neural Netw.* **1988**, *1*, 593–605. [CrossRef]
32. Palau, A.; Velo, E.; Puigjaner, L. Use of neural networks and expert systems to control a gas/solid sorption. *Int. J. Refrig.* **1999**, *22*, 59–66. [CrossRef]
33. Membrain Neuronale Netze Editor und Simulator. Available online: https://membrain-nn.de/index.htm (accessed on 16 October 2020).
34. Popko, A.; Jakubowski, M.; Wawer, R. Membrain Neural Network for Visual Pattern Recognition. *Sci. Adv.* **2013**, *7*, 54–59. [CrossRef]
35. Woernle, I.A. Anwendbarkeit Künstlicher Neuronaler Netze zur Untergrundbewertung in der Oberflächennahen Geothermie. Ph.D. Thesis, Universität Fridericiana zu Karlsruhe, Karlsruhe, Germany, 2008.

© 2020 by the authors. Licensee MDPI, Basel, Switzerland. This article is an open access article distributed under the terms and conditions of the Creative Commons Attribution (CC BY) license (http://creativecommons.org/licenses/by/4.0/).

Article

Minimal Invasive Diagnostic Capabilities and Effectiveness of CFRP-Patches Repairs in Long-Term Operated Metals

Grzegorz Lesiuk [1],*, Bruno A. S. Pedrosa [2], Anna Zięty [1], Wojciech Błażejewski [1], Jose A. F. O. Correia [3], Abilio M. P. De Jesus [3] and Cristiano Fragassa [4]

[1] Department of Mechanics, Materials and Biomedical Engineering, Faculty of Mechanical Engineering, Wroclaw University of Science and Technology, Smoluchowskiego 25, PL-50-370 Wrocław, Poland; anna.ziety@pwr.edu.pl (A.Z.); wojciech.blazejewski@pwr.edu.pl (W.B.)
[2] ISISE, Departamento de Engenharia Civil, Universidade de Coimbra, Rua Luís Reis Santos, Pólo II, 3030-788 Coimbra, Portugal; bruno.pedrosa@uc.pt
[3] Faculty of Engineering, University of Porto, Rua Dr Roberto Frias, 4200-465 Porto, Portugal; jacorreia@inegi.up.pt (J.A.F.O.C.); ajesus@fe.up.pt (A.M.P.D.J.)
[4] Department of Industrial Engineering, University of Bologna, Viale Risorgimento 6, 40133 Bologna, Italy; cristiano.fragassa@unibo.it
* Correspondence: Grzegorz.Lesiuk@pwr.edu.pl; Tel.: +48-71-320-39-19

Received: 12 May 2020; Accepted: 17 July 2020; Published: 21 July 2020

Abstract: The paper deals with the subject of diagnostics and the quick repairs of long-term operated metallic materials. Special attention was paid to historical materials, where the structure (e.g., puddle iron) is different from modern structural steels. In such materials, the processes of microstructural degradation occur as a result of several decades of exposure, which could overpass 100 years. In some cases, their intensity can be potentially catastrophic. For this reason, the search for minimally invasive diagnostic methods is ongoing. In this paper, corrosion and fracture toughness tests were conducted, and the results of these studies were presented for two material states: post-operated and normalized (as a state "restoring" virgin state). Moreover, through the use of modern numerical methods, composite crack-resistant patches have been designed to reduce the stress intensity factors under cyclic loads. As a result, fatigue lifetime was extended (propagation phase) by more than 300%.

Keywords: extended finite element method (xFEM); polarization curve; long-term operated metals; hybrid materials; fatigue crack growth; stress intensity factors (SIF)

1. Introduction

During the long-term operation of metallic materials and structures, the problem of their degradation becomes an issue, which could be serious when referring to bridge structures erected at the turn of the 19th to 20th centuries. This is the case of puddle iron, typically used in 19th-century metallic bridge structures, which is more susceptible to structural degradation processes than the old mild steel from the early 20th century. It is worth noting that in structural engineering at the beginning of the 20th century, both types of metallic materials were largely used. Nowadays, the maintenance and diagnostic of these old bridge structures, erected using puddle iron or old mild steel, is still a vital topic [1,2].

The puddle iron that was used as a constructional material is strongly influenced by local material flaws, which significantly limited the repairing techniques excluding (for instance) welding. Typically, the yield strength of puddle iron is about 250 ÷ 310 MPa, showing at a total elongation of 7 ÷ 25% [1–4]. Puddle iron has been characterized by variations in elongation, yield, and ultimate tensile strengths for the rolling direction as well as in the perpendicular direction. Therefore, puddle iron is often characterized by significant anisotropy of mechanical properties. In the rolling direction, they are significantly higher than in the direction perpendicular to it. This effect is a result of the former metallurgy process (puddling). However, it is more likely that this effect is strongly assisted by the degradation processes taking place in old steels [3–9] consisting, among others, of the decomposition of pearlite to ferrite and carbide, nitride, and carbide separations inside grains and at grain borders.

In order to illustrate the structure types and features indicating the degradation progress, all mentioned symptoms are documented in Figure 1, where the microstructures of steel from the Pomorski Środkowy and Północny Bridges located in Wrocław, Poland (Środkowy bridge erected in 1861–puddle iron, Północny bridge erected in 1930–mild steel) are shown. Figure 1a–d shows the typical microstructure of 19th-century puddle iron with numerous slags and nonmetallic inclusions (A) with brittle precipitations inside the ferrite grains (B) and a thick envelope of Fe_3C_{III} on the grain boundaries (C). The enlarged ferrite grains with degradation symptoms, brittle precipitations inside ferrite grains, are noticeable in Figure 1d. Due to the lack of a contemporary equivalent of the material from the 19th century, the normalization process is used as a heat treatment procedure simulating the original state of the material. However, if the degradation processes are very advanced, normalization does not remove all of the degradation products. This is a clear sign that the old material has been disposed of and this example is shown in Figure 1e.

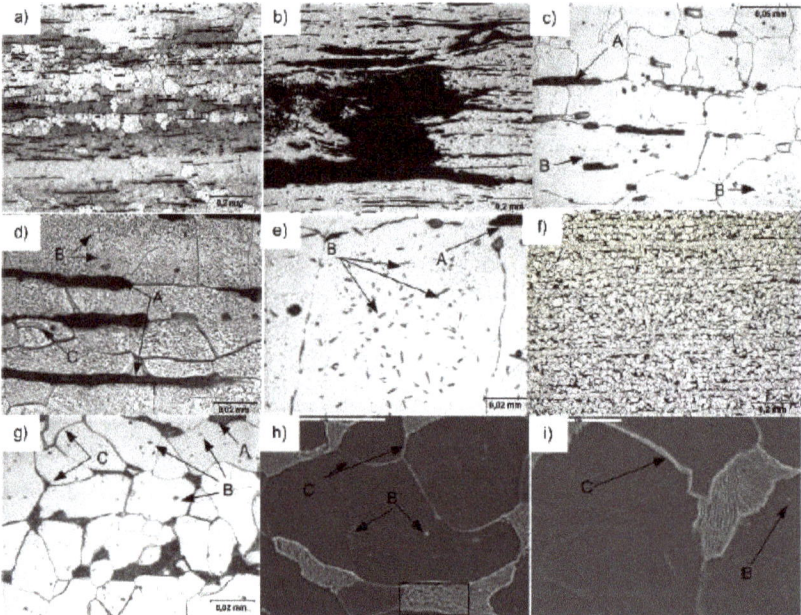

Figure 1. A typical microstructure of the long term-operated late 19th and early 20th century steels (described in the text). (**a**) typical microstructure of puddle iron, (**b**) enlarged non-metallic inclusion in puddle iron, (**c**) magnified microstructure of puddle iron, (**d**) magnified microstructure of puddle iron with microstructural degradation symptoms, (**e**) enlarged ferrite grain with degradation symptoms–puddle iron, (**f**) typical microstructure of early 20th century mild steel, (**g**) enlarged ferrite-pearlite microstructure of early 20th century mild steel, (**h**) SEM (Scanning Electron Microscope) image of early 20th century mild steel misstructure with small precipitations inside ferrite grains and degraded pearlites areas, (**i**) SEM (Scanning Electron Microscope) image of early 20th century mild steel misstructure without degradation symptoms.

The structural steels from the early 20th century (shown in Figure 1f–h) have a typical ferritic-pearlitic (A–pearlite areas) microstructure that is close to that of modern mild steels with small amounts of the degradation processes manifested in (B) brittle precipitations inside ferrite grains and degenerated pearlite areas (marked by frame). However, in the case of the presented early 20th-century mild steel, after normalization, the degradation symptoms disappeared (see Figure 1i). This type of structure leads to strain localization and makes the problem of fracture site prediction in the material more complex [6].

The presented microstructural features in old metallic materials also lead to significant changes in the fracture mechanisms visible in fractograms. Figure 2 illustrates the fracture surfaces of the specimens after Charpy testing for puddle iron and early 20th-century mild steel (the same materials as used for metallographic analysis from Figure 1).

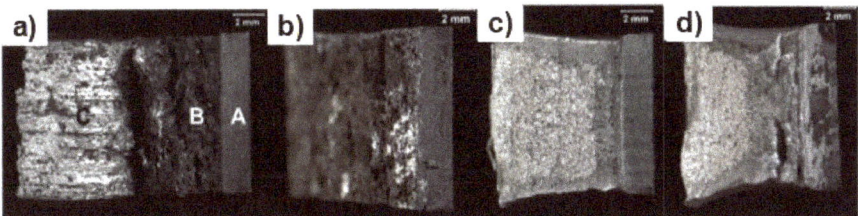

Figure 2. Fracture surfaces of the specimens after the Charpy impact test (+20 °C): (**a**) puddle iron in post-operated state (A–notch zone, B–crack propagation area, C–the plastic zone from the bottom part of the Charpy specimen), (**b**) puddle iron after heat treatment (normalized), (**c**) early 20th-century mild steel, post-operated state, (**d**) early 20th century mild steel, normalized state.

As shown in Figure 2, the differences in ductility were noticeable for both long-term operated materials. As reported in several papers devoted to the degradation of long-term operated materials, usually during maintenance service, two crucial issues have been raised [3–9]:

- How can we detect damage caused by material degradation using non-destructive or minimally invasive inspection at the operational level?
- How long can structures, mostly old steel structures, be operated under fatigue loads, and how can their service life be effectively extended temporarily until further repairs and renovations?

This article presents selected aspects of the strategy addressing the questions above in relation to historic long-term operated bridge steels

2. Materials and Methods

Scientific investigations were conducted on 19th-century metallic materials extracted from the structural elements of a bridge built in the 19th-century in Bayonne, France (marked as "A" and subjected to fatigue crack growth tests. In 2013, it was demolished and replaced by a new bridge. The original one was mainly composed of puddle iron components through five spans (47 m + 3 × 60 m + 47 m) and a total length of 274 m. The main beams were formed with trusses in a Saint Andrew's cross shape that were 5.5 m high. After the disassembly of the bridge, several structural elements were preserved in order to perform scientific investigations. For corrosion and fracture toughness tests, we selected puddle iron (marked as "B") gained from the renovated 19th century Main Railway Station in Wrocław, Poland, as both materials had a similar chemical composition (reported in Table 1).

One group of materials was selected to test the degradation of the material and its potential impact on the susceptibility to brittle cracking. For this purpose, the crack resistance expressed as a critical integral *J* value was tested according to ASTM E1820 [10]. The main aim of the conducted works was to check if it would be possible to detect the material degradation phenomenon on the basis of corrosion tests based on electrochemical parameters. In this group of samples, one part of the material was tested in the post-operative state and the other after heat treatment (normalization) to simulate the initial state. The relevance of such analysis has been widely confirmed by the authors' works concerning destructive mechanical tests [3–5,7,9], following the assumption of degradation theory [3,11,12]. For corrosion and fracture toughness tests, samples from the hall of the Wrocław Main Railway Station were used, resembling the materials published in the work (taken from other beams) [4].

Table 1. Chemical composition (% by weight) of the tested material compared with typical puddle irons and old mild steels.

Material	C	Mn	Si	P	S
puddle iron for fatigue crack growth rate (bridge from Bayonne, France) marked as "A"	<0.01	<0.02	0.28	0.41	0.054
Puddle iron from Main Railway Station (Wrocław, Poland) marked as "B"	0.02	0.03	0.13	0.29	0.048
typical values for puddle irons	<0.8	0.4	n/a	<0.6	<0.04
typical values for old mild steels	<0.15	0.2 ÷ 0.5	Variable	<0.06	<0.15

The second important contribution of this work was to develop a strategy for strengthening cyclically loaded structural members: crack repair with the use of composite materials. In this part of the study, the influence of the microstructural degradation phenomenon was not explicitly accounted. The problem of the degradation mechanism impact on fatigue lifetime reduction has been discussed in previous works by the authors [3–5]. However, it should be assumed that the reinforcement will work equally effectively in the post-operational state of the structure as well as after normalization, and the key elements of the research are the numerical analyses that allow for the correct design of the process to retard the development of fatigue-fracture by reducing the crack-like defect stress intensity factors (SIF). After the numerical part, fatigue crack growth laboratory tests were carried out on "hybrid" modified CT (Compact Tension)–metal–puddle iron (with comparable properties to the metal used in corrosion tests from the object tested in the work [13]) with adhesive and carbon fiber reinforced polymer (CFRP).

2.1. Corrosion Tests and Electrochemical Indicators of the Degradation Processes Activity

In order to perform corrosion tests, samples were carefully extracted using electro discharging machining (EDM) from parts of the steel structure. Special attention was paid to the extraction procedure in order not to lead to temperature changes. Before the main electrochemical measurement, each cylindrical sample (Figure 3) was degreased in acetone in an ultrasound bath for five minutes.

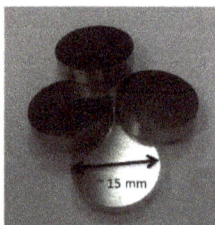

Figure 3. The geometry of the samples for the corrosion test.

Afterward, they were exposed for approximately 60 min in a corrosive medium of 3.5% NaCl solution. During this time, samples reached stability in the new environment. Samples prepared in this way were tested in a three-electrode polarization measuring system. The fully automated test stand (Figure 4) consisted of a measuring vessel, the ATLAS 0531–Electrochemical UNIT & IMPEDANCE ANALYSER (ATLAS 0531 ELEKTROCHEMICAL UNIT & IMPEDANCE ANALYSER, Atlas-Sollich, Gdansk, Poland) potentiostat, and the computer drivers. The counter electrode was made of austenitic stainless steel, while the reference electrode was used as the electrode, a saturated Ag/AgCl. During the open circuit, potential measuring polarization curves were registered.

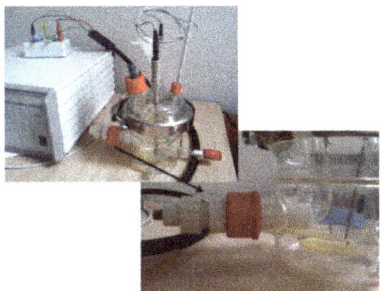

Figure 4. The electrochemical test stand used in the experimental campaign.

Moreover, the test was performed in the same polarity in the direction of the anode solution at a rate of $dE/dt = 1$ mV/s. The initial value of the potential was determined based on a stabilized value of the open circuit potential (E_0). All relevant parameters of corrosion current density (i_{corr}), corrosion potential (E_{corr}), Tafel coefficients ba and bc, and polarization resistance (R_p) were determined by the Stern method.

2.2. Fracture Toughness Tests

The tests were carried out using a MTS809 test machine, in accordance with ASTM E1820 [10], on the SEN (B) type three-point bend specimens with thickness $B = 10$ mm and width $W = 20$ mm. Mechanical notches were cut to a length of $a = 10.7$ mm using the EDM. Before the proper test, the fatigue pre-cracking procedure was involved according to ASTM E1820 [10]. For crack length calculation, an elastic-unloading compliance procedure was involved. Based on the registered signals of force, deflection, crack mouth opening displacement (CMOD), the J-R curves were evaluated, allowing the determination of the fracture toughness parameters expressed as critical values of the J integral.

2.3. Concept of the CFRP (Carbon Fiber Reinforced Polymer) Patches Strengthening and Numerical Analysis

The fatigue crack growth process is generally related to the severity of the stress distribution around the crack, which can be determined by the stress intensity factor. It is expected that using CFRP patches bonded to the material can contribute to decreasing the severity of the stresses around the crack tip, leading to an extension in fatigue life. Recent investigations have been performed in order to estimate the contribution of implementing the CFRP patch in CT specimens through numerical analysis. In recent considerations [14], different configurations of CFRP patches were studied (full-face and two strips) and numerical stress intensity factors were evaluated using X-FEM (eXtended Finite Element Method). However, in these studies, the adhesive was not modeled, and the interface between the CFRP and CT specimen was considered as perfectly tied, which means that there was no relative displacement in the interface. The crack in the CT specimens was imposed by node separation (X-FEM approach), and SIFs computed using the modified virtual crack closure technique (VCCT) for several imposed crack increments. In the present paper, numerical simulations were conducted to assess the values of the stress intensity factor for both strengthened and non-strengthened scenarios, as presented in Figure 5. The models were composed of a compact tension specimen with geometry defined by W (width) equal to 50 mm and B (thickness) equal to 8.5 mm, a patch of CFRP, and an adhesive to establish the connection.

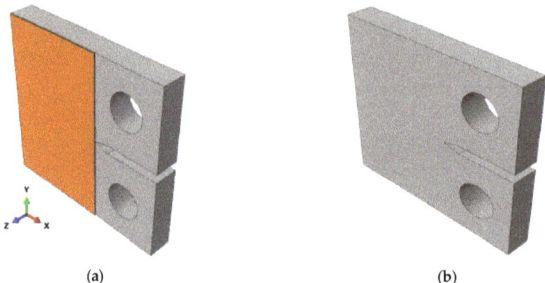

(a)   (b)

Figure 5. Numerical models: (a) Strengthened compact tension (CT) specimen; (b) Plain steel CT specimen.

The metallic component was modeled with 3D finite elements with eight nodes, C3D8; the CFRP component was created using 3D shell elements with four nodes, S4; and the adhesive was modeled using 3D cohesive finite elements with eight nodes, COH3D8. Boundary conditions were set as pinned in the upper pin, $U_x = U_y = U_z = 0$, and as $U_x = U_z = 0$ for the lower pin. Loading was applied in the lower pin with a value equal to −7500 N in the y-direction. Boundary conditions and loads (Figure 6) were applied in reference nodes that are linked to the holes' surfaces by rigid links (kinematic coupling).

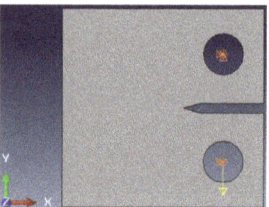

Figure 6. Boundary conditions and loads in the models.

Since the main objective of this study was to evaluate the stress intensity factor, an elastic fracture mechanics parameter and all the materials were set to linear elastic. In this sense, the metallic component was modeled with $E = 185$ GPa (typical average value within puddle iron [15]) and $\nu = 0.3$. The mechanical properties of the CFRP component were determined using the parameters for Sika® CarboDur® E-1014 (Sika Poland Sp. z o.o., Warszawa, Poland), as presented in Table 2. Direction 1 is the direction of the fibers and is aligned with direction y in Figures 5 and 6. The Young modulus in direction 2 (correlated with x in Figures 5 and 6) was considered as 10% of the main direction. The values of the shear moduli were defined using the data published by Naboulsi and Mall [16].

Table 2. Mechanical properties of CFRP (Carbon Fiber Reinforced Polymer) [17].

E_1 [GPa]	E_2 [GPa]	ν_{12}	G_{12} [GPa]	G_{13} [GPa]	G_{23} [GPa]
170	17	0.17	7.24	7.24	4.94

For the adhesive component, the material law was defined with a cohesive mixed-mode damage model, as presented in Figure 7. It is characterized by a linear relation between stresses, t, and relative displacements, δ. The maximum value of stresses $t_{u,i}$ ($i =$ I, II) defines the damage initiation and the energy release rate, J_{ic}, is the parameter that gives the complete failure. In the case of mixed-mode behavior, damage initiation is based on the following quadratic stress criterion:

$$\left(\frac{t_I}{t_{u,I}}\right)^2 + \left(\frac{t_{II}}{t_{u,II}}\right)^2 = 1 \quad (1)$$

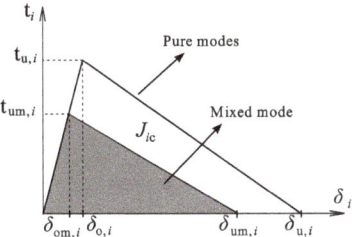

Figure 7. Bilinear cohesive zone model for pure modes (I and II) and mixed-mode (I + II).

For the crack propagation, the following linear energetic criterion was used:

$$\frac{J_I}{J_{IC}} + \frac{J_{II}}{J_{IIC}} = 1 \quad (2)$$

The parameters of the model that was adopted in the analysis are described in Table 3.

Table 3. Mechanical properties of adhesive [18].

E [GPa]	ν [-]	$t_{u,I}$ [MPa]	$t_{u,II}$ [MPa]	J_{IC} [N/mm]	J_{IIC} [N/mm]
11.5	0.3	30	18	0.35	1.1

A total of four analyses were conducted in order to compute the stress intensity factor for different values of crack length: 15.5 mm, 18.5 mm, 21.5 mm, and 24.5 mm. The evaluation of this parameter was done using the modified virtual crack closure technique (VCCT) [19].

Recent studies have shown that as fatigue crack grows on a subtract, a debonded area is created between the subtract and reinforcement [20]. The definition of this region is presented in Figure 8. The parameter b is obtained by Equation (3), where a refers to the total crack length, u_0 is the distance between the loading axis and the CFRP patch (12.5 mm), and r_p is the plastic zone ahead of the crack tip, which can be calculated from Equation (4).

$$b = a - a_0 + r_p \quad (3)$$

$$r_p = \frac{1}{\pi}\left(\frac{K_{max}}{f_y}\right)^2 \quad (4)$$

The application of the modified VCCT allows for the computation of stress intensity factors in all points through-thickness of the crack tip. The validation of the method and the influence of the mesh were assessed by comparing the values of the stress intensity factors obtained for the non-strengthened model with the analytical solution presented in the ASTM E 647 standard for fatigue crack growth tests [21]:

$$\Delta K = \frac{\Delta P}{B\sqrt{W}} \frac{(2+\alpha)}{(1-\alpha)^{1.5}}[0.886 + 4.64\alpha - 13.32\alpha^2 + 14.72\alpha^3 - 5.6\alpha^4] \quad (5)$$

where α is the ratio a/W with a as the corresponding crack length measured during the test; ΔP is the applied force range; B is the thickness of the specimen; and W is the width.

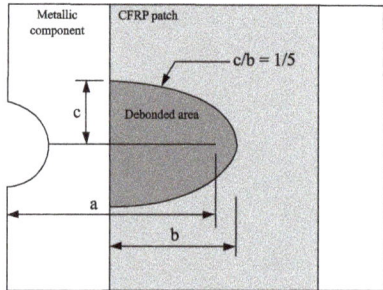

Figure 8. Definition of the debonded area.

The values of the stress intensity factor in the nodes through the thickness of the CT specimen (all points are shown in Figure 9) can be observed in Figure 10. In these graphics, node 1 is the node in the surface of the CT specimen, which is not reinforced with the CFRP patch, and node 29 is the opposite node situated in the face where the CFRP patch is bonded (see Figure 9). It is observed that the application of the CFRP patch produces a reduction on the stress intensity factor in all nodes of the crack tip.

Figure 9. Mesh of the bare steel CT specimen: identification of node number through crack thickness.

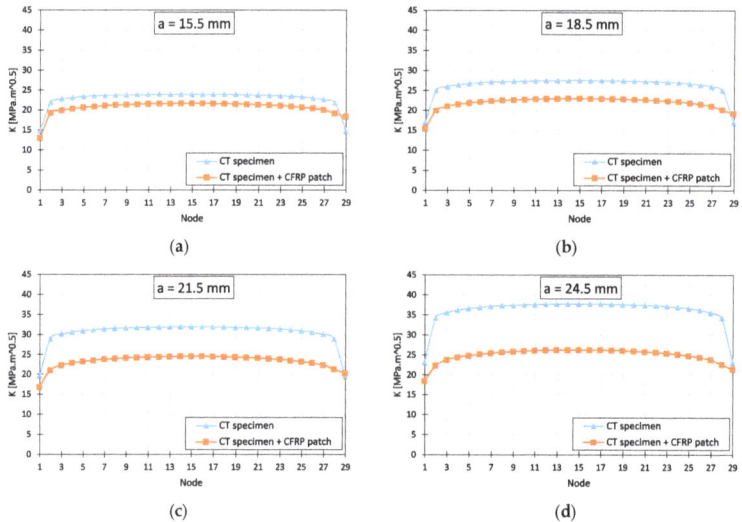

Figure 10. Numerical stress intensity factors for the strengthened and non-strengthened CT specimens, (**a**) variations of stress intensity factors (SIF) for crack length 15.5 mm, (**b**) variations of stress intensity factors (SIF) for crack length 18.5 mm, (**c**) variations of stress intensity factors (SIF) for crack length 21.5 mm, (**d**) variations of stress intensity factors (SIF) for crack length 24.5 mm.

Another important aspect that can be concluded from these graphics is that the difference between the stress intensity factors of strengthened and non-strengthened solutions increases for higher values of crack length. In fact, as presented in Table 4, the reduction on the stress intensity factor increased from 9% for a crack length equal to 15.5 mm to 32.3% for a crack length equal to 24.5 mm. In fact, the reduction of stress intensity factor is marginal for short cracks because the stiffness of the steel plate is much higher than that of the CFRP strips. However, the effectiveness of the patch reinforcement is maximized for long cracks because the patch covers a larger amount of the crack and its plastic zone. On average, the use of the CFRP patch led to a reduction in the stress intensity factor by 21%. Figure 11 and Table 4 present the evolution of the stress intensity factor with the crack length. It is shown that the numerical computation of stress intensity factors is in line with the analytical values and the difference varied between 0.9 and 2.7%.

Table 4. Comparison of final stress intensity factors.

Crack Length	CT Specimen			CT Specimen + CFRP Patch	
	K [MPa*m$^{0.5}$]		Dif. (%)	K [MPa*m$^{0.5}$]	Dif. (%)
a [mm]	Analytical	Numerical VCCT		Numerical VCCT	
15.5	22.8	23.0	0.9%	20.7	9.0%
18.5	26.6	26.3	0.9%	22.0	17.1%
21.5	31.1	30.5	2.1%	23.4	24.8%
24.5	37.0	36.0	2.7%	25.0	32.3%

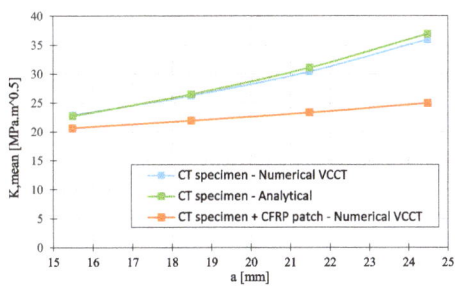

Figure 11. Comparison of final stress intensity factors.

3. Experimental Results and Discussion

3.1. Corrosion Resistance

Corrosion tests were performed for materials gained from the Main Railway Station (Wrocław, Poland, mechanical tests reported in [4]). Specimens marked as B1P mean that it is a material (puddle iron) from the Main Railway Station (Wrocław, Poland) in the post operated state, B1N are materials (puddle iron) from the Main Railway Station (Wrocław, Poland) after normalization. All electrochemical parameters for such material are shown in Figure 12. Analyzing the obtained values of E_0 and E_{corr} for the first group, their increase was observed. As presented in Figure 12, the parameters of samples after the normalizing process are increasing from about 10 mV for E_0 to almost 20 mV for values of E_{corr}. Moreover, Rp value also indicates a significant increase. According to the literature rule, when values of a potential and a resistant rise, the value of current should decrease.

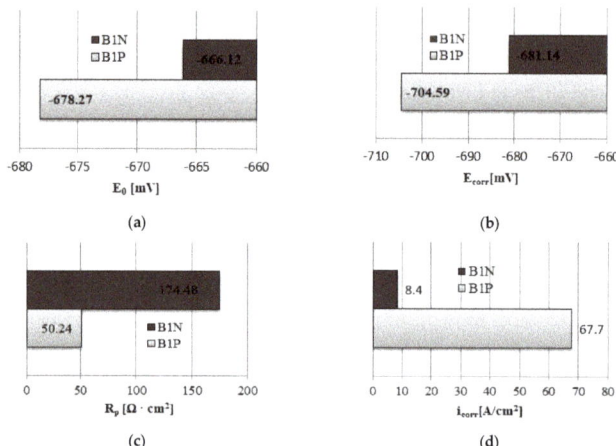

Figure 12. Results of the main parameters of electrochemical tests: (a) open circuit potential, (b) corrosion potential, (c) polarization resistance, (d) corrosion current density.

A higher corrosive resistance for samples with normalized conditions was also observed on the obtained polarization curves (Figure 13).

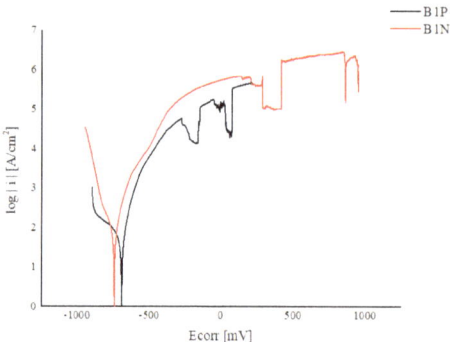

Figure 13. Polarization curves with the results of electrochemical tests for the B1P and B1N samples, where B1P is a post-operated state and B1N is a specimen after normalization.

3.2. Fracture Toughness Analysis: J Integral

According to the ASTM E1820 [10] recommendations, the J-R curves were drawn in Figure 14 for material in a post operated state and after heat treatment-normalization. The values of engineering J-integral $J_{0.2}$ were also determined. In Figure 14, there are marked horizontal lines corresponding with the J integral for the crack onset of 0.2 mm. However, because the thickness of 19th-century structural elements does not usually exceed 18 mm (in our case 10 mm), the plane strain condition cannot be achieved. In the presented results, the critical J_C values can be treated only (!) as a substitute (conditional) material resistant to cracking. The analysis of the experimental data showed that in the normalized state, the integral value of J_c was estimated on level J_c = 104 N/mm, and in the post-operated state at J_c = 87 N/mm. Comparing it with modern mild, low carbon steels, these values were far below the critical J integral, for example, S235/355 grade steel is J_{IC} = 320–360 N/mm. By also analyzing the archival data [22–25] for a wide group of degraded materials (including pipeline steels), it is very likely that the applied corrosion testing method should find a non-invasive application in the detection of degradation damage. This shift in polarity curves and the change in the value of electrochemical parameters of steel including 19th-century puddle iron could be a good indicator in the assessment of material exploitation level.

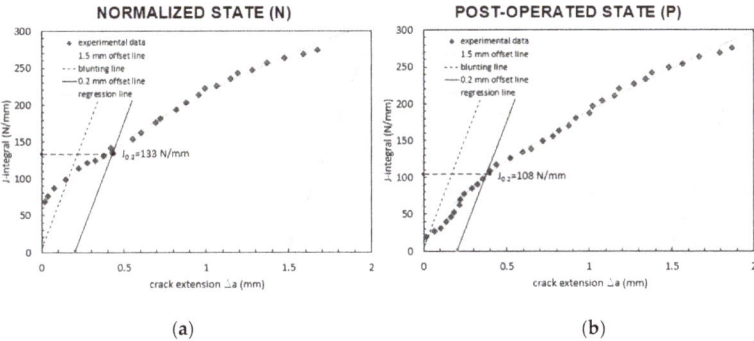

Figure 14. J-R curves for puddle iron after normalization heat treatment (a) and in the post-operated state (b).

Figure 14 clearly demonstrates the higher fracture resistance for the normalized state than for the post-operated state. After normalization, the critical J_c parameter increased by almost 20% and $J_{0.2}$ was −23% compared to the post-operated state.

3.3. Fatigue Crack Growth Test Results

The experimental campaign followed a numerical analysis of SIF and the influence of the CFRP patch on the SIF reduction. The geometry of CT specimens was defined as in the previous part, by W (width) = 50 mm and B (thickness) = 8.5 mm, as presented in Figure 15. The notch was created using electrical discharge machining. A preliminary FCG (Fatigue Crack Growth Rate) test produced the pre-crack under ΔK control with a maximum value of 15 MPa*m$^{0.5}$. To evaluate the influence of the mean stress effect, for non-strengthened specimens, different values of the stress ratio were used: 0.05, 0.1, and 0.7. During the tests of the fully metallic specimens, the crack length was monitored using the compliance method prescribed in the standard. The crack opening displacement and the applied force were also recorded in order to assess the effective stress intensity factor. Strengthened CT specimens were tested with the geometry and components described in Figure 15. The CFRP patch was glued in one face of the specimen with the fibers oriented perpendicular to the crack growth direction (parallel to the load application line). These tests were conducted under constant amplitude loading with a maximum value of 7500 N and a frequency of 10 Hz.

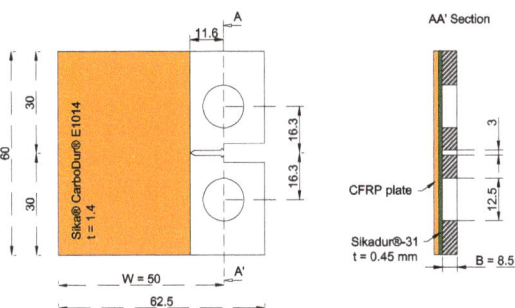

Figure 15. Geometry of CT specimens strengthened with the CFRP patch (all dimensions in mm).

The elaboration of FCG tests enabled us to compute the material parameters related to the crack growth resistance. In Figure 16, it is possible to observe the data obtained from FCG tests on non-strengthened specimens. It can be stated that the stress ratio influences the growth behavior since the data obtained for the stress ratio equal to 0.7 were significantly distinct from the data obtained for stress ratio equal to 0.05 and 0.1. While in the first case the constants C and m were 5.63×10^{-45} and 3.61, for the second case, the constants assumed the values 2.28×10^{-22} and 6.07. The following graphic represents the evolution of the crack growth curves using applied and effective stress intensity factors. It was observed that all the tests performed converged for one line when the effective stress intensity factors were used, computed using the compliance information.

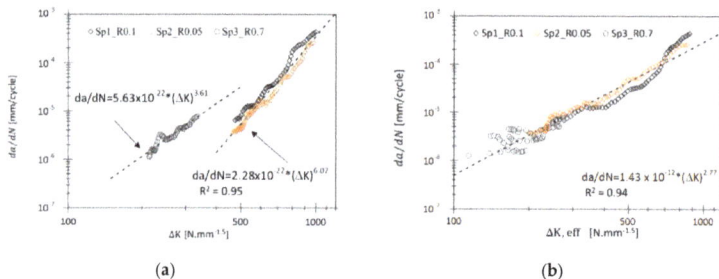

Figure 16. Fatigue crack growth data for the CT specimens: (a) use of applied stress intensity factors; (b) use of effective stress intensity factors.

In the case of the CT specimens strengthened with CFRP patches, the crack length was monitored using a microscope in the face with no patch. Figure 17 presents the evolution of the crack growth throughout the test performed under $R = 0.1$ and $F_{max} = 7500$ N. Comparing the data obtained from the FCG tests on strengthened specimens with the a-N curve defined by the material parameters determined with FCG tests on non-strengthened CT specimens, it is possible to state that the implementation of a CFRP patch led to a fatigue life extension of 318%.

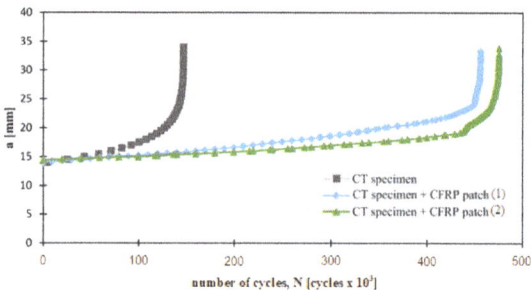

Figure 17. Crack growth evolution for strengthened and non-strengthened solutions ($R = 0.1$ and $F_{max} = 7500$ N).

4. Conclusions

The paper presents the diagnostic role of corrosion testing in the detection of degradation phenomena in the long-term operated metals. It also focuses on the capabilities of repairing metallic structures through the CFRP application. The following conclusions and observations can be made based on the performed investigations:

- shifting of polarization curves and change of electrochemical parameters are a good indicator to evaluate the operating condition of metals in the presented case puddle iron (19th century);

- in the case of puddle iron (and early mild steels), comparative analysis in the post-operated and normalized state—in the absence of material in the virgin state—is a reasonable solution to confirm the presence of microstructural degradation phenomena;
- application of modern adhesive solutions and CFRP patches allow effectively decelerating the development of fatigue cracking, which was confirmed numerically and experimentally. A prerequisite is to design the hybrid joint (metal + composite) correctly using numerical methods;
- further work will aim to establish the relationship between the fracture mechanism and diagnostic methods of structural elements subjected to multi-axial stress; and
- the future linking of changes in the electrochemical parameters of steel to the progress of degradation is also a further condition for the development of a diagnostic system for long term-operated materials, which requires even more experimental work. However, this study has shown the possible direction of further research.

Author Contributions: Conceptualization, G.L. and B.A.S.P.; Methodology, W.B. and J.A.F.O.C.; Validation, G.L. and C.F.; Formal analysis, A.M.P.D.J.; Investigation, G.L. and A.Z.; Resources, C.F.; Data curation, G.L. and A.Z.; Writing—original draft preparation, G.L. and B.P.; Writing—review and editing, G.L., J.A.F.O.C., and C.F.; Visualization, A.Z. and B.P.; Supervision, G.L. and C.F.; Project administration, G.L.; Funding acquisition, G.L. All authors have read and agreed to the published version of the manuscript.

Funding: The project was co-financed by the Polish National Agency for Academic Exchange grant number PPN/BUA/2019/1/00086.

Conflicts of Interest: The authors declare no conflict of interest.

References

1. Hołowaty, J. Toughness tests on steels from old railway bridges. *Procedia Struct. Integr.* **2017**, *5*, 1043–1050. [CrossRef]
2. Sieber, L.; Stroetmann, R. The brittle fracture behaviour of old mild steels. *Procedia Struct. Integr.* **2017**, *5*, 1019–1026. [CrossRef]
3. Lesiuk, G.; Correia, J.A.F.O.; Krechkovska, H.V.; Pekalski, G.; Jesus, A.M.P.; de Student, O. Degradation Theory of Long Term Operated Materials. In *Structural Integrity Series*; Springer Nature: Heidelberg, Germany, 2021; Volume 15, in press.
4. Lesiuk, G.; Szata, M.; Bocian, M. The mechanical properties and the microstructural degradation effect in an old low carbon steels after 100-years operating time. *Arch. Civ. Mech. Eng.* **2015**, *15*, 786–797. [CrossRef]
5. Lesiuk, G.; Szata, M. Aspects of structural degradation in steels of old bridges by means of fatigue crack propagation. *Mater. Sci.* **2011**, *47*, 82. [CrossRef]
6. Balokhonov, R.; Romanova, V. On the problem of strain localisation and fracture site prediction in materials with irregular geometry of interfaces. *Facta Univ. Ser. Mech. Eng.* **2019**, *17*, 169–180.
7. Krechkovska, H.; Student, O.; Lesiuk, G.; Correia, J. Features of the microstructural and mechanical degradation of long term operated mild steel. *Int. J. Struct. Integr.* **2018**, *9*, 296–306. [CrossRef]
8. Zvirko, O.I.; Mytsyk, A.B.; Tsyrulnyk, O.T.; Gabetta, G.; Nykyforchyn, H.M. Corrosion degradation of steel of long-Term operated gas pipeline elbow with large-scale delamination. *Mater. Sci.* **2017**, *52*, 861–865. [CrossRef]
9. Szata, M.; Lesiuk, G.; Pękalski, G. Assessment of degrading processes progress in the old brigde steel in terms of fracture mechanics-Part one-The investigation of possibilities. *Logistyka* **2009**. bwmeta1.element.baztech-article-BUS6-0035-0025. Available online: http://yadda.icm.edu.pl/yadda/element/bwmeta1.element.baztech-article-BUS6-0035-0025 (accessed on 12 May 2020).
10. ASTM E1820; *Standard Test Method for Measurement of Fracture Toughness*; American Society for Testing and Materials: West Conshohocken, PA, USA.
11. Dudziński, W.; Konat, Ł.; Pękalski, G. Modern constructional steels. In *Maintenance Strategy of Surface Mining Machines and Facilities with High Technical Degradation Levels*; Wrocław, D.D., Ed.; Publishing House of Wrocław University of Technology: Wrocław, Poland, 2013; pp. 346–366.
12. Konat, Ł.; Pękalski, G. Overview of Materials Testing of Brown-Coal Mining Machines (Years 1985–2017). In *Mining Machines and Earth-Moving Equipment*; Springer: Heidelberg, Germany, 2020; pp. 21–58.

13. Lesiuk, G.; Sire, S.; Ragueneau, M.; Correia, J.A.F.O.; Pedrosa, B.A.S.; Jesus, A.M.P. Mean Stress Effect and Fatigue Crack Closure in Material from Old Bridge Erected in the Late 19th Century. *Procedia Struct. Integr.* **2019**, *17*, 198–205. [CrossRef]
14. Lesiuk, G.; Katkowski, M.; Correia, J.; De Jesus, A.; Blazejewski, W. Fatigue Crack Growth Rate in CFRP Reinforced Constructional Old Steel. *Int. J. Struct. Integr.* **2018**, *9*, 381–395. [CrossRef]
15. Lesiuk, G.; Kucharski, P.; Correia, J.; De Jesus, A.; Rebelo, C.; Simões da Silva, L. Mixed Mode (I+II) Fatigue Crack Growth in Puddle Iron. *Eng. Fract. Mech.* **2017**, *185*, 175–192. [CrossRef]
16. Naboulsi, S.; Mall, S. Modelling of a Cracked Metallic Structure with Bonded Composite Patch Using the Three Layer Technique. *Compos. Struct.* **1996**, *35*, 295–308. [CrossRef]
17. *Sika Product Data Sheet: Sika® CarboDur® E-1014*; Sika Sp. z. o.o.: Warsaw, Poland, 2019.
18. Dourado, N.; Pereira, F.A.M.; de Moura, M.F.S.F.; Morais, J.J.L. Repairing Wood Beams under Bending Using Carbon-Epoxy Composites. *Eng. Struct.* **2012**, *34*, 342–350. [CrossRef]
19. Krueger, R. Virtual Crack Closure Technique: History, Approach, and Applications. *Appl. Mech. Rev.* **2004**, 109–143. [CrossRef]
20. Karbhari, V. *Rehabilitation of Metallic Civil Infrastructure Using Fiber-Reinforced Polymer (FRP) Composites*; Woodhead, P., Ed.; Elsevier: Cambridge, UK, 2014.
21. ASTM E 647; *Standard Test Method for Measurement of Fatigue Crack Growth Rates*; American Society for Testing and Materials: West Conshohocken, PA, USA, 2015.
22. Gallegos Mayorga, L.; Sire, S.; Ragueneau, M.; Plu, B. Understanding the Behaviour of Wrought-Iron Riveted Assemblies: Manufacture and Testing in France. In *Proceedings of the Institution of Civil Engineers-Engineering History and Heritage*; ICE Publishing: London, UK, 2017; Volume 170, pp. 67–79.
23. Zvirko, O.; Nykyforchyn, H.; Szata, M.; Kutnyi, A.; Lesiuk, G. Corrosion degradation of old structures steels. In *XII International Conference "Problems of Corrosion and Corrosion Protection of Structural Materials"*; Corrosion: Lviv, Ukraine, 2014.
24. Tsyrul'nyk, O.T.; Kret, N.V.; Voloshyn, V.A.; Zvirko, O.I. A procedure of laboratory degradation of structural steels. *Mater. Sci.* **2018**, *53*, 674–683. [CrossRef]
25. Zvirko, O.; Zagórski, A. Corrosion and electrochemical properties of the steel of exploited oil tanks in bottom water. *Mater. Sci.* **2008**, *44*, 126–132. [CrossRef]

 © 2020 by the authors. Licensee MDPI, Basel, Switzerland. This article is an open access article distributed under the terms and conditions of the Creative Commons Attribution (CC BY) license (http://creativecommons.org/licenses/by/4.0/).

Article

Influence of Hot Forging Parameters on a Low Carbon Continuous Cooling Bainitic Steel Microstructure

Antonio Carlos de Figueiredo Silveira [1,*], William Lemos Bevilaqua [1], Vinicius Waechter Dias [1], Pedro José de Castro [1], Jeremy Epp [2] and Alexandre da Silva Rocha [1]

1. Post-Graduation Program in Mining, Metallurgical and Materials Engineering, Federal University of Rio Grande do Sul, Avenida Bento Gonçalves, 9500, 91509-900 Porto Alegre, Brazil; william.bevilaqua@ufrgs.br (W.L.B.); vinicius.waechter@ufrgs.br (V.W.D.); pedro.castro@ufrgs.br (P.J.d.C.); alexandre.rocha@ufrgs.br (A.d.S.R.)
2. Leibniz-Institut für Werkstofforientierte Technologien, Badgasteiner Street 3, 28359 Bremen, Germany; epp@iwt-bremen.de
* Correspondence: figueiredo.silveira@ufrgs.br; Tel.: +55-(53)-99977-0780

Received: 13 April 2020; Accepted: 4 May 2020; Published: 6 May 2020

Abstract: Thermomechanical processing of low carbon bainitic steels is used to obtain a bainitic microstructure with good strength and toughness by continuous cooling after forging without the need of further heat treating, hence reducing manufacturing costs. However, hot forging parameters can significantly influence the microstructure in the forged material. A series of heat treating and forging experiments was carried out to analyze the effect of austenitizing time and temperature on the grain growth and the effect of forging temperature on the Prior Austenite Grain Size (PAGS) and continuously cooled microstructure. The forged microstructures were characterized by optical microscopy, microhardness tests, and X-ray diffraction. The results indicate that at 1200 °C austenitizing temperature abnormal grain growth takes place. Forging temperature significantly affects the PAGS and the subsequently formed microstructure. At high forging temperature (1200 °C), an almost fully bainitic microstructure was obtained. As the forging temperature was reduced to 1100 and 1000 °C, the PAGS refined, while the polygonal ferrite faction increased and the amount of retained austenite decreased. Further evaluations showed that a decrease in the forging temperature results in a higher carbon concentration in solution in the retained austenite leading to a stabilization effect.

Keywords: thermomechanical processing; grain growth; forging; retained austenite; bainitic microstructure

1. Introduction

In the last decade, the application of advanced bainitic steels in thermomechanical processing (TMP) has gained significant importance [1]. Caballero et al. [2] presented a comprehensive study of continuously cooled bainitic steels for automotive structural parts. Other authors [3–5] have also studied the effect of different thermomechanical routes on the microstructure and mechanical properties of bainitic steels. The implementation of TMP in forged components is complex. Different cooling rates between the center and surface of the component can generate a heterogeneous microstructure [6,7], restricting the TMP processing window. Hence, a significant number of articles focus on developing steels with a chemical composition that enables them to obtain a bainitic microstructure using a broader cooling range [8,9]. The use of continuous cooling can reduce the manufacturing costs by replacing long isothermal treatments or even subsequent conventional quenching and tempering, therefore decreasing significantly the thermal cycles and energy consumption [10].

Thermomechanical processing comprehends controlled steps of austenitizing, plastic deformation, and cooling. Thus, understanding how the process parameters affect recrystallization and the final microstructure is crucial to ensure a well-defined processing window of forged components with suitable mechanical properties. For instance, the use of an excessive high austenitizing temperature can lead to strong grain coarsening and abnormal grain growth [11]. During the hot deformation, temperature in combination with strain and strain rate affects directly the recrystallization.

Yang et al. [12] investigated the workability of a low carbon bainitic steel through processing map analysis and showed that the amount of strain significantly affected the suitable hot deformations parameters, also known as stable regions in a processing map. At lower deformation degree (0.2–0.4 true strain), higher values of strain rate can be applied (5–10 s^{-1}) without compromising the recrystallized microstructure homogeneity. However, at higher levels of deformation (0.6–0.8), lower values of strain rate must be applied (0.001–0.016 s^{-1}) to avoid incomplete recrystallized regions, which can result in poor mechanical properties. The Prior Austenite Grain Size (PAGS) significantly influences the packet size and growth orientation of bainite [13–16]. PAGS also affects the formation of other phases, such as Polygonal Ferrite (PF) [15,17,18] and martensite [19]. Therefore, non-homogeneous austenitic grain size can lead to variations on the final microstructure after the cooling step. Consequently, the TMP of bainitic steels requires further investigation to avoid undesirable microstructures related to grain coarsening, incomplete recrystallization, and heterogeneous microstructure in the forged component.

Most of the work carried out on bainitic transformation has been concentrated on isothermal forging conditions. Therefore, there is a lack of information regarding the thermomechanical processing of bainitic steel in conditions similar to industrial hot forging. In this article, we discuss the effect of austenitizing and forging parameters on the growth and recrystallization of PAGS and continuously cooled microstructure of a low carbon bainitic steel through optical microscopy, microhardness, and X-ray diffraction analysis.

2. Materials and Methods

For this work, hot rolled bars of a DIN 18MnCrSiMo6-4 (HSX 130) steel from Swisstec (Swiss Steel), Emmenbrücke, Switzerland, with 43 mm of diameter were employed. The bars were directly air-cooled after hot rolling. Table 1 gives the chemical composition of the steel, and the as-received microstructure is shown in Figure 1.

Table 1. Chemical composition (in ma. %).

C	Si	Mn	S	Ni	Cr	Cu	Mo	Al	Ti	N	Fe
0.18	1.19	1.42	0.015	0.063	1.17	0.10	0.27	0.005	0.004	0.01	Balance

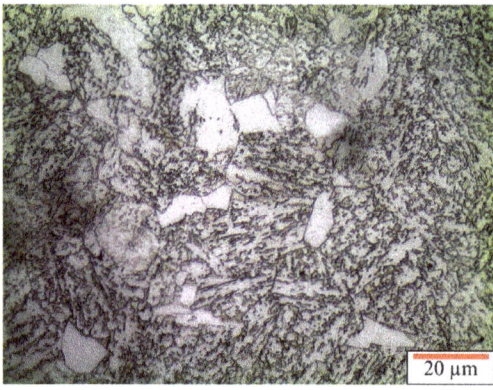

Figure 1. Optical microscopy of the DIN 18MnCrSiMo6-4 as-received microstructure.

2.1. Grain Growth Evaluation on the Austenitizing Step for the Hot Forging

To analyze the effect of temperature and time on the grain growth, quarter-circle specimens with 21.5-mm radius and 10-mm thickness were machined from the steel bars. The small size was adopted to avoid significant temperature gradients during heating. The experiments consisted in an austenitizing step at temperatures of 905, 1000, 1100, and 1200 °C with different holding times of 0, 10, 20, and 40 min followed by rapid cooling to room temperature by water quenching to retain PAGS. The temperature of the sample at its core was monitored by inserted type K thermocouples. The holding time of 0 indicates the minimum time for the sample's core to achieve the austenitizing temperature i.e. minimum soaking time. According to the material composition, 905 °C is the minimum temperature for a complete austenitization (Ac3). By using this temperature and the minimum soaking time, the resulted PAGS was estimated as the austenitic grain size of the material as received. For the remaining temperatures, by increasing the holding time from 0 to 40 min, both effects of temperature and time on the grain growth could be evaluated.

2.2. Hot Forging Experiments and Microstructure Evaluation

For the hot forging experiments, billets with 38-mm diameter and 54-mm height were machined from hot rolled steel bars. Forging was carried out in a 40-tonf (356-kN) hydraulic press equipped with a load cell and Linear Variable Differential Transformer (LVDT) to acquire force versus displacement curves during forging. Flat dies manufactured from hardened and tempered AISI H13 steel were used for the upsetting. Graphite water solution was applied to the surfaces of the dies as a forging lubricant. To simulate temperature losses of an ongoing industrial process, the dies were pre-heated by a resistance coil attached to the lower die (stationary); then, both dies were kept in contact for 1 h before forging to achieve temperatures around 250 and 180 °C for lower and upper die, respectively. The temperature losses between air and billet during transportation from the furnace to the press were quantified by type K thermocouples in the billet. The samples were submitted to a 60% height reduction. Figure 2a shows the thermomechanical routes. Three different routes were applied, with heating temperatures of 1000, 1100, and 1200 °C, always with a holding time of 10 min after the sample's core had reached the heating temperature set in the furnace (minimum soaking time). For each route, one sample was water quenched for PAGS analysis, and another one was cooled in calm air to characterize the continuous cooled microstructure. Figure 2b shows a macrograph of a forged sample as an example, after a height reduction of 60%. Due to the friction and thermal losses to the dies, the top and bottom regions of the samples presented were excluded for PAGS and phase quantification; only the center areas (red rectangle) were used in the analysis.

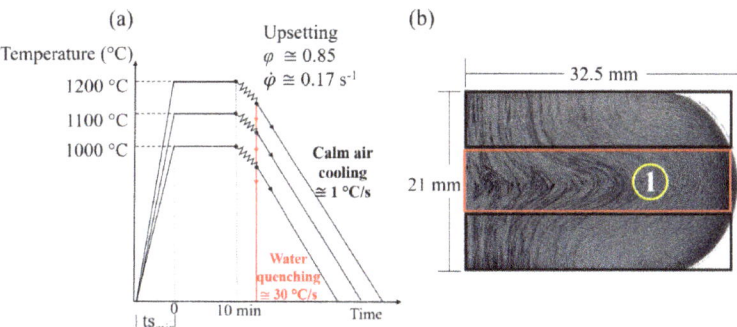

Figure 2. (a) Thermomechanical routes. ts_{min}, minimum soaking time. (b) Half-section macrograph of the forged sample. The red rectangle indicates the analyzed area. The yellow circle labeled 1 indicates the position from where the micrographs shown in this work were taken.

2.3. Critical Strain

Based on the true flow stress curves from the hot forging experiments, the onset of dynamic recrystallization (DRX) can be determined by the double-differentiation technique proposed by Poliak and Jonas [20]. Equation (1) shows the basic equation for the strain hardening effect, where θ is the strain hardening rate, σ is the flow stress, and ϕ is the true strain.

$$\theta = \partial\sigma/\partial\phi \qquad (1)$$

The Poliak and Jonas approach identifies the beginning of the DRX as an inflection point of the strain hardening rate. This inflection point is more easily seen as a minimum on the $-(\partial\theta/\partial\sigma)$ vs. σ plot [21]. This minimum value corresponds to critical stress, which is equivalent to a critical strain, where strain hardening rates reach a minimum value indicating the recrystallization start acting point.

2.4. Metallurgical Characterization

Metallographic samples parallel to the compression direction from the forged billets and from the hot rolled bars (as-received microstructure) were prepared following standard procedures [22] and etched by immersion in a Nital 2% solution for 10 s. For PAGS characterization, a saturated aqueous picric acid solution was used. The etchant consists of wetting agent (42 mL), distilled water (58 mL), and picric acid (2.3 g). The samples were etched by swabbing for 5 min.

The circular intercept procedure was adopted for the PAGS quantification [23]. From 6 to 12 images were used to achieve a good statistical significance, and the evaluations were performed with the Omnimet 9.8 Software from Buehler, IL, USA [24]. The ferrite quantification was carried out with the ImageJ Software after the binarization of the images from optical microscopy (OM). Vickers microhardness was measured using an INSIZE ISH-TDV 1000 microhardness tester according to ASTM standard E384 [25]. For each forged sample, thirty measurements were performed with a load of 1 kgf for 10 s at different testing positions of the samples to obtain a mean hardness value. Further tests were carried out with a lower load of 0.1 kgf applied for 10 s to evaluate the hardness of each microconstituent of the microstructure. Three measurements were performed per microconstituent. For all hardness tests, a distance larger than the recommended indentations spacing was adopted to assure no interference between measurements.

The Retained Austenite (RA) fraction was quantified using X-ray Diffraction (XRD) analysis. Before the analysis, all samples were electropolished to remove approximately 100 μm in order to eliminate the possible effects of mechanical preparation. The measurements were performed with a GE-Analytical X-ray MZ VI E Diffractometer with Cr-Kα radiation with a wavelength of 2.2897 Å. A 2θ range of 60–164° was adopted to obtain the {111}, {200}, and {220} austenite peaks and {110}, {200}, and {211} ferrite peaks, with a 0.05° step size. The retained austenite (RA) phase quantification was done with the Rietveld refinement software TOPAS version 4.2 (Bruker AXS, Karlsruhe, Germany) [26].

3. Results and Discussion

3.1. Grain Growth

Figure 3a shows the evolution of PAGS as a function of temperature and time. PAGS of the as-received material (indicated as the gray dashed line), was found to be 24 μm. At 1000 °C, a slight growth was observed at 40 min of holding time (≈4 μm) in comparison to the initial grain size. At 1100 °C, no growth was noticed at time $t = 0$ min and 10 min. A sharper growth was observed at 20 min, followed by a stabilization of the grain size at $t = 40$ min. At 1200 °C and $t = 0$ min, the grain showed an increase of ≈20 μm with the highest growth between 10 and 20 min and a stabilization of the grain size at 40 min, similar to 1100 °C. However, at 1200 °C, PAGS had a significant data dispersion (indicated by the error bars). This dispersion happened because larger grains than those of the matrix are present in the microstructure, as indicated in highlighted in red in Figure 3b.

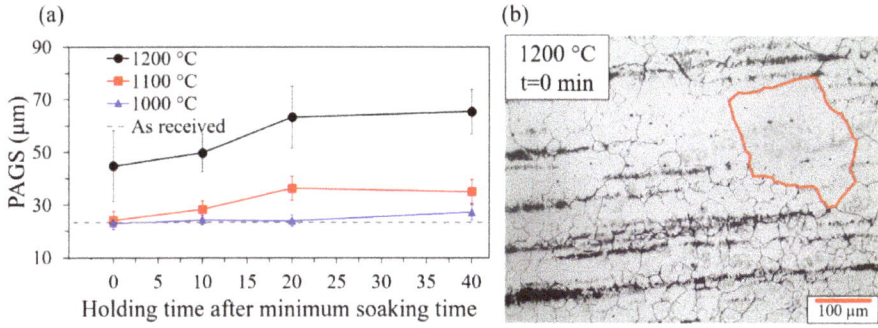

Figure 3. (a) Effect of temperature and holding time on austenitic grain; and (b) PAGS microstructure after austenitizing at 1200 °C with 0 min of holding time, indicating the presence of abnormal grain. Highlighted grain has a mean size of 200 µm and the matrix 45 µm.

This heterogeneity in the grain size indicates that abnormal grain growth took place at 1200 °C; however, this affect was not observed at lower temperatures. The abnormal grain growth may be associated with the dissolution of precipitates formed by microalloying elements such as Ti, N, and Al present in the steel [27]. Based on the chemical composition of the steel and solubility equations for precipitates in austenite [28], the temperature for total dissolution can be calculated for carbides such as TiC (907 °C) and nitrides such as TiN (1339 °C) and AlN (1004 °C). These precipitates inhibit the growth of the austenitic grains due to pinning effect [29]. The presence of impurities such as S also promotes the formation of sulfites such as MnS (1701 °C), which can act as a barrier to the grain growth as well [30]. However, as the temperature rises, these precipitates can be totally or partially dissolved in the austenite or even coalesce, as may be the case with MnS bands. Hence, the pressure exerted by the pin particles decreases, leading to excessive grain growth, resulting in a nonhomogeneous microstructure [11,31–33].

3.2. Effect of the Hot Deformation Temperature

3.2.1. Austenitic Grain Refinement

Figure 4 shows the morphology of the PAGS immediately after the hot deformation step. The deformation at 1000 and 1100 °C (Figure 4a,b, respectively) resulted in similar refinement, achieving PAGS approximately 50% smaller than the material in its initial state. The resulting PAGS of the forging at 1200 °C showed a similar size to the material in its initial state, indicating no refinement of the austenitic grain by forging at this temperature (Figure 4c).

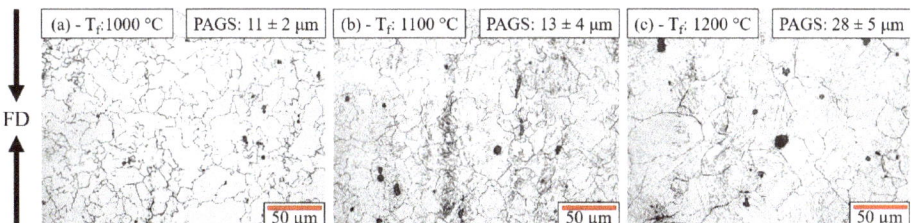

Figure 4. Optical microscopy of the PAGS after hot deformation with 60% of reduction at (a) 1000 °C; (b) 1100 °C; and (c) 1000 °C T_f, forging temperature; FD, forging direction.

3.2.2. Critical Strain

Figure 5a shows the flow stress curves from the hot forging experiments indicating the critical strain for the onset of DRX, while Figure 5b shows the $-(\partial \theta / \partial \sigma)$ versus σ, where the minimum values correspond to the critical stress for the beginning of DRX.

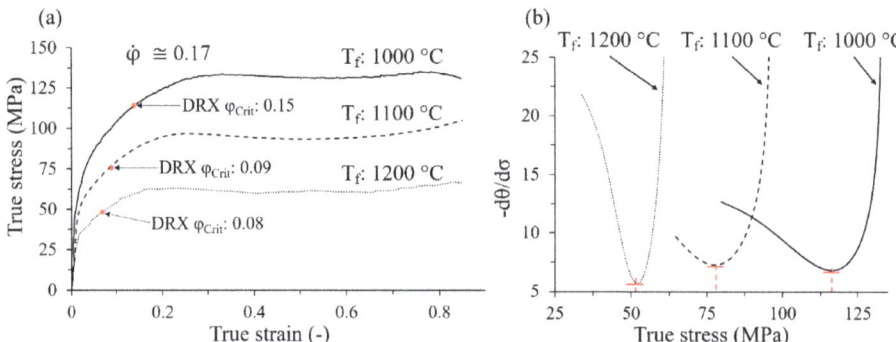

Figure 5. (a) Flow stress curves of the hot forging experiments with the critical strain for the onset of the DRX; and (b) $-(\partial \theta / \partial \sigma)$ versus σ curve, where the minimum value corresponds to the critical stress.

Figure 5 shows that, as the forging temperature increases, the required strain necessary to initiate the DRX decreases. The hot forging at 1200 °C presented a smaller critical strain for the initiation of the DRX, and, as the forging temperature decreases to 1100 and 1000 °C, a higher degree of deformation was required to start the recrystallization. Although an increase in temperature decreased the critical strain for recrystallization (due to an increase of the internal energy), larger initial grain sizes had the opposite effect by decreasing grain boundary area, which are preferred nucleation sites [34].

A comparison between the flow stresses curves (Figure 5a) and PAGS (Figure 4) from the forging experiments show that an increase in the forging temperature reduces the flow stress and results in a larger PAGS. This is expected since PAGS during forging is associated to the strain rate and flow stress shape, which is controlled by the temperature [35]. Moreover, increasing the forging temperature means more time during cooling at temperatures where grain growth occurs, leading to a larger PAGS when the phase transformation starts. Even though recrystallization was observed at 1200 °C, the microstructure was subjected to longer periods at temperatures where grain growth occurred. As a result, the austenite grains grew sufficiently, reaching a similar size to PAGS of the as-received microstructure.

3.2.3. Continuously Cooled Microstructure

Figure 6 shows the microstructures and overall hardness of the samples forged and directly cooled by calm-air to room temperature in comparison with the as-received material. Figure 6a shows the as-received microstructure composed mainly of Granular Bainite (GB), which is characterized by Bainitic Ferrite (BF) (white regions) between the blocky RA and martensite/austenite (M/A) shown as the darker regions in the figure. The Polygonal Ferrite (PF) is revealed as a massive white block. The hot forging at 1200 °C (Figure 6b) resulted in a GB microstructure with a small portion of PF. As the forging temperature decreased to 1100 °C, as indicated in Figure 6c, the fraction of PF increased. Figure 6d shows the resulting microstructure after forging at 1000 °C composed by a high fraction of PF, small fractions of GB, and an additional microconstituent highlighted in yellow (MC). The microhardness tests results for the microconstituents are 301, 407, and 495 HV 0.1 for PF, GB, and MC, respectively. The latter presents a similar morphology to martensite [36]. Moreover, its hardness is approximately 90 HV 0.1 higher than the GB, indicating that this may indeed be martensite.

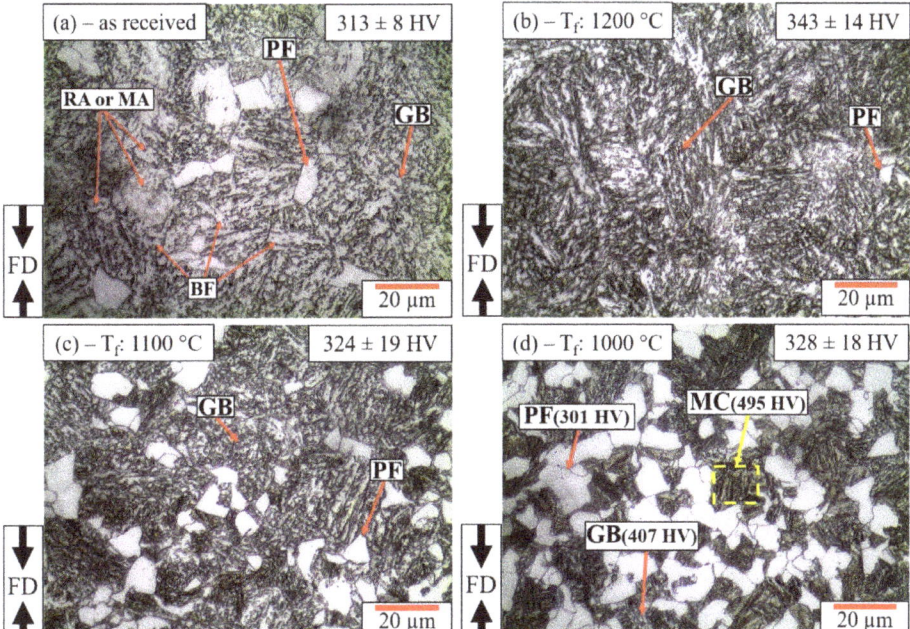

Figure 6. Optical microscopy of (**a**) as-received condition and forged samples cooled in calm air: (**b**) forged at 1200 °C; (**c**) forged at 1100 °C; and (**d**) forged at 1000 °C. PF, polygonal ferrite; GB, granular bainite; MC, third microconstituent; RA, retained austenite; MA, martensite/austenite; BF, bainitic Ferrite; FD, forging direction.

Deformation can significantly affect the PF formation [37]. The increase dislocation density caused by the plastic deformation leads to more unstable austenite. Usually, this instability is countered by recovery and recrystallization. The latter refines the austenite grain, increasing the total area of the austenitic grain boundaries. The boundaries are the preferred sites for the ferrite nucleation during the cooling [15,38,39]. Thus, a higher quantity of PF can be formed from a microstructure with smaller PAGS. Moreover, the increase of defects caused by the plastic deformation also elevates the temperature of austenite–ferrite transformation A3. Therefore, hot forging at the lowest temperature of 1000 °C not only increased the number of possible nucleation sites but also favored the beginning of the transformation is closer to the A3 temperature of the steel [37].

The polygonal ferrite formation during cooling can affect the bainite formation in two different ways. As reported by Quidort and Brechet [40], the formation of a small fraction of ferrite at grain boundaries (less than 10%) increases the number of nucleation sites. Since the grain boundaries are also the preferred site for bainite nucleation, the presence of ferrite may accelerate the bainitic reaction. However, Zhu et al. [33] reported that, as the ferrite nucleates and grows, the carbon and other alloys diffuse to the remaining austenite (partitioning process). However, due to the diffusion limitation during the cooling, this enrichment is not homogeneous, resulting in an austenite–ferrite interface with a higher concentration of these elements and regions distant from the interfaces with a chemical composition closer to the initial composition shown in Table 1. This chemical heterogeneity in the austenite would inhibit the bainitic transformation at the interfaces (higher carbon and alloy content), while facilitating the martensite reaction at the center regions of the remained austenite with lower carbon and alloy content. Lambert et al. [41] also found martensite at the center of austenitic grain, indicating a similar chemical variation in the austenite and, by Transmission Electron Microscope (TEM), identified dislocations in the interior of the austenite which could initiate the martensite formation. This indicates that the microconstituent highlighted in yellow in Figure 6d could indeed be martensite.

Figure 7 shows the polygonal ferrite obtained by OM quantification, and the RA fraction obtained by the XRD analysis of the forged samples. Forging at 1000 °C resulted in 30% of PF, 6% of RA, and a small parcel of GB. As temperature increased, the PF fraction decreased, and RA and GB increased. At 1200 °C, the PF reduced to almost zero, resulting in a GB microstructure with 10% of RA. RA in the microstructure is mainly related to the carbide suppression caused by Si in the steel, resulting in a carbide free bainitic microstructure with stable RA at room temperature due to the carbon partitioning [42]. This result indicates that RA stability is closely related to the GB presence in the microstructure. With the PF formation at lower forging temperatures, a higher RA stability should be expected, increasing the RA fraction. However, the lowest fraction of RA is found in the microstructure with higher quantities of PF. This low fraction is possibly related to martensite formation at the austenite regions with lower carbon and alloying elements content. With the increase of forging temperature, a higher fraction of GB is formed, favoring the formation of RA during bainitic transformation.

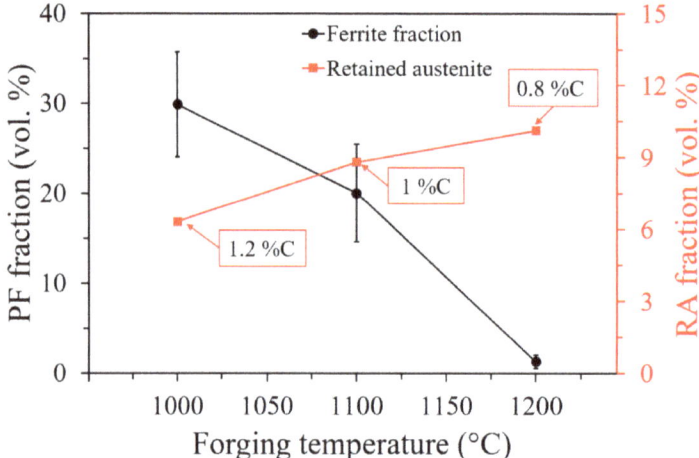

Figure 7. The effect of the forging temperature on the PF and RA fractions and carbon content of the austenite.

Based on the retained austenite lattice parameter obtained from the TOPAS Rietveld method, the average carbon content of the RA was calculated according to Dyson and Holmes's equation [43]. Figure 7 shows that, at 1000 °C, RA reached 1.2 ma. % of carbon; as the forging temperature increased, the carbon concentration reduced to 1 ma. % and 0.8 ma. % at 1100 and 1200 °C, respectively. This result shows that, despite the lowest fraction of RA at 1000 °C, the resulting austenite is highly carbon enriched. According to Xiong et al. [44], this heterogeneous chemical composition favors the formation of a high carbon blocky RA at the austenite/ferrite interface. At the same time, the austenite regions with lower carbon would transform to martensite, as explained above. At higher forging temperature, since lower quantities of PF forms and no martensite appears, the RA is enriched only by the GB formation.

3.2.4. Summary of the Microstructure Characterization

In this study, the effect of austenitizing temperature and forging temperature on the microstructure of bainitic steel was investigated using stress–strain curves and microstructure analysis. Table 2 shows the main results of thermomechanical processed samples at different forging temperatures. The characterization shows that the forging temperature resulted in different PAGS and significantly affected the phases fraction and the RA carbon content in solution. This means that, by changing the forging temperature, a range of microstructures can be obtained without modifying the cooling parameters. At 1000 °C, the smallest PAGS and the highest fraction of PF were achieved with the presence of a third microconstituent with superior hardness. Moreover, the lowest fraction of RA and GB were obtained. Forging at 1100 °C resulted in similar PAGS refinement with a reduction in the PF fraction and an increase in RA and GB fractions. At 1200 °C, PAGS shows similar size with the material as received, and the microstructure obtained is almost entirely bainitic with small fractions of PF and higher quantities of RA.

Table 2. Summary of the microstructure characterization.

Forging Temperature	PAGS (μm)	Microstructure	PF Fraction (vol. %)	RA (vol. %)	C_{RA} Content (ma. %)	Overall Hardness (HV 1)
As received	24 ± 3.2	GB + PF	3.5 ± 1.8	10.5	0.8	313 ± 8
1000 °C	11 ± 1.7	PF + MC + GB	30.0 ± 5.8	6.3	1.2	328 ± 18
1100 °C	13 ± 3.0	GB + PF	20.0 ± 5.4	8.8	1.0	324 ± 19
1200 °C	28 ± 5.3	GB + PF	1.3 ± 0.8	10.1	0.8	343 ± 14

4. Conclusions

- Heat treating experiments with different austenitizing temperatures and holding times were carried out to evaluate the prior austenite grain size. The prior austenite grain size results show that an austenitizing temperature of 1200 °C promoted a significant grain growth and the presence of abnormal grain growth, however, abnormal grain growth was not identified at 1100 and 1000 °C.
- The temperature also affected the prior austenite grain size after the hot forging, which affected the phase fractions in the final continuous cooling microstructure, especially the polygonal ferrite formation.
- X-ray diffraction analysis showed that the fraction of retained austenite was reduced as the forging temperature decreased due to the formation of higher quantities of polygonal ferrite. Moreover, the carbon content increased in austenite due to the growth of polygonal ferrite during cooling.
- Hot forging at 1000 °C promoted the formation of 30% of polygonal ferrite, resulting in a chemical heterogeneity of the remaining austenite, leading to martensite formation at the austenite regions with lower carbon and alloying elements.
- The granular bainite formation and retained austenite fraction are favored in the coarser austenitic microstructure obtained at higher hot forging temperatures.

Author Contributions: Conceptualization, A.C.d.F.S.; methodology, A.C.d.F.S. and P.J.d.C.; validation, A.C.d.F.S. and P.J.d.C.; formal analysis, A.C.d.F.S. and J.E.; investigation, A.C.d.F.S., P.J.d.C., and W.L.B.; data curation, A.C.d.F.S.; writing—original draft preparation, A.C.d.F.S.; writing—review and editing, A.d.S.R., W.L.B., V.W.D., P.J.d.C., and J.E.; visualization, A.C.d.F.S.; and Funding acquisition, A.d.S.R. and J.E. All authors have read and agreed to the published version of the manuscript.

Funding: This research was funded in part by the Coordenação de Aperfeiçoamento de Pessoal de Nível Superior—Brasil (CAPES)—Finance Code 001 and n° 1844/2017, Conselho Nacional de Desenvolvimento Científico e Tecnológico (proc 167149/2017-2 and Pq-2018, proc 308773/2018-7) and the Deutsche Forschungsgemeinschaft_DFG (ZO 140/21-1).

Acknowledgments: The authors express their gratitude to the following laboratories: Metal Forming Laboratory (LdTM), Foundry Laboratory (LAFUN), and Institut für Werkstofforientierte Technologien (IWT) for the investigation support and the company Swiss Steel for the material donation. Antonio Carlos de Figueiredo Silveira, William Lemos Bevilaqua, Vinicius Waechter Dias and Pedro José de thank the Coordenação de Aperfeiçoamento de Pessoal de Nível Superior (Brazil) (CAPES) and Conselho Nacional de Desenvolvimento Científico e Tecnológico (Brazil) (CnPQ) for the master and doctorate scholarships.

Conflicts of Interest: The authors declare no conflict of interest.

References

1. Sourmail, T.; Smanio, V.; Ziegler, C.; Heuer, V.; Kuntz, M.; Caballero, F.G.; Garcia-Mateo, C.; Cornide, J.; Elvira, R.; Leiro, A.; et al. *Novel Nanostructured Bainitic Steel Grades to Answer the Need for High-Performance Steel Components*; European Commission: Brussels, Belgium, 2013. [CrossRef]
2. Caballero, F.G.; Garcia-Mateo, C.; Cornide, J.; Allain, S.; Puerta, J.; Crouvizer, M.; Mastrorillo, T.; Jantzen, L.; Vuorinen, E.; Lindgren, L.E.; et al. New Advanced Ultra High Strength Bainitic Steels: Ductility and Formability. *Res. Fund Coal Steel Eur. Commision* **2013**, 1–124. [CrossRef]
3. Cao, J.; Yan, J.; Zhang, J.; Yu, T. Effects of thermomechanical processing on microstructure and properties of bainitic work hardening steel. *Mater. Sci. Eng. A* **2015**, *639*, 192–197. [CrossRef]
4. Liang, X.; Deardo, A.J. A study of the influence of thermomechanical controlled processing on the microstructure of bainite in high strength plate steel. *Metall. Mater. Trans. A Phys. Metall. Mater. Sci.* **2014**, *45*, 5173–5184. [CrossRef]
5. Zhao, L.; Qian, L.; Liu, S.; Zhou, Q.; Meng, J.; Zheng, C.; Zhang, F. Producing superfine low-carbon bainitic structure through a new combined thermo-mechanical process. *J. Alloys Compd.* **2016**, *685*, 300–303. [CrossRef]
6. Sourmail, T. Bainite and Superbainite in Long Products and Forged Applications. *HTM J. Heat Treat. Mater.* **2017**, *72*, 371–378. [CrossRef]
7. Bleck, W.; Bambach, M.; Wirths, V.; Stieben, A. Microalloyed Engineering Steels with Improved Performance—An Overview. *HTM J. Heat Treat. Mater.* **2017**, *72*, 346–354. [CrossRef]
8. Fang, H.S.; Chun, F.E.; Zheng, Y.K.; Yang, Z.G.; Bai, B.Z. Creation of Air-Cooled Mn Series Bainitic Steels. *J. Iron Steel Res. Int.* **2008**, *15*, 1–9. [CrossRef]
9. Hasler, S.; Roelofs, H.; Lembke, M.; Caballero, F.G. New air cooled steels with outstanding impact toughness. In Proceedings of the 3nd International Conferente On Steels in Cars and Trucks, Salzburg, Austria, 5–9 June 2011; p. 9.
10. ASM. *Handbook. Volume 14: Forming and Forging*; ASM International: Materials Park, OH, USA, 1993; Volume 14, ISBN 0-87170-007-7. [CrossRef]
11. Fernández, J.; Illescas, S.; Guilemany, J.M. Effect of microalloying elements on the austenitic grain growth in a low carbon HSLA steel. *Mater. Lett.* **2007**, *61*, 2389–2392. [CrossRef]
12. Yang, Z.; Zhang, F.; Zheng, C.; Zhang, M.; Lv, B.; Qu, L. Study on hot deformation behaviour and processing maps of low carbon bainitic steel. *Mater. Des.* **2015**, *66*, 258–266. [CrossRef]
13. Rancel, L.; Gómez, M.; Medina, S.F. Influence of Microalloying Elements (Nb, V, Ti) on Yield Strength in Bainitic Steels. *Steel Res. Int.* **2008**, *79*, 947–953. [CrossRef]
14. Zhao, H.; Wynne, B.P.; Palmiere, E.J. Effect of austenite grain size on the bainitic ferrite morphology and grain refinement of a pipeline steel after continuous cooling. *Mater. Charact.* **2017**, *123*, 128–136. [CrossRef]
15. Li, X.; Xia, D.; Wang, X.; Wang, X.; Shang, C. Effect of austenite grain size and accelerated cooling start temperature on the transformation behaviors of multi-phase steel. *Sci. China Technol. Sci.* **2013**, *56*, 66–70. [CrossRef]
16. Rancel, L.; Gómez, M.; Medina, S.F.; Gutierrez, I. Measurement of bainite packet size and its influence on cleavage fracture in a medium carbon bainitic steel. *Mater. Sci. Eng. A* **2011**, *530*, 21–27. [CrossRef]
17. Capdevila, C.; Caballero, F.G. Carlos García de Andrés Austenite Grain Size Effects on Isothermal Allotriomorphic Ferrite Formation in 0.37C-1.45Mn-0.11V Microalloyed Steel. *Mater. Trans.* **2003**, *44*, 1087–1095. [CrossRef]
18. Ferry, M.; Thompson, M.; Manohar, P.A. Decomposition of coarse grained austenite during accelerated cooling of C-Mn steels. *ISIJ Int.* **2002**, *42*, 86–93. [CrossRef]
19. Celada-Casero, C.; Sietsma, J.; Santofimia, M.J. The role of the austenite grain size in the martensitic transformation in low carbon steels. *Mater. Des.* **2019**, *167*. [CrossRef]
20. Poliak, E.I.; Jonas, J.J. A one-parameter approach to determining the critical conditions for the initiation of dynamic recrystallization. *Acta Mater.* **1996**, *44*, 127–136. [CrossRef]
21. McQueen, H.J.; Yue, S.; Ryan, N.D.; Fry, E. Hot working characteristics of steels in austenitic state. *J. Mater. Process. Technol.* **1995**, *53*, 293–310. [CrossRef]
22. ASM International (Ed.) *Volume 9: Metallography and Microstructure*, 9th ed.; ASM International: Materials Park, OH, USA, 2004; Volume 9, ISBN 0871707063.

23. ASTM Standard. *E112-12:Standard Test Methods for Determining Average Grain Size*; ASTM International: Materials Park, OH, USA, 2012. [CrossRef]
24. Voort, G. Vander Introduction to quantitative metallography. *Buehler TechNotes* **2015**, *1*, 5.
25. ASTM International. *E384 Standard Test Method for Microindentation Hardness of Materials*; ASTM International: Materials Park, OH, USA, 2017; pp. 1–40. [CrossRef]
26. Dinnebier, R.E.; Leineweber, A.; Evans, J.S.O. *Rietveld Refinement—Practical Powder Diffraction Pattern Analysis Using TOPAS*; De Gruyter, Ed.; De Gruyter: Berlin, Germany; Boston, MA, USA, 2018; ISBN 9783110456219. [CrossRef]
27. Razzak, M.A.; Perez, M.; Sourmail, T.; Cazottes, S.; Frotey, M. Preventing Abnormal Grain Growth of Austenite in Low Alloy Steels. *ISIJ Int.* **2014**, *54*, 1927–1934. [CrossRef]
28. Totten, G.E.; Xie, L.; Funatani, K. *Handbook of Mechanical Alloy Design*; CRC Press: New York, NY, USA, 2003; ISBN 978-0-8247-4308-6. [CrossRef]
29. Zener, C. Theory of Growth of Spherical Precipitates from Solid Solution. *J. Appl. Phys.* **1949**, *20*, 950. [CrossRef]
30. Guo, L.; Roelofs, H.; Lembke, M.I.; Bhadeshia, H.K.D.H. Effect of manganese sulphide particle shape on the pinning of grain boundary. *Mater. Sci. Technol.* **2017**, *33*, 1013–1018. [CrossRef]
31. Totten, G.E. *Steel Heat Treatment*; Taylor & Francis Group: Abingdon, UK, 2006; pp. 1–820. ISBN 978-0-8493-8455-4.
32. Li, M.; Li, J.; Qiu, D.; Zheng, Q.; Wang, G.; Zhang, M.X. Crystallographic study of grain refinement in low and medium carbon steels. *Philos. Mag.* **2016**, *96*, 1556–1578. [CrossRef]
33. Zhu, K.; Chen, H.; Masse, J.P.; Bouaziz, O.; Gachet, G. The effect of prior ferrite formation on bainite and martensite transformation kinetics in advanced high-strength steels. *Acta Mater.* **2013**, *61*, 6025–6036. [CrossRef]
34. Humphreys, J.; Rohrer, G.S.; Rollett, A. Recrystallization and Related Annealing Phenomena. In *Recrystallization and Related Annealing Phenomena*, 2nd ed.; Elsevier: Amsterdam, The Netherlands, 2017; pp. 1–734. ISBN 978-0-08-098235-9.
35. Derby, B. The dependence of grain size on stress during dynamic recrystallisation. *Acta Metall. Mater.* **1991**, *39*, 955–962. [CrossRef]
36. Krauss, G. *Steels: Processing, Structure, and Performance*, 2nd ed.; ASM, I.H.C., Ed.; ASM International: Materials Park, OH, USA, 2005.
37. Sietsma, J. 14—Nucleation and growth during the austenite-to-ferrite phase transformation in steels after plastic deformation. In *Phase Transformations in Steels*; Woodhead Publishing Limited: Sawstone, Cambridge, UK, 2012; Volume 1, pp. 505–526. ISBN 9781845699703. [CrossRef]
38. Aranda, M.M.; Kim, B.; Rementeria, R.; Capdevila, C.; De Andrés, C.G. Effect of prior austenite grain size on pearlite transformation in A hypoeutectoid Fe-C-Mn steel. *Metall. Mater. Trans. A Phys. Metall. Mater. Sci.* **2014**, *45*, 1778–1786. [CrossRef]
39. Lan, H.F.; Du, L.X.; Liu, X.H. Microstructure and mechanical properties of a low carbon bainitic steel. *Steel Res. Int.* **2013**, *84*, 352–361. [CrossRef]
40. Quidort, D.; Brechet, Y.J.M. Isothermal growth kinetics of bainite in 0.5% C steels. *Acta Mater.* **2001**, *49*, 4161–4170. [CrossRef]
41. Lambert, A.; Drillet, J.; Gourgues, A.F.; Sturel, T.; Pineau, A. Microstructure of martensite-austenite constituents in heat affected zones of high strength low alloy steel welds in relation to toughness properties. *Sci. Technol. Weld. Join.* **2000**, *5*, 168–173. [CrossRef]
42. Caballero, F.G. Carbide-free bainite in steels. In *Phase Transformations in Steels*; Elsevier: Amsterdam, The Netherlands, 2012; Volume 1, pp. 436–467. ISBN 9781845699703. [CrossRef]
43. Dyson, D.J. Holmes B effect of alloying additions on the lattice parameter of austenite. *J. Iron Steel Inst.* **1970**, *208*, 469–474.
44. Xiong, X.C.; Chen, B.; Huang, M.X.; Wang, J.F.; Wang, L. The effect of morphology on the stability of retained austenite in a quenched and partitioned steel. *Scr. Mater.* **2013**, *68*, 321–324. [CrossRef]

© 2020 by the authors. Licensee MDPI, Basel, Switzerland. This article is an open access article distributed under the terms and conditions of the Creative Commons Attribution (CC BY) license (http://creativecommons.org/licenses/by/4.0/).

Article

A Comprehensive Study into the Boltless Connections of Racking Systems

Rodoljub Vujanac [1], Nenad Miloradović [1,*], Snežana Vulović [2] and Ana Pavlović [3]

[1] Faculty of Engineering, University of Kragujevac, Sestre Janjić 6, 34000 Kragujevac, Serbia; vujanac@kg.ac.rs
[2] Department of Technical-Technological Sciences, Institute of Information Technologies, University of Kragujevac, Jovana Cvijića bb, 34000 Kragujevac, Serbia; vsneza@kg.ac.rs
[3] Department of Industrial Engineering, University of Bologna, Viale Risorgimento 2, 40136 Bologna, Italy; ana.pavlovic@unibo.it
* Correspondence: mnenad@kg.ac.rs; Tel.: +381-34-335-990

Received: 2 February 2020; Accepted: 18 February 2020; Published: 20 February 2020

Abstract: In practice, structures of pallet racks are characterized by very wide options of beam-to-column connections. The up to date part of the standard Eurocode 3 considers details for the design of connections. However, experimental determination of the joint properties in steel pallet racks is the most reliable process, since it takes into account an inability to develop a general analytical model for the design of these connections. In this paper, a test procedure for the behavior of beam-to-column connections is presented and the results are analyzed according to the procedure defined in the relevant design codes. With aim to avoid expensive experiments to determine structural properties of different types of connections, a polynomial model and a corresponding numerical model were developed to be used for simulating the experiment. After verification, the developed analytical and numerical model can be applied for investigation of various combinations of beam-to-column connections.

Keywords: pallet rack; moment-rotation curve; connection; experiment; numerical analysis

1. Introduction

Racking systems play a key role in satisfying today's manufacturing and distribution needs that are determined by competitive markets. When choosing storage equipment, an engineer is faced with a wide variety of options. Racking systems, ranging from selective/adjustable racks, double-deep, drive-in or drive-through configurations, to live pallet storage, push-back and mobile storage systems, are all conventional pallet racking configurations. All these different types of racks vary slightly in their structure and functioning. They are self-sustaining thin-walled steel constructions, with the ability to carry significant vertical and lateral loads. Racking systems are designed as easy-to-install structures and this means that connections must be easily detachable in order to allow the users to change the layout according to their needs. Thus, bolted and welded connections do not qualify for these purposes. The design and development of connections between parts of the spacious pallet racking system are very important due to carrying the capacity and profitability of a steel structure. Cold-formed, boltless, semi-rigid connections between the beams and columns of a frame pallet structure offer cost savings from materials and from the costs of manufacturing and assembly, which are the main reason for their wide application. Nevertheless, pallet rack structures are prone to structural failure due to lateral loads e.g., seismic loads due to semi-rigid connections between the beams and columns. For this reason, special European standards and regulations give guidelines for structural design requirements to all types of adjustable pallet racking systems, especially for the self-sustaining warehouses, fabricated from steel members subject to seismic actions. The modern technical practice treats connections according to the European Eurocode 3 standard [1]. The study on joint rigidity dates back to the beginning of the 1990s, including both the experimental study and the analytical approach. However, the studies on joints in cold-formed steel structures, particularly those of pallet rack systems, are only a few decades old.

A simple design approach which ensures the stability of pallet rack structures and includes the influence of the form of the moment-rotation characteristics on the type of stability of the system was described by Lewis in 1991 [2].

In order to determine the parameters governing an efficient beam-end-connector design, Markazi et al. [3] performed tests on four different types of beam-end-connectors. Research presented in reference [4] implies that the required ductility does not depend on the stiffness of the connector. A comparison between results of an elastic 3D linear analysis of the connector and corresponding experimental results is presented and discussed in reference [5].

The research presented by Bernuzzi et al. [6] in 2001 points out the impracticality of analytical tools in the prediction of the stiffness and strength of connectors due to wide variations in the beam-end connectors and the fact that major international codes for rack design demand the conducting of experiments in order to determine the properties of connectors.

Using the cantilever and double cantilever test set-ups, Bajoria and Talikoti [7] conducted experiments to determine the flexibility of the beam-to-column connectors of conventional pallet racking systems. For the verification of results, a full-scale frame test was conducted. The double cantilever set-up was found to be superior to the conventional single cantilever test because the shear-to-moment ratio in an actual frame is better presented by this test. In addition, both tests together with the full-scale test were subjected to non-linear finite element analyses.

Prabha et al. [8] proposed two analytical models for the calculation of the stiffness of cold formed boltless semi-rigid pallet rack connections: the polynomial model based on the Frye-Morris method and the power model. It was established that the polynomial model predicts the initial stiffness of the tested connections reasonably well and that it is useful in linear design space, while the power model can predict the ultimate capacity of the connection.

The results of the experimental tests conducted on double-sided semi-rigid beam-to-column joints as predominant joints in typical pallet racking systems were analyzed by Krolin 2014. A comparison between the experimentally obtained stiffness and the bending moment of the double-sided and the single-sided joints was presented in reference [9].

Large displacements, geometrical properties and material nonlinearities were taken into account in a 3D non-linear finite element model developed in Shah et al. [10] in 2016. The model was verified by comparing the numerical data with experimental data; good agreement between the two sets of results was obtained.

In order to predict the initial rotational stiffness of the beam-to-column connections used in cold-formed steel racks, Zhao et al. [11] developed a corresponding mechanical model in 2017. The model was verified by experiments and the obtained results showed good agreement between the initial rotational stiffness given by the model and that recorded in the experimental results. The main factors influencing the observed initial rotational stiffness of the connections that were included in the model were also discussed in the paper.

In 2018, Gausella et al. [12] presented the results of monotonic and cyclic tests carried out on four different types of industrial rack joints. The experimental results from the cyclic tests enable the moment-rotation curves of joints to be accurately defined, confirming that the industrial rack joints are significantly different from traditional joints used in steel framed buildings due to the pinching in hysteresis loops. The curves obtained in the cyclic tests can also be used for reliable modeling of joints in the analysis of seismic behavior of steel pallet racks.

The behavior of the beam-to-column connections and the column bases has a major influence on the stability of rack structures [2–12]. Complex design details such as different mechanical devices used in beam-to-column connections (tabs and hooks without bolts and welds) do not allow the flexural behavior of the beam-to-column connections of steel storage racks to be easily predicted.

The properties of the beam-to-column connections can be determined only through experiments because it is currently impossible to develop a general analytical model. In this paper, with aim to define moment-rotation curve (M-Φ curve) a test procedure for the behavior of beam-to-column connections is presented and the results are analyzed according to the procedure (cantilever and/or portal test method) defined in the current design codes for steel pallet racks, FEM (European Materials Handling Federation) [13] and European standard EN 15512 [14]. Since the experiments are too expensive, in order to reduce the costs of determination of joint properties, this paper presents a polynomial model as well as numerical model developed for the simulation of the experiment. After the verification of the models by comparing the simulation results with the available experimental results, the proposed models can be applied to various combinations of beam-to-column connections. The model makes it possible to determine characteristics of connections. The determined structural properties can be used for the comprehensive study of the racking structure and for the analysis of each element by following the proceedings from the code [13] and standard [14]. European Standard EN 16681 [15] deals with all the relevant and specific seismic design issues for racking systems, based on the criteria defined in EN 1998-1, Eurocode 8 [16]. While the basic technical description of an earthquake is the same for all structures, the general principles and technical requirements applicable for conventional steel structures have to be adapted for racking systems, in order to take the peculiarities of racking to achieve the requested safety level into account [15].

2. Configuration of a Pallet Racking

A typical selective pallet rack configuration is shown in Figure 1. The side frames and horizontal beams, usually made of thin-walled cold-formed profiles, form a spatial frame structure of the pallet racking system. The horizontal and vertical bracing system of frames provides the rack stability in the cross-aisle direction. The beam-to-column connectors as a special part are welded to the beams or otherwise formed as an integral part of the beam. They have special devices like tabs, stud or hooks engaged in the perforations of the column. In this way, through the stiffness of the beam-to column connection, the stability of the rack in the direction of the corridor is ensured.

In general, starting from the traditional assumption of the ideal connections among the elements in the joint, connections are classified as rigid or elastic. Nevertheless, the modern practice and experiments have confirmed behavior of some joints between ideal characteristics. Thus, a new division of the joints arose on:

- simple or elastic joints,
- semi-rigid joints and.
- continuous or rigid joints.

The new semi-rigid joint between the main racking elements provides completely specific behavior of spatial racking structure. Such behavior of joint in thin walled structures of rack is caused by deformation of the special devices on a beam-end connector, destruction of the upright perforation and distortion of the column walls. In practice, there are different types and designs for these connections, which are characteristic of different producers of racks [3].

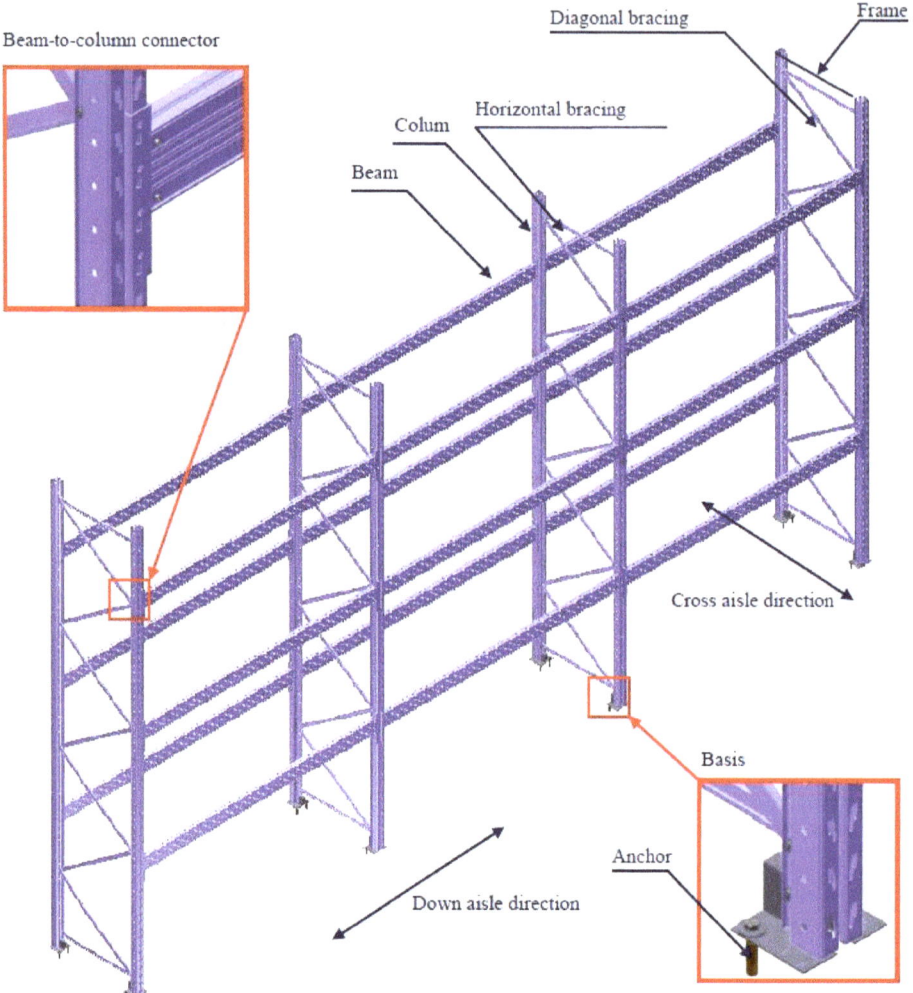

Figure 1. Parts of the racking system.

3. Analytical Approach and Experimental Study on Beam-to-Column Connections

The diagram shown in Figure 2 defines correlation between the bending moment at the connecting point, $M_{j,Ed}$, and the relative rotation of the joint, Φ_{Ed}. This M-Φ curve (M-Φ characteristic) can be reliably determined in several ways: through an experiment, by using semi-empirical expressions developed for different connections or by using numerical methods or the recommendations from FEM codes and Eurocodes. Sometimes the real M-Φ curve includes some initial deviations due to the various effects such as insufficient alignment of the elements in the assembly or mistakes in production and installation.

The outcome can be the significant initial rotation and this mast be taken into account when deriving the M-Φ curve.

Three zones with their boundaries (1, 2 and 3, respectively) corresponding to the rigid, semi-rigid or simple joints with their structural properties that can be determined by using the M-Φ curve are shown in Figure 2:

- bending strength, $M_{j,Rd}$,
- rotational stiffness, S_j, and
- rotational capacity, Φ_{Cd}.

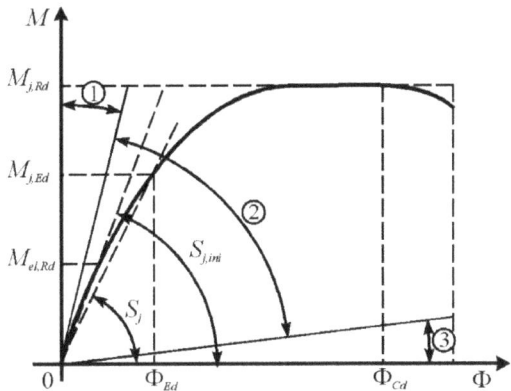

Figure 2. Moment-rotation characteristics.

3.1. Cantilever Test

The purpose of the test is to determine the stiffness and the bending strength of the beam-end connector [13,14]. The behavior of the beam-to-column connection is influenced by both members in the joint with many of their characteristics. Some of the factors, which must be taken into account during the analysis of joint behavior, are:

- the column profile,
- the thickness of the column wall,
- the beam profile,
- the thickness of the beam walls,
- the position of the connector on the beam,
- the way of connection between the connector and the beam,
- the connector type and
- the characteristics of the materials for all elements in the connection.

The combinations of the factors of the beam-to-column connection mentioned above that occur within the pallet racking system should be considered separately.

According to the standard procedure defined in [13] and [14] for each beam-end connector column joint, a minimum of three identic tests should be done in order to statistically interpret obtained results.

3.1.1. Experiment Set-up

Figure 3 shows cantilever bending test arrangement. Within the very rigid testing frame as shown in the Figure 3, a short part of the racking column is fitted whose length should satisfy the following condition:

$$h_c < c + 2 \cdot b \qquad (1)$$

During the experiment, the column should not come in the contact with the testing frame outside this distance. Connection between short piece of the beam and stiff column is obtained over a beam-end connector. The beam is secured from disassembling by means of a beam locks during the test.

A special part of the experiment settings is lateral guides, which prevent lateral movement and twisting of the beam end. However, these guides provide the beam end to move freely in the direction of the applied force.

The force is applied at a distance of 400 mm far from the perforated face of the column by a loading jack that is at least 750 mm long between the support at the testing frame and beam level according to the test set up. The rotation shall be determined by either of the following:

- two gauges C_1 and C_2 as shown in Figure 3 bearing onto a plate fixed to the beam near to the connector, but far from it in order to allow for connector distortion, or
- by an inclinometer connected to the beam close to the connector.

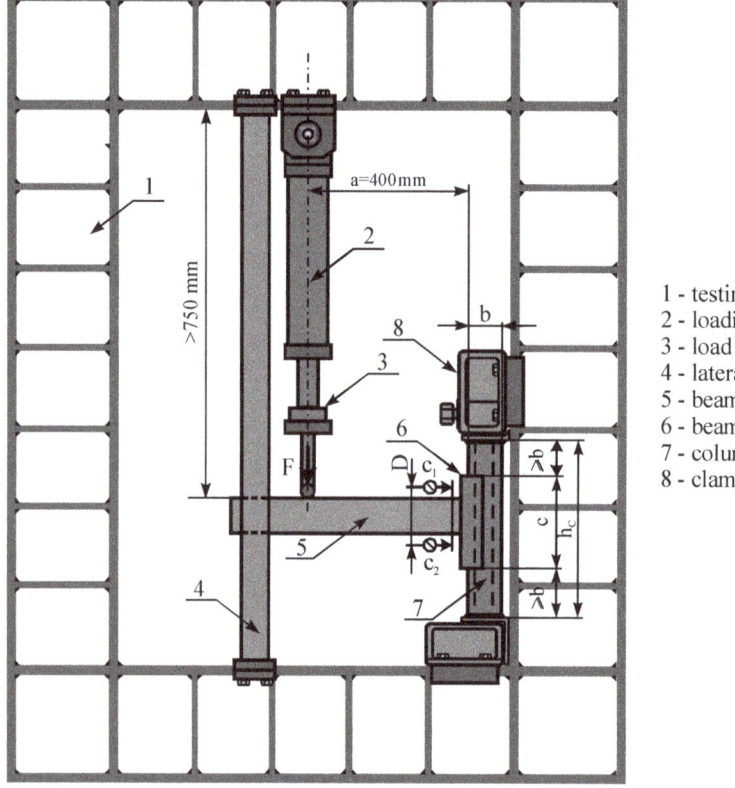

Figure 3. Cantilever test set-up.

A complete procedure for cantilever bending test is defined in code [13] and standard [14].

Table 1 shows four combinations of tested samples with different size of column and beam wall thickness. Materials with their standard properties used for elements of the connection are S350 GD Z 200 UNI EN 10326 for the column and S320 GD Z 200 UNI EN 10326 for the beam.

Table 1. Joint combinations of columns and beams.

No.	Joint (Column-Beam)	Column Wall Thickness, mm	Thickness of the Beam Wall, mm	Height of Beam Profile, mm	Experimental Sample
1	S80ML-R100L	1.5	1	100	G-5, G-6, G-7, G-8, (G-9)
2	S80ML-R120L	1.5	1	120	H-5, H-6, H-7, H-8, (H-9)
3	S80ML-R140ML	1.5	1	140	I-5, I-6, I-7, I-8, (I-9)
4	S80M-R120M	2	1.25	120	A-5, A-6, A-7, A-8, (A-10)

Dimensions of all parts in the connection and their position necessary for the experiment performing are shown in Figure 4. Table 2 shows values of all dimensions shown in Figure 4 for the five samples of three tested joints made by the same producer.

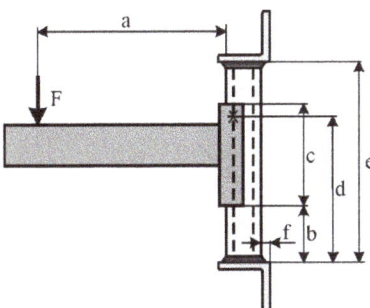

Figure 4. Arrangement of the parts of the sample.

Table 2. Dimensions of the elements in connection.

Joint	Sample	a, mm	b, mm	c, mm	d, mm	h_c, mm	e, mm
S80ML-R100L	G-5	400	119	215	311	454	17.0
	G-6	400	120	215	312	454	19.1
	G-7	400	118	214	311	454	19.9
	G-8	400	118	214	311	454	19.3
	G-9	400	119	215	140	454	18.5
S80M-R120M	A-5	400	120	214	311	455	23.2
	A-6	400	120	215	311	454	23.0
	A-7	400	120	215	311	455	23.6
	A-8	400	120	215	311	455	23.9
	A-10	400	120	214	143	455	23.5
S80ML-R140L	I-5	400	120	290	389	530	19.7
	I-6	400	120	290	387	530	19.2
	I-7	400	120	290	387	529	20.1
	I-8	400	120	289	387	529	18.8
	I-9	400	120	290	141	529	18.9

The disposition of measuring devices is shown in Figure 5. In Table 3, dimensions of the position of all devices for measuring displacement are given. Bending tests were performed on five samples of each joint as shown in Tables 2 and 3. Each sample consists of the short part of the beam with beam-end connector connected to the short part of the racking column and secured from the disassembling by a beam lock. The samples of each joint marked from 5 to 8 are subjected to the force which generates positive bending moment under normal operating conditions, while the fifth sample marked as 9 is loaded in such a way that the applied force generates a negative bending moment which endeavors to separate the connection.

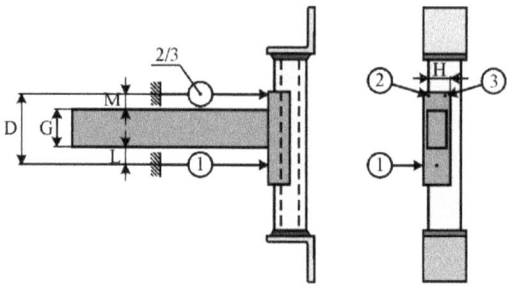

Figure 5. Disposition of the measuring equipment.

Table 3. Dimensions of the position of measuring equipment.

Joint	Sample	G, Mm	H, mm	L, mm	M, mm
S80ML-R100L	G-5	99.5	69.2	47	12
	G-6	99.5	69.3	55	19
	G-7	99.5	69.3	56	18
	G-8	99.6	69.4	55	19
	G-9	99.6	69.4	49	33
S80M-R120M	A-5	119.9	69.5	35	18
	A-6	119.9	69.7	35	18
	A-7	119.9	69.3	27	19
	A-8	119.6	69.4	27	18
	A-10	119.9	69.5	30	30
S80ML-R140L	I-5	139.1	69.0	54	18
	I-6	139.5	69.2	54	17
	I-7	139.0	69.2	55	16
	I-8	139.0	69.1	54	17
	I-9	139.3	69.3	46	35

3.1.2. Experiment Procedure

Within the performed tests, the applied force acting in the downward direction parallel to the beam-end connector causes shear. If tests in the upward direction show the results for stiffness and strength, which are less than 50% of the values measured in these tests, then the actual figures will be measured to be used in the design. The design of the connectors should use the mean value for the stiffness and strength obtained from the values for the right and left connectors.

The load, F, must be slowly increased until the moment at the connector reaches a value equal to 10% of the failure moment in order to mount the components. After assembling of the parts, load should be removed and displacement transducers reset. Then, the gradual increase of the load F should be applied until the maximum is reached and the connection breaks.

For each test, the moment, M, and the rotation, Φ, should be plotted using the following relations [13,14,17]:

$$M = a \cdot F, \tag{2}$$

and

$$\Phi = \frac{\delta_2 - \delta_1}{D}, \tag{3}$$

where:
- a—the length at which the force F acts,
- D—distance between the displacement transducers on the opposite sides of the beam,
- δ_1, δ_2—displacement measured by gauges C_1 and C_2.

Finally, the connection rotation, Φ, is determined according to the expression:

$$\Phi = \frac{\frac{\delta_2+\delta_3}{2} - \delta_1}{L+G+M}, \quad (4)$$

where:

- δ_3 is the displacement measured by gauge C_3,
- L, G and M are dimensions of gauges position as shown in Figure 5.

Cantilever bending tests on beam-end connectors up to their collapse under normal operating conditions (bending moment is conventionally defined positive) were performed on the racking elements described previously. Initial loading-unloading cycles for the assembly and fitting of the connected parts were provided, up to the maximum load level of F_0, after which the load was increased incrementally until it reached the value of the failure load, F_{ti}. Table 4 shows the maximum measured values of the achieved force, F, for each sample, with corresponding failure moments, M_{ti}, calculated according to formula (2). Duration of each test, t, is also given in Table 4 for each sample.

Table 4. Values of the obtained force and moment.

Joint	Sample	F_0, kN	F_{ti}, kN	t, s	M_{ti}, kNm
S80ML-R100L	G-5	0.762	4.168	496	1.667
	G-6	0.393	4.087	577	1.635
	G-7	0.402	4.093	575	1.637
	G-8	0.394	4.084	495	1.634
	G-9	−0.179	−2.486	512	−0.994
S80M-R120M	A-5	0.581	5.678	340	2.271
	A-6	0.578	5.458	381	2.183
	A-7	0.563	5.708	407	2.283
	A-8	0.554	5.741	436	2.2.96
	A-10	−0.542	−3.516	412	−1.407
S80ML-R140L	I-5	0.516	6.042	375	2.417
	I-6	0.604	6.029	383	2.412
	I-7	0.618	6.132	383	2.453
	I-8	0.600	6.270	400	2.508
	I-9	−0.305	−3.267	348	−1.307

3.1.3. Test Results

The maximum observed moment seen in Figure 6 is the failure moment, M_{ti}. The mean value, M_m, of the individual test results is:

$$M_m = \frac{1}{n} \cdot \sum_{i=1}^{n} M_{ti}. \quad (5)$$

For each tested joint, the characteristic failure moment, M_k, can be determined according to procedure defined in [13] for the derivation of characteristics values as following:

$$M_k = M_m - k_s \cdot s, \quad (6)$$

in which:

k_s is the coefficient given in [13], which depends on the number of tests (for $n = 4$, $k_s = 2.68$),
s is the standard deviation of the adjusted test results according to the following expression [13]:

$$s = \sqrt{\frac{1}{(n-1)} \sum_{i=1}^{n} (M_{ti} - M_m)^2}. \quad (7)$$

The design moment, M_{Rd}, for the connection is as follows:

$$M_{Rd} = \eta \cdot \frac{M_k}{\gamma_M},\qquad(8)$$

in which:

γ_M is the partial safety factor for connections, [1,13],

η is variable moment reduction factor selected by the designer ≤ 1.

It is permissible to choose any value of the design moment less than or equal to the allowable maximum in order to optimize the possibly conflicting requirements for stiffness and strength. Thus, by reducing the design strength, it is possible to achieve a greater design stiffness.

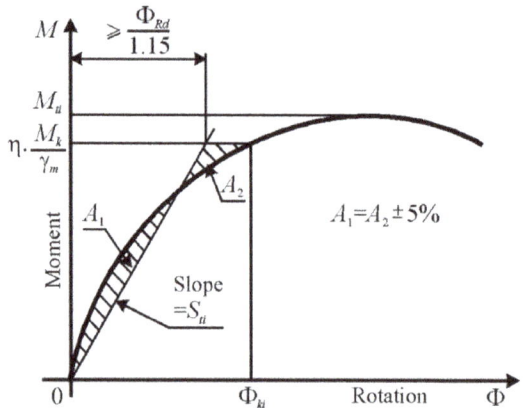

Figure 6. Derivation of moment-rotation relationship.

The rotational stiffness of the connector, S_{ti}, is the slope of the line going through the origin and forming the equal areas between the straight line and the experimental curve below the design moment, M_{Rd}, (Figure 6), under condition [15]:

$$S_{ti} \leq 1.15 \cdot \frac{M_{Rd}}{\Phi_{ki}}.\qquad(9)$$

The design value of the connector stiffness, S_d, should be taken as the average value, S_m, as shown in Table 5, where:

$$S_m = \frac{1}{n}\sum_{i=1}^{n} S_{ti}.\qquad(10)$$

Table 5. Obtained experimental results.

Joint	Sample	M_{ti}, kNm	M_m, kNm	M_k, kNm	M_{Rd}, kNm	S_{ti}, kNm/rad	S_m, kNm/rad
S80ML-R100L	G-5	1.667	1.643	1.601	1.455	32.65	37.20
	G-6	1.635				41.07	
	G-7	1.637				35.28	
	G-8	1.634				39.79	
S80M-R120M	A-5	2.271	2.258	2.121	1.928	46.22	44.15
	A-6	2.183				43.05	
	A-7	2.283				43.43	
	A-8	2.296				43.90	
S80ML-R140L	I-5	2.417	2.448	2.329	2.117	60.26	63.42
	I-6	2.412				64.30	
	I-7	2.453				62.15	
	I-8	2.508				66.98	

3.2. Frye-Morris Polynomial Model

The Frye-Morris method [8] proposes a non-dimensional polynomial model for determining the moment-rotation characteristic of a single connection; the model is generated by replacing the numerical values of its individual parameters in a standardized connection. The parameters used to determine the equation can be: the thickness of the wall of the column, t_u, the beam height, d_b and the thickness of the wall of the beam profile, t_b. The standardized link is then given by the equation:

$$\Phi_r = C_1(K \cdot M) + C_2(K \cdot M)^3 + C_3(K \cdot M)^5, \qquad (11)$$

where:

- Φ_r—relative rotation in rad,
- M—moment of rotation in Nmm,
- K—coefficient that scales the ordinate of curves,
- C_1, C_2, C_3—constants for curve fitting.

The coefficient K which scales the ordinates of the curves taking into account the numerical value of the individual connection parameters is calculated according to:

$$K - \prod_{j=1}^{m} q_j^{a_j}, \qquad (12)$$

where:

- q_j—numerical value of j parameter,
- a_j—exponent that shows the effect of the numerical value of the j parameter on the moment-rotation relation,
- m—number of parameters j.

The determination of the exponent a_j in Equation (12) is performed on the basis of the pair of experimentally obtained moment-rotation curves for two identical joints, but in which the parameter q_j is not included.

The relationship between the moments M_1 and M_2 for connections 1 and 2 at rotation Φ is assumed in the form:

$$\frac{M_1}{M_2} = \left(\frac{q_{j1}}{q_{j2}}\right)^{a_j}, \qquad (13)$$

where q_{j1} and q_{j2} are the values of the parameters q_j for connections 1 and 2, respectively.

From relation (13), the coefficient a_j can be expressed according to:

$$a_j = \frac{\log(M_1/M_2)}{\log(q_{j2}/q_{j1})}. \tag{14}$$

Expression (14) is used to calculate the values of a_j corresponding to different rotations for each combination of experimental curves. When the mean value is calculated for all "m" exponents a_j, they are applied to a standardized moment-rotation diagram. Finally, the curve fitting is done to generate a standardized moment-rotation connection.

The mean value of a_1 for variable column thickness is −0.126, the mean value of a_2 for variable heights of the beam is −2.981 and the mean value of a_3 for the variable thickness of the beam is −0.121. Therefore, the standardized coefficient K is expressed as:

$$K = t_u^{-0.126} \cdot d_b^{-2.981} \cdot t_b^{-0.121}. \tag{15}$$

Constants for curve fitting obtained for all connectors are shown in Table 6. They are calculated using a procedure developed in Microsoft Excel [17].

Table 6. Constants for curve fitting.

Joint	Sample	C_1	C_2	C_3
S80ML-R100L	G-5	34.10479	0.0200	0.0002
	G-6	28.20721	0.0201	0.0003
	G-7	30.61559	0.0701	0.0006
	G-8	27.61125	0.0144	0.0002
S80M-R120M	A-5	44.65244	0.0159	0.0001
	A-6	46.20446	0.0875	0.0005
	A-7	47.30988	0.0133	0.0008
	A-8	47.27306	0.0125	0.0006
S80ML-R140L	I-5	41.19487	0.0225	0.0004
	I-6	38.47692	0.0326	0.0006
	I-7	41.81794	0.0847	0.0007
	I-8	36.31579	0.0223	0.0003

The mean values of the coefficients are:

$$C_1 = 43.693;\ C_2 = 0.0393;\ C_3 = 0.000435. \tag{16}$$

The Frye-Morris equation for the observed structure is:

$$\Phi_r = 43.693(K \cdot M) + 0.0393(K \cdot M)^3 + 0.000435(K \cdot M)^5. \tag{17}$$

4. Numerical Analysis of Beam-to-Column Connection

4.1. Finite Element Model of Cantilever Test

As a state-of-the-art method in the field of structural analysis, the Finite Element Method is commonly addressed to provide accurate and reliable predictions of structural deformation and stress states. Recent developments [18] enable high computational efficiency of finite element models even if nonlinear effects are involved. Finite element models for the cantilever test were generated in Femap with the NX Nastran software version 2019 sold by software company Siemens Digital Industries Software from Plano, TX, USA, based on the data given in the tables in chapter 3.1.1 provided by the producer of the equipment. The numerical analysis was conducted using the elasto-plastic material model with kinematic reinforcement made in the LS-Dyna software version R.9.0.1. developed by

Livermore Software Technology Corporation (LSTC) from Livermore, CA, USA. The elements of the tested samples, shown in Figure 3 are modeled with finite elements as following:

- Column with 41,356 3D 8-nodal finite elements, Figure 7a.
- Beam with 150,540 3D 8-nodal finite elements, Figure 7b.
- Beam-end connector with 63,777 3D 8-nodal finite elements, Figure 7c.
- Screws for joint beam-end connector and beam with 3395 3D 8-nodal finite elements, Figure 7d.
- Load transfer plate with 4136 shell elements, Figure 7e.
- 1D finite elements, i.e., rods were used for load modeling, Figure 7f.
- Surface-to-surface contact elements were used for the connected parts in samples: the column-beam end connector, beam end connector-beam and beam parts for blocking the lateral movement.

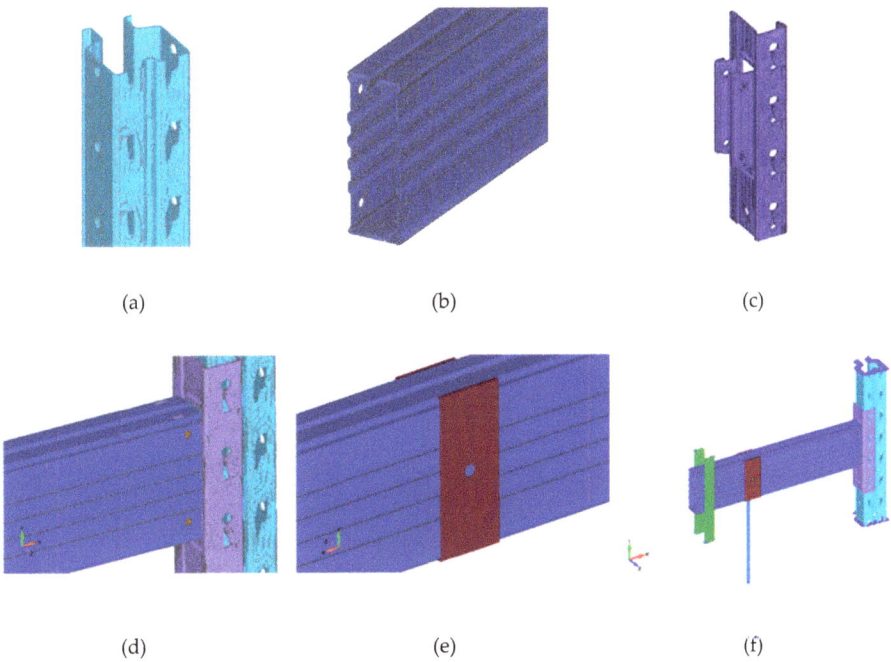

Figure 7. Finite element model of elements of cantilever test: (**a**) Column; (**b**) Beam; (**c**) Beam-end connector; (**d**) Screw; (**e**) Loading transfer plate; (**f**) Boundary conditions and the applied load.

As shown in Figure 7f, the corresponding movement acts according to the diagram shown in Figure 8 along the direction of the rod. Actually, the value of the movement at the end of the rod on which the load acts is calculated using the experimental data based on dependence between the angle of rotation and the corresponding force value, i.e., the bending moment.

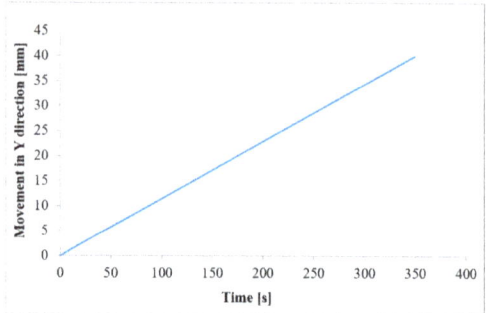

Figure 8. Diagram of the displacement.

Figure 9 shows the experimentally obtained M-Φ curves for the four tested samples of the S80ML-R140L joint, the curve generated by using the analytical polynomial model of the Frye-Morris method, and the curve generated using a finite element model of the tested joint.

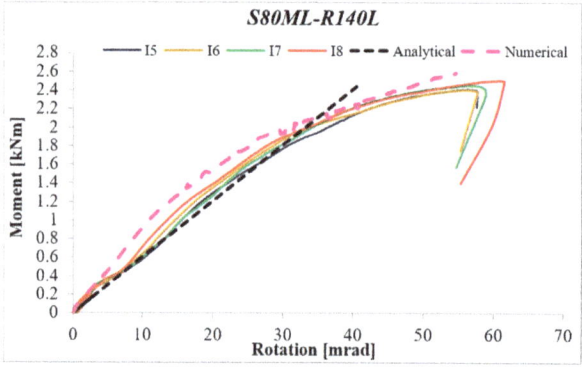

Figure 9. Moment-rotation curves for the S80ML-R140L joint.

Very good agreement between the experimental results and the proposed polynomial model of the initial part of the moment-rotation capacity curve can be observed in Figure 9. The Frye-Morris method-based analytical polynomial model contains the standardization coefficient K, which involves three dimension parameters: column wall thickness, beam profile height and beam profile wall thickness. This constant is evaluated using the experimental results. The numerical analysis made by the LS-Dyna software showed that the finite element model developed using the test results was the best fit for experimental behavior. Figure 10 shows the deformation of one of the tested samples and displacement fields obtained by the numerical model, which shows good agreement. That is why the validated finite element model can be used in further parametric studies.

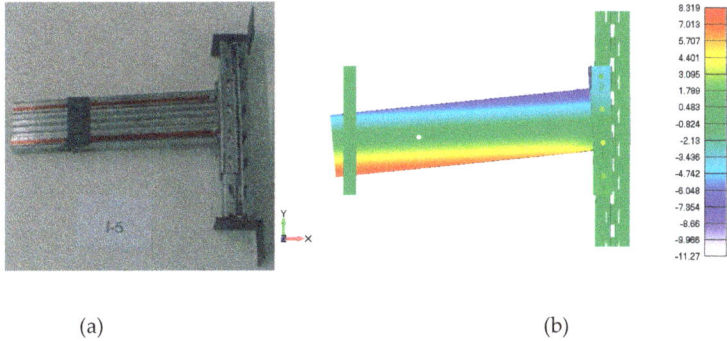

(a) (b)

Figure 10. Comparison of the results of the tested joints S80ML-R140L: (**a**) Deformation of the tested sample (**b**) Displacement field in the x direction obtained by the numerical model.

Figure 11 shows the M-Φ curves for the S80M-R140M joint obtained by the numerical model and the analytical application of expression (14). Using the described methodology for determination of the rotational stiffness of the connection according to references [13,15], defined in chapter 3.1.3, a value of 74.65 kNm/rad was obtained for the observed joint, as shown in Figure 11. The use of the numerical model only has its limitations because the applied maximum moment resistance is the mean value in determination of the rotational stiffness [15]. However, despite this limitation, developed finite elements model can be applied to determine the structural properties of the beam-to-column connection.

The numerical model provides a detailed further investigation of each constituting part as well as of the structure as a whole. In addition, the validated finite element model can be used for a parametric analysis and identification of the effects of various parameters on the overall performance of the observed beam-to-column connection. Further parametric studies should analyze the influence of the number of tabs or hooks, the column thickness and the connector depth. The design of the beam-to-column connector and the efficiency of the accompanying members (beam, column) determine the moment-rotation characteristics of the joint.

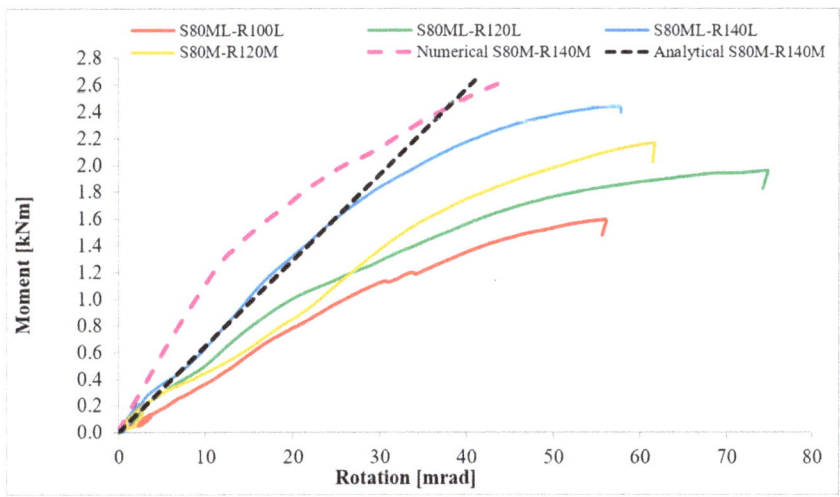

Figure 11. Comparative average moment-rotation curves for the tested samples and numerical model.

4.2. Influence of the Column and Beam Wall Thickness on the Behaviour of the Connection

Comparing the average moment-rotation characteristics for tested samples (shown as green and orange curves in Figure 11) with the simultaneous change of the thickness of the wall of the column profile from 1.5 mm to 2.0 mm and the thickness of the wall of the beam profile from 1.00 mm to the 1.25 mm, with the same other parameters of the joints (height of the beam profile and beam-end connector), the bending strength is increased by 10%. The rotational stiffness remains almost unchanged. Further analysis of the behavior of the joint S80M-R140M numerically obtained, given as pink curve in Figure 11, with the experimentally curve of the joint S80ML-R140L, shows increasing of the bending strength by 13%, while rotational stiffness is increased by 15% after taking in consideration the limitations of the numerical model. From this analysis, it follows that the wall thickness of the column and beam profile has a dominant influence on the bending strength of the connection.

4.3. Influence of the Height of the Beam Profile and Beam-End Connector on the Behaviour of the Connection

Observing the joints with the same column wall thickness of 1.5 mm and the same height of the beam-end connector, with three "teeth" of 215 mm height, after changing only the height of the beam profile with the wall thickness of 1.00 mm from 100 mm up to 120 mm as shown in Figure 11 on red and green curves, the bending strength and rotational stiffness increases proportionally by 20%. However, by changing the height of the beam profile from 100 mm to 140 mm, the bending strength increases proportionally by 40%, while the rotational stiffness increases by 70% as shown in Figure 11 with red and blue curves and the data given in Table 5. The beam-end connector with three teeth in height had to be changed with a four-tooth connector from 215 mm to 290 mm. The conclusion is that dominant influence on the rotational stiffness has a beam-end connector with its parameters.

5. Conclusions

Better cognition of behavior of the beam-to-column connection based on the M-Φ characteristics is of great importance for comprehensive analysis of the joint in the structure and its influence on the whole spacious construction of racking system. Design codes like the FEM [13] or EN 15512 standard codes [14] demand experimental testing for which they supply the testing protocols with marginal differences for predicting the moment-rotation M-Φ behavior of any pallet rack beam-to-column connection. However, the prescribed experimental testing is expensive, so a possible solution may be found in the development of a particular uniform M-Φ relationship for each type of connection in terms of parameters by using analytical prediction or finite element modeling. The experimental testing of the beam-to-column connection was the subject of a large number of recent studies. However, few studies have considered the behavior of this connection numerically.

The Frye-Morris method-based analytical polynomial model predicts reasonably well the initial stiffness of the tested connections, but it cannot capture the overall strength of the connection. Analytical models have not given satisfactory results so far, while the usage of numerical method in combination with experimental testing provides very useful results. By simply changing the model of the experiment, which is possible due to modern computer technology, characteristics of the connections of various combinations of elements in the joint can be examined and determined. In this way, rough approximations of real characteristics and their introduction into the calculation are avoided.

Further research in the field of semi-rigid connections of the pallet rack elements will certainly refer to the possibility of improving the connection. This can be achieved, for example, by increasing the load capacity, i.e., by enabling an additional, multiple contacts between the parts of the beam-end connector and the column. Possible problems that may occur in this case, such as the weakening of the column because of multiple perforations, should be previously analyzed and solved using the proposed numerical model. Numerical analysis enables rapid and optimized construction without the need for expensive experiments.

Author Contributions: Conceptualization, R.V. and S.V.; methodology, R.V.; software, R.V. and S.V.; validation, S.V and A.P.; formal analysis, R.V., N.M. and S.V.; investigation, R.V. and A.P.; resources, R.V. and S.V.; writing—original draft preparation, R.V. and N.M; writing—review and editing, R.V. and N.M.; visualization, R.V. and S.V; supervision, R.V. All authors have read and agreed to the published version of the manuscript.

Funding: This research received no external funding.

Conflicts of Interest: The authors declare no conflict of interest.

References

1. *Eurocode 3: EN 1993-1-1: 2005, Design of Steel Structures, Part 1-1: General Rules and Rules for Buildings, Part 1-3: General Rules—Supplementary Rules for Cold Formed Thin Gauge Members and Sheeting, Part 1-8: Design of Joints*; European Union: Brussels, Belgium, 2005.
2. Lewis, G.M. Stability of rack structures. *Thin Walled Struct.* **1991**, *12*, 163–174. [CrossRef]
3. Markazi, F.D.; Beale, R.G.; Godley, M.H.R. Experimental Analysis of Semi-Rigid Boltless Connectors. *Thin Walled Struct.* **1997**, *28*, 57–87. [CrossRef]
4. Godley, M.H.R. Plastic design of pallet rack beams. *Thin Walled Struct.* **1997**, *29*, 175–188. [CrossRef]
5. Markazi, F.D.; Beale, R.G.; Godley, M.H.R. Numerical modeling of semi-rigid boltless connectors. *Thin Walled Struct.* **2001**, *79*, 2391–2402. [CrossRef]
6. Bernuzzi, C.; Castiglioni, C.A. Experimental analysis on the cyclic behavior of beam-to-column joints in steel storage pallet racks. *Thin Walled Struct.* **2001**, *39*, 841–859. [CrossRef]
7. Bajoria, K.M.; Talikoti, R.S. Determination of flexibility of beam-to-column connectors used in thin walled cold-formed steel pallet racking systems. *Thin Walled Struct.* **2006**, *44*, 372–380. [CrossRef]
8. Prabha, P.; Marimuthu, V.; Saravanan, M.; Jayachandran, S.A. Evaluation of connection flexibility in cold formed steel racks. *J. Constr. Steel Res.* **2010**, *66*, 863–872. [CrossRef]
9. Krol, P.A.; Papadopoulos-Wozniak, M.; Wojt, J. Experimental tests on semi-rigid, hooking-type beam-to-column double-sided joints in sway-frame structural pallet racking systems. *Procedia Eng.* **2014**, *91*, 238–243. [CrossRef]
10. Shah, S.N.R.; Sulong, R.N.H.; Khan, R.; Jumaat, M.Z.; Shariati, M. Behavior of industrial steel rack connections. *Mech. Syst. Signal Process.* **2016**, *70–71*, 725–740. [CrossRef]
11. Zhao, X.; Dai, L.; Wang, T.; Sivakumaran, K.S.; Chen, Y. A theoretical model for the rotational stiffness of storage rack beam-to-column connections. *J. Constr. Steel Res.* **2017**, *133*, 269–281. [CrossRef]
12. Gusella, F.; Lavacchini, G.; Orlando, M. Monotonic and cyclic tests on beam-column joints of industrial pallet racks. *J. Constr. Steel Res.* **2018**, *140*, 92–107. [CrossRef]
13. *FEM 10.2.02: Racking Design Code, The Design of Static Steel Pallet Racking*; European Racking Federation—FEM Racking and Shelving Product Group: Brussels, Belgium, 2000.
14. *European Standard EN 15512: Steel Static Storage Systems—Adjustable Pallet Racking Systems—Principles for Structural Design*; European Committee for Standardization: Brussels, Belgium, 2009.
15. *European Standard EN 16681:2016: Steel Static Storage Systems—Adjustable Pallet Racking Systems—Principles for Seismic Design*; European Committee for Standardization: Brussels, Belgium, 2016.
16. *Eurocode 8: EN 1998-1: Design of Structures for Earthquake Resistance—Part 1: General Rules, Seismic Actions and Rules for Buildings*; European Union: Brussels, Belgium, 2004.
17. Vujanac, R.; Vulovic, S.; Disic, A.; Miloradovic, N. Numerical analysis of beam-to-column connection of pallet racks. In Proceedings of the IOP Conference Series: Materials Science and Engineering, The 10th International Symposium Machine and Industrial Design in Mechanical Engineering (KOD 2018), Novi Sad, Serbia, 6–8 June 2018; Volume 393, pp. 1–11. [CrossRef]
18. Nguyen, V.A.; Zehn, M.; Marinković, D. An efficient co-rotational FEM formulation using a projector matrix. *Facta Univ. Ser. Mech. Eng.* **2016**, *14*, 227–240. [CrossRef]

© 2020 by the authors. Licensee MDPI, Basel, Switzerland. This article is an open access article distributed under the terms and conditions of the Creative Commons Attribution (CC BY) license (http://creativecommons.org/licenses/by/4.0/).

Article

Prediction and Analysis of Tensile Properties of Austenitic Stainless Steel Using Artificial Neural Network

Yuxuan Wang [1,2], Xuebang Wu [1,*], Xiangyan Li [1], Zhuoming Xie [1], Rui Liu [1], Wei Liu [1], Yange Zhang [1], Yichun Xu [1] and Changsong Liu [1,*]

[1] Key Laboratory of Materials Physics, Institute of Solid State Physics, Chinese Academy of Sciences, P.O. Box 1129, Hefei 230031, China; wyx811@mail.ustc.edu.cn (Y.W.); xiangyanli@issp.ac.cn (X.L.); zmxie@issp.ac.cn (Z.X.); liurui@issp.ac.cn (R.L.); wliu@issp.ac.cn (W.L.); yangezhang@issp.ac.cn (Y.Z.); xuyichun@issp.ac.cn (Y.X.)
[2] Department of Materials Science and Engineering, University of Science and Technology of China, Hefei 230036, China
* Correspondence: xbwu@issp.ac.cn (X.W.); csliu@issp.ac.cn (C.L.)

Received: 14 January 2020; Accepted: 7 February 2020; Published: 10 February 2020

Abstract: Predicting mechanical properties of metals from big data is of great importance to materials engineering. The present work aims at applying artificial neural network (ANN) models to predict the tensile properties including yield strength (YS) and ultimate tensile strength (UTS) on austenitic stainless steel as a function of chemical composition, heat treatment and test temperature. The developed models have good prediction performance for YS and UTS, with R values over 0.93. The models were also tested to verify the reliability and accuracy in the context of metallurgical principles and other data published in the literature. In addition, the mean impact value analysis was conducted to quantitatively examine the relative significance of each input variable for the improvement of prediction performance. The trained models can be used as a guideline for the preparation and development of new austenitic stainless steels with the required tensile properties.

Keywords: austenitic stainless steel; tensile properties; artificial neural network; MIV analysis

---

1. Introduction

Metallic materials are widely used in daily life, especially a variety of steel that have a very long history of research. It is known that there are many variables that can affect the properties of steels such as strength [1]. Strength is the ability of a material to resist plastic deformation or fracture, and the strength properties such as tensile strength and plasticity of steels are usually dependent on chemical composition. In addition, the heat treatments, such as annealing, tempering and quenching, can effectively control the microstructure, grain size and defects, which are all closely related to the tensile properties of steels. In addition, the tensile properties are also affected by the service conditions such as working temperature and irradiation environment [2]. To discover the inherent mechanism of how these variables affect the steels, researchers could only get the preliminary influence trends through the continuous experiments via changing a part of variables for a long time. A formula obtained with the traditional linear fitting regression is generally difficult to capture the exact correlations between relative variables and corresponding properties of steels, because they often have complex nonlinear relationships [3].

With the vigorous development of computer technology, machine learning sprang up and became a powerful method for finding the patterns in high-dimensional data [4]. Machine learning is actually an efficient statistical analysis method to capture the linear or nonlinear internal relationships by learning from empirical data [5]. The common machine learning methods include artificial neural network (ANN) [6,7], support vector machine (SVM) [8,9], decision tree (DT) [10] and so on. Nowadays, machine learning prompts data science and analytics to become a significant tool to find the desired causal relations in the material research [11], and results in developing a new field termed as "Materials Informatics" [12,13] in recent years. Machine learning has been rapidly used in the fields of metals [14–18], as well as polymers [19], semiconductors [20,21], which fully demonstrates its powerful universality.

Meanwhile, in order to meet the needs of machine learning for big data [22], many experimental data were collected and established databases such as MatNavi [23], MatWeb [24] and Matmatch [25]. MatNavi contains a large amount of data about the fatigue and creep properties of various steels, which have been already used for machine learning model establishment and research. Agrawal et al. [26] proved the practicality of machine learning for fatigue strength research with the Fatigue Data Sheet. Sourmail et al. [27] correctly captured the important influence trends using the established models with ANN based on the Creep Data Sheet. Besides the fatigue and creep properties, a few machine learning models have been established to obtain the correlations between the tensile properties of steels and the important variables. Guo et al. [28] used the ANN model to well characterize the relationships between the mechanical properties of maraging steels and composition, processing and working conditions. Fragassa et al. [29] chose the metallographic factors as the input features and designed three kinds of machine learning methods to model the mechanical properties of cast iron.

However, for austenitic stainless steel (ASS), more attention is paid on the creep, fatigue and corrosion resistance [30–32]. Besides the corrosion resistance, the mechanical properties such as strength of ASS are also important for their application and have drawn much attention in past decades. The tensile properties of ASS have been extensively studied both experimentally and theoretically. Sivaprasad et al. [33] developed an artificial neural network model to correlate alloy composition and test temperature to tensile properties of 15Cr-15Ni-2.2Mo-Ti modified ASS. Desu et al. [34] used test temperature and strain rates as descriptors to predict the tensile properties of ASS 304L and 316L using the ANN model. However, there are no general machine learning models to correlate chemical composition, heat processes and service conditions to tensile properties of ASS, and clarify how each variable affects the tensile properties of ASS.

In this work, we proposed a machine learning method using ANN to predict the tensile properties of ASS with the chemical composition, solution treatment conditions (heat processes) and test temperature (service condition) as descriptors. The models established by partial data in the database have high predictive accuracy for the remaining data and some new data outside the database. We also calculated the impact degrees of each variable with the mean impact value (MIV) method and predicted the influence trends of several important variables on tensile properties. Our results conform to the previous metallurgical theories, and the established models can guide us for further research and development of new ASS with the expected tensile properties.

2. The Database and ANN Model

2.1. Information of the Database

Our study is based on the tensile test data of some classical types of ASS including SUS 304, SUS 316, SUS 321, SUS 347 and NCF 800H, which is referenced to the Creep Data Sheet of Steel (No.4B, 5B, 6B, 14B, 15B, 26B, 27B, 28B, 32A, 42 and 45) from NIMS MatNavi and BSCC High Temperature Data from The British Steelmakers Creep Committee [35], and collected by the Material Algorithm Project (MAP) of University of Cambridge [36].

The original data contains 1916 samples, of which 1107 samples unfortunately lack the necessary information, so the remaining 809 samples are selected for further research. The data has the following characteristics: (1) Every sample has two kinds of tensile properties as output, yield strength (YS) and ultimate tensile strength (UTS); (2) The data contains 20 variables selected as input: chemical composition, solution treatment conditions and test temperature, as shown in Table 1. The chemical composition includes some common elements such as carbon (C), nickel (Ni), chromium (Cr), Nitrogen (N) and the microalloying additions such as titanium (Ti), vanadium (V) and niobium (Nb). The parameters of solution treatment conditions on ASS are temperature and time. It should be noted that after the solution treatment, in order to stabilize the austenite structure and prevent the carbide precipitation at room temperature, it is generally required to perform the rapid water quenching of the materials. Here, the samples that are water-quenched in the database are set as the label 1 and the air-cooled ones without water quenching are set as the label 0. These features are related to the physical metallurgy which is important for modeling and property predictions of ASS [37]. There are some other features in the original database, such as the type of melting, grain size and the form of products, but the data of them are incomplete or they have a lower correlation with tensile properties. So they are not considered in this study. Further information about the input variables is given in Tables A1 and A2 of Appendix A.

Table 1. The input variables of austenitic stainless steel in this research.

Number	Variables	Number	Variables
1	Chromium (Cr, wt%)	11	Carbon (C, wt%)
2	Nickel (Ni, wt%)	12	Boron (B, wt%)
3	Molybdenum (Mo, wt%)	13	Phosphorus (P, wt%)
4	Manganese (Mn, wt%)	14	Sulfur (S, wt%)
5	Silicon (Si, wt%)	15	Cobalt (Co, wt%)
6	Niobium (Nb, wt%)	16	Aluminum (Al, wt%)
7	Titanium (Ti, wt%)	17	Solution treatment temperature (T_s, K)
8	Vanadium (V, wt%)	18	Solution treatment time (t_s, s)
9	Copper (Cu, wt%)	19	Water-quenched or Air-quenched
10	Nitrogen (N, wt%)	20	Test temperature (T_t, K)

2.2. Division of Data and Pre-Processing

The data are randomly divided into two groups. Eight-three percent of the data (674 samples) are selected as the training set for the model establishment while the remaining 17% (135 samples) are employed as the testing set for accuracy verification of models, which makes the training/testing ratio close to 5/1.

Deeply understanding and pre-processing the data with appropriate normalization before modeling is one of the most vital steps of effective data mining [26]. The input data generally have more than one dimension and each variable has different size of range, so it is necessary to make each variable normalized within the range from 0 to 1 using the following equation firstly to improve the accuracy and efficiency of calculation and prediction:

$$x_n = (x - x_{min})/(x_{Max} - x_{min}), \tag{1}$$

where x_n is the normalized value of the corresponding x, x_{Max} and x_{min} are the maximum and minimum values of x respectively.

2.3. ANN Model Development

Artificial neural network (ANN) is a flexible model for non-linear statistical analysis and it can be used for both data classification and regression calculation. It looks like a box that links input data and output data together via a set of non-linear functions. More details of this method can be found elsewhere [38], but it is necessary to have a brief introduction on the main features of ANN.

A simple three-layer feedforward network is competent for general works, such as the one shown in Figure 1. As the name suggests, it consists of three layers: input, hidden and output. The transfer function in the second layer can be any kind of non-linear function as long as it is continuous and differentiable, such as the hyperbolic tangent function tanh (Equation (2)), which can effectively capture the interaction between the inputs and map many functions of practical interest [39]. The transfer function in the third layer is usually linear (Equation (3)):

$$h_i = \tanh\left(\sum_j w_{ij}^{(1)} x_j + \theta_i^{(1)}\right), \tag{2}$$

$$y = \sum_i w_{ij}^{(2)} h_i + \theta^{(2)}, \tag{3}$$

where x_i are inputs, w_i the weights which determine the strength of the transfer function and the biases h_i the analogous just like the constant in linear regression. The number of neurons in the hidden layer determines the complexity of the model and intensely influences the effect of modeling. How to determine it will be introduced later.

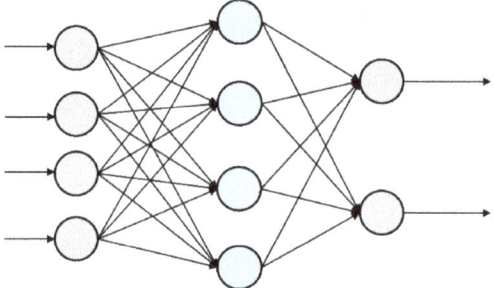

Figure 1. Structure of a three-layer feedforward neural network.

We generally take the following 6 steps for the ANN model development: (1) determine the input and output variables and collect the data; (2) pre-process the data such as normalizing; (3) divide the original database into the training set and testing set; (4) use the training set for modeling; (5) test the established model with the testing set; (6) use the model for further simulation and prediction.

There are many types of network models and what we use in this work is the one based on the back propagation learning algorithm which is called BPNN [40,41]. The traditional algorithms of BPNN usually make the model inaccurate and reduce the efficiency of calculation. Therefore, some kinds of new algorithms start to be widely used with the technological advancement. What we select is one of the most popular algorithms with good generalization called Bayesian regularization (TRAINBR) [42]. The transfer functions we select are the hyperbolic tangent sigmoid function (TANSIG) in the second layer and the linear function (PURELIN) in the third layer. The combination of these functions can meet the basic modeling requirement. To implement the ANN modeling, we use the Neural Network Toolbox (nntool) of MATLAB (R2018b edition) on PC, and TRAINBR, TANSIG, PURELIN are the MATLAB commands of nntool. Some detailed information can be viewed in the manual of MATLAB nntool [43]. In order to describe the modeling process more intuitively, a schematic diagram of the model is illustrated in Figure 2.

As mentioned above, a three-layer network with one hidden layer is found to be sufficient for this study. A suitable neural network usually needs to have a high accuracy and good correction on both the training and testing set. However, it is difficult to select the best architecture of network. Although the model has a very good description of the training set, the accuracy of prediction could be very poor for other new data in the testing set. This phenomenon is generally called overfitting.

Since we have already determined 20 variables as the input and 2 properties as the output, the number of units in the hidden layer will greatly change the architecture and performance of the networks. We determine the optimal number of units in the hidden layer by comparing the predictive accuracy of different networks on the testing set. Here, we use root mean square error (RMSE) and correlation coefficient (R) as the error statistical parameters. The RMSE can accurately measure the deviation between original values and predicted ones, and the R is able to provide information on the strength of correlation between them. They are calculated using the following equation:

$$\text{RMSE} = \sqrt{\frac{1}{n}\sum_{i=1}^{n}\left(f(x_i)-y_i\right)^2}, \tag{4}$$

$$R = \frac{\sum_{i=1}^{n}\left(f(x_i) - \overline{f(x)}\right)\left(y_i - \overline{y}\right)}{\sqrt{\sum_{i=1}^{n}\left(f(x_i) - \overline{f(x)}\right)^2}\sqrt{\sum_{i=1}^{n}\left(y_i - \overline{y}\right)^2}} \tag{5}$$

where n is total number of data, y_i the original values, $f(x_i)$ the predicted ones, $\overline{f(x)} = \frac{1}{n}\sum_{i=1}^{n}f(x_i)$ and $\overline{y} = \frac{1}{n}\sum_{i=1}^{n}y_i$, respectively [26]. Once the predicted values and the original ones have a small deviation, a strong correlation is found with a small value of RMSE and R close to 1.

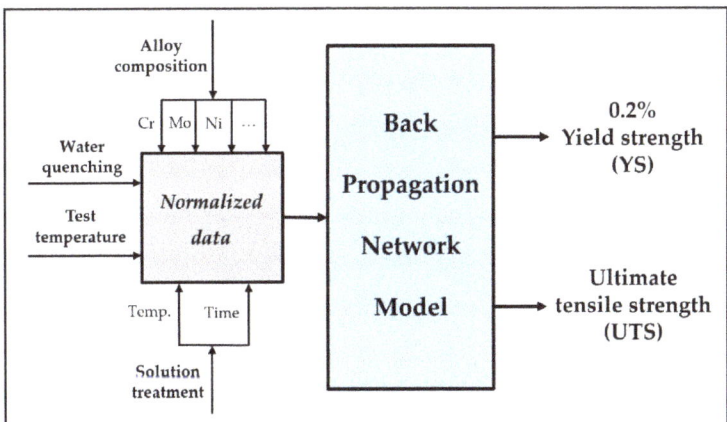

Figure 2. Schematic diagram of Back Propagation Neural Network (BPNN) model for predicting two tensile properties of austenitic stainless steels in this study.

Figure 3a–c shows the values of RMSE and R for YS, UTS and YS/UTS ratio, respectively, with different numbers of the hidden units of the training and testing set. The increase in number of hidden units means the more complexity of the model. It is obviously seen that both for YS and UTS, R increases and RMSE decreases in the training set, indicating that the more complicated model has better prediction for the training set. However, a high degree of complexity probably causes the overfitting and makes the model not suitable for the unseen data in the testing set. As shown in Figure 3, the prediction for the testing set first becomes better and then gets slightly worse when the number of units in the hidden layer increases. Hence, we set the optimum number of units by the RMSE and R of the testing set in this study and it is clearly found to be 8 for YS and 11 for UTS in Figure 3. The architectures of the models we use in the following work are [20-8-1] for YS and [20-11-1] for UTS. The YS/UTS ratio is also an important parameter of tensile properties that determines the reliability of metallic materials [44]. We also establish the network for the YS/UTS ratio and find the similar phenomenon of prediction from the observation of Figure 3c. Its optimum architecture is [20-6-1].

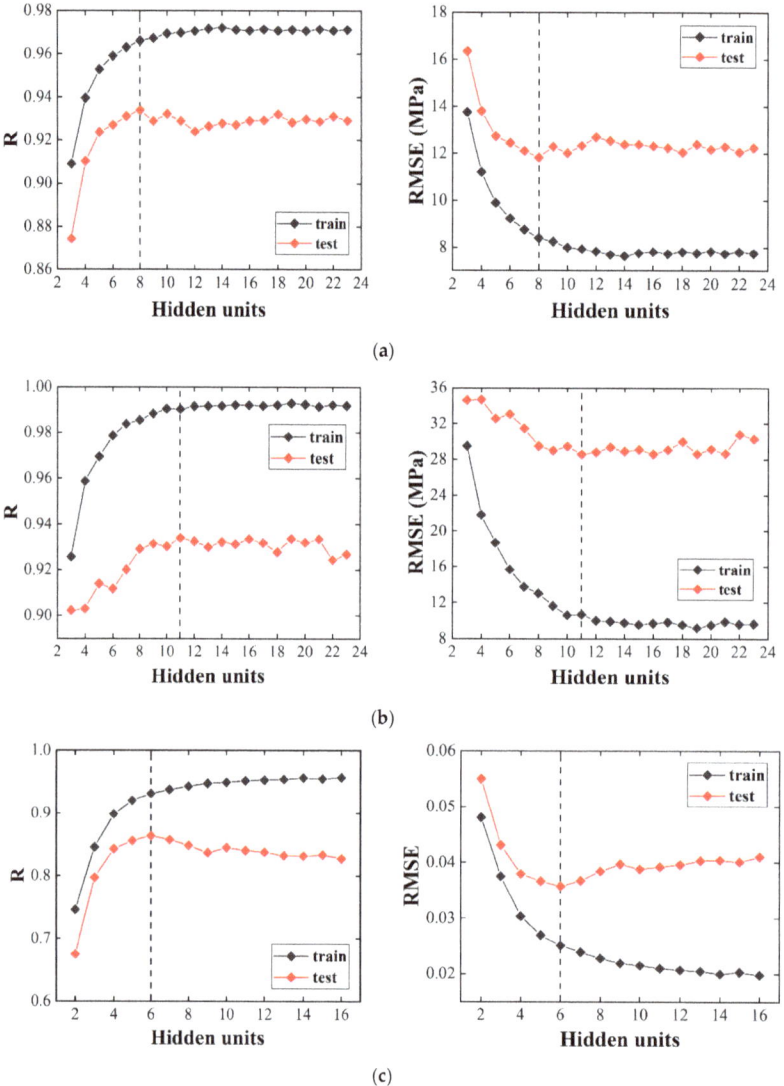

Figure 3. The number of units in the hidden layers influences the prediction of models on the training set and testing set: (**a**) yield strength (YS), (**b**) ultimate tensile strength (UTS) and (**c**) YS/UTS ratio. The black and red dots represent correlation coefficient (R) and root mean square error (RMSE) on the training set and testing set, respectively. We use the dotted lines to mark the selection of the optimal number of hidden units in every figure.

3. Results and Discussion

3.1. Model Performance and Validation

In order to verify the accuracy of the established models, besides the RMSE and R, another goodness-of-fit statistical parameter is the mean absolute percentage error (MAPE) [26], which is used to measure the error between the predicted and original values, and it particularly considers the ratio of error to the original values. The smaller the MAPE value, the higher the predictive accuracy. The value of MAPE is calculated as following equation:

$$\text{MAPE} = \frac{1}{n}\sum_{i=1}^{n}\left|\frac{f(x_i)-y_i}{y_i}\right|. \tag{6}$$

Table 2 shows the values of MAPE, RMSE and R of ANN models for three tensile properties, YS, UTS and YS/UTS, of the training and testing sets with the optimal hidden units. As it can be seen that the predictions of the training set are generally better than those of the testing set, which could be expected because the model is better for predicting the known data than the unseen data. However, for the testing set, the values of MAPE are less than 6% and of R are above 0.86 for the properties YS, UTS and YS/UTS, indicating a good prediction performance of the models. It is worth noting that the value of R for UTS is nearly 0.99, indicating that the present model has a more accurate description and prediction for UTS than YS and YS/UTS. Figure 4 shows the original and predicted values of UTS, YS and YS/UTS for the training and testing sets. Good performance of the models for UTS, YS and YS/UTS is observed for both the training and testing sets.

Table 2. Statistical parameters for the training and testing set with the optimal hidden units.

Tensile Properties	Hidden Units	Training (83% of the Data)			Testing (17% of the Data)		
		MAPE (%)	RMSE (MPa)	R	MAPE (%)	RMSE (MPa)	R
YS	8	4.09	8.40	0.97	5.21	11.82	0.93
UTS	11	1.76	10.69	0.99	3.56	28.54	0.93
YS / UTS	6	4.80	0.025 [1]	0.93	5.91	0.035 [1]	0.86

[1] There is no unit for the predicted RMSE of YS/UTS ratio.

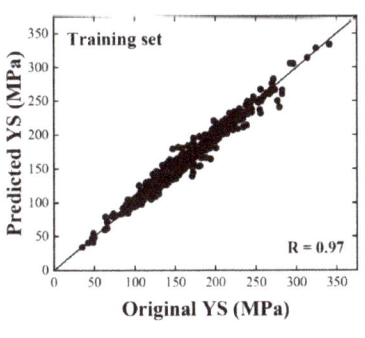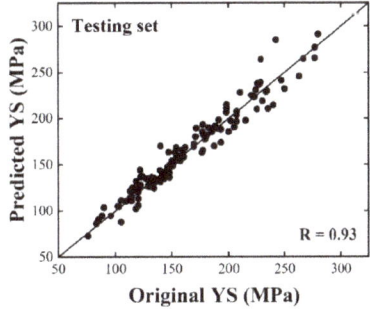

(a)

Figure 4. Cont.

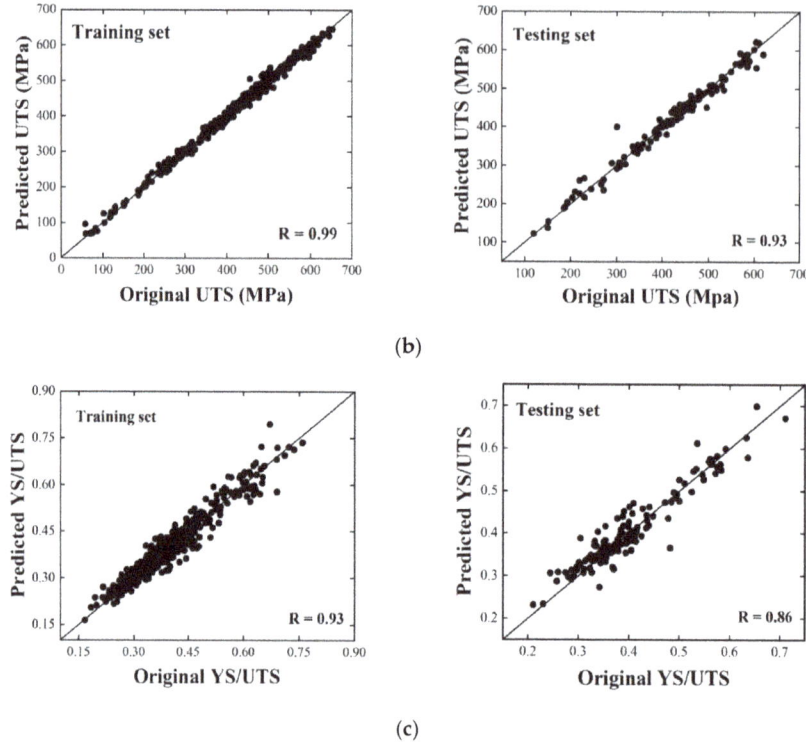

Figure 4. Comparison between the predicted and original values on the training set and testing set for: (**a**) YS, (**b**) UTS and (**c**) YS/UTS ratio. X-axis is the original value of tensile properties, Y-axis is the predicted one and the equation of black line in these figures is y = x (The more points concentrated near the black line, the more accurate the predicted values).

In order to verify the ability of established models to predict the unseen data outside the database, some new data of tensile properties of ASS in the Fatigue Data Sheet from MatNavi are tested. The details of these data are listed in Table A3 of Appendix A. The prediction results are shown in Figure 5 and the statistical parameters are presented in Table 3. It can be seen from Figure 5 and Table 3 that the models have good performance for UTS and YS with R ≥ 0.95. This means that the models have a good ability to predict the unknown data. Moreover, the models have a relative better prediction for UTS than YS with these new data, which is similar to the result for the original data mentioned above.

Table 3. Statistical parameters of predicted results for the new data.

Tensile Properties	MAPE (%)	RMSE (MPa)	R
YS	9.61	5.27	0.95
UTS	4.16	6.06	0.97

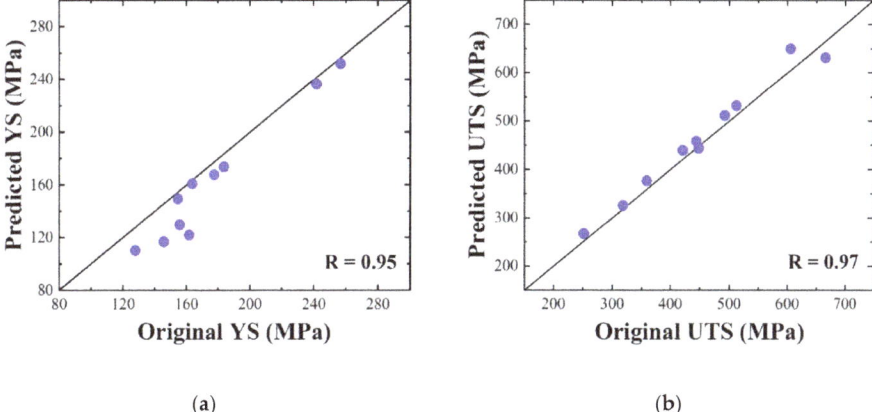

Figure 5. Comparison between the predicted and original values of the unseen data outside the database for: (**a**) YS and (**b**) UTS.

3.2. Feature Analysis

3.2.1. Mean Impact Value

Using machine learning cannot only make accurate predictions, but also further analyze the effect of each single variable on the corresponding properties. The mean impact value (MIV) method has been widely used for quantitative feature analysis in the machine learning application to explore the relative importance of each input variable for the improvement of the prediction performance [45]. The algorithm process of MIV is as follow:

1. Building two new datasets by varying the magnitude of one of variables by ±10% for the original training set;
2. Inputting two new datasets as the simulation samples to the model and obtaining two predicted results;
3. Calculating the difference value of these two predicted results, called the impact value (IV);
4. According to the amount of samples in the original training set, calculating the average value of IV, that is MIV of each variable;
5. Repeating the above steps in turn to get the MIV of each independent variable. It should be noted that the value of MIV indicates the positive or negative effect as well as the intensity of influence.

The MIV values of each variable are calculated and shown in Figure 6. MIV results show that test temperature and Ni content are two most important factors for both YS and UTS. Generally, the tensile properties of steels largely depend on the test temperatures and experimentally, the properties often need to be carried out at different test temperatures. Moreover, for the UTS, besides the test temperature, the importance of Ni and Cr contents is in good agreement with the traditional metallurgical theories and the engineering practice. In ASS steels, Ni and Cr are intentionally added in large quantities into ASS to improve the tensile properties and high-temperature oxidation resistance. In addition, Ti and Mo contents and temperature of solution treatment are also strong indicators of YS, while Cr and Mn contents are highly related to UTS. Note that the MIV value of test temperature for UTS is much greater than that for YS, which means that the test temperature has a much stronger effect on UTS than YS.

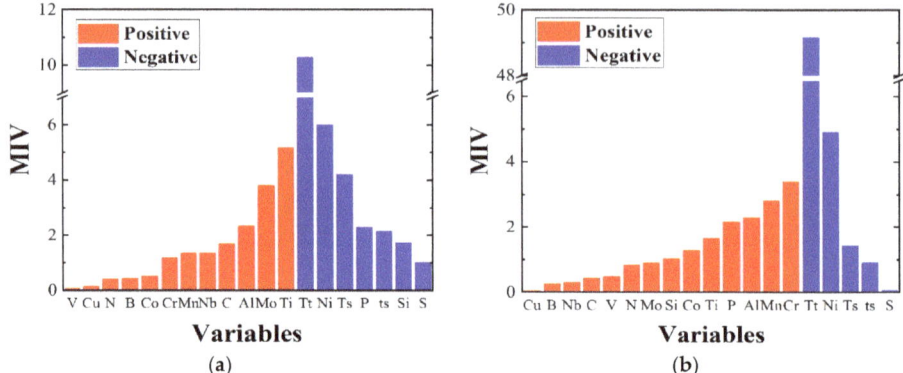

Figure 6. The absolute values of mean impact value (MIV) of each variable on two tensile properties: (a) YS and (b) UTS. The red and blue columns represent the positive and negative correlation, respectively (The higher the MIV value, the greater the effect on the corresponding tensile properties).

The information about the positive and negative correlation between the properties and all features is also shown in Figure 6. The elements Cr, V, Ti, Mo, Nb, C and Mn appear positively correlated with YS and UTS, which is consistent with the characteristics of previous theories of precipitation strengthening and solid solution strengthening [2,46]. The elements Cr, V, Ti, Mo and Nb could form strong carbides or nitrides that play a role of second phase precipitation strengthening. The elements like C, N and Mn could form interstitial solid solution that lead to the solid solution strengthening. In addition, it is worth mentioning that S content, test temperature and solution treatment time all exhibit the negative influences on YS and UTS.

3.2.2. Influence Trends of Variables

As discussed above, the present models show good prediction performance for the tensile properties YS and UTS, so they can be used to predict the influence trends of some important variables on the properties. Here, the effects of some typical elements and treatment conditions are examined. To investigate the effect of each variable, a new dataset is built and the values of this variable are set from the maximum to minimum and the values of other variables are set to the average of the original database, which are listed in Table A1 of Appendix A.

Figure 7 shows the effect of typical elements C, Cr, Ni, Ti, Nb, and V on the tensile properties UTS and YS. It can be observed that both UTS and YS generally show positive correlations with the contents of C, Cr, Ti, Nb and V, while they exhibit negative correlation with the Ni content. The results are in good agreement with the above MIV analyses that C and Cr are added to stabilize the austenite and to form interstitial solid solution or carbides dispersed in the matrix [2,46,47]. Note that for YS, it shows a firstly decreasing trend when the C content is less than 0.05%. This result is probably due to the lack of sufficient amount of data that limits the accuracy of the model. Moreover, the reduced value is not large and the overall tendency is still increasing, so the model has a reasonable prediction result for the effect of C content on YS within a certain error tolerance. The microalloying elements Ti, Nb and V could form the stable second carbides to prevent austenite grain coarsening, hindering the dislocation motion and then strongly strengthening the mechanical properties of ASS. Previous experimental results show that Nb and Ti can improve the tensile properties of ASS [48]. As shown in Figure 7c, Ni is a deliberate element which has a strong negative correlation with tensile strengths and makes ASS have good plasticity and ductility for subsequent processing [49].

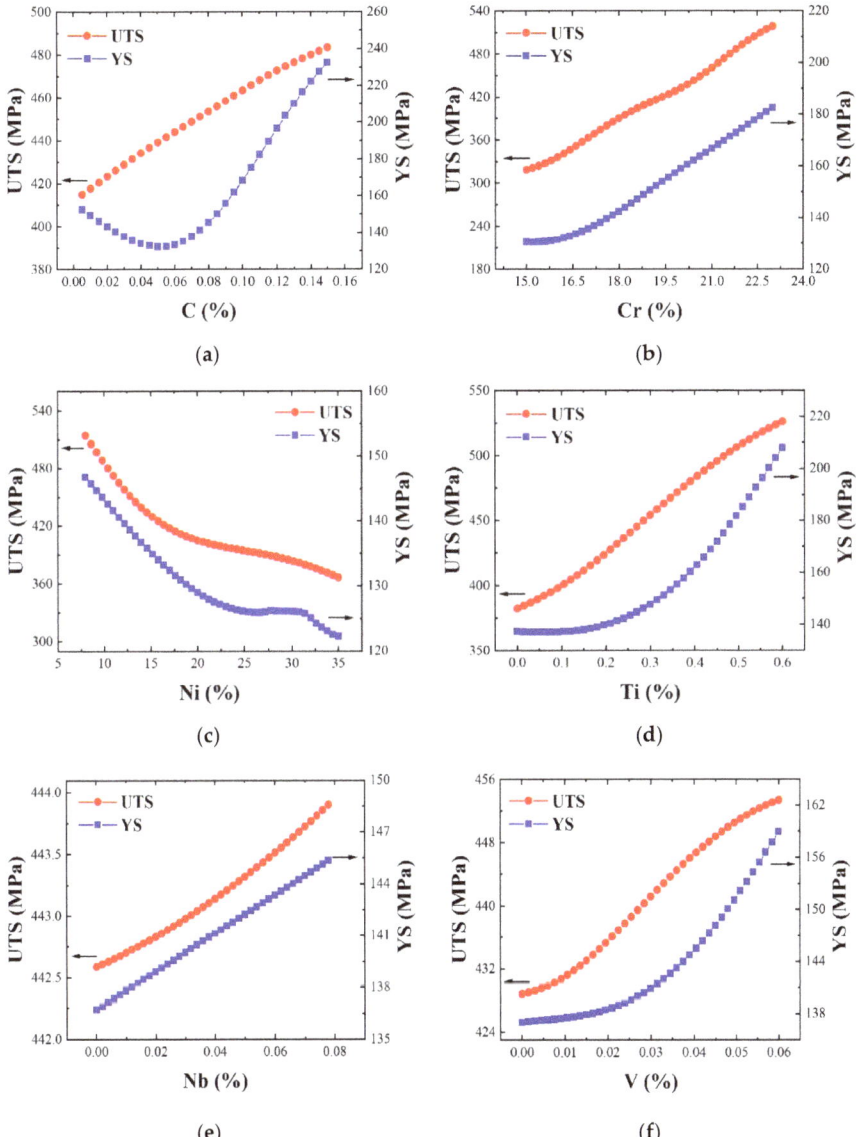

Figure 7. The predicted influence trends of (**a**) carbon, (**b**) chromium, (**c**) nickel, (**d**) titanium, (**e**) niobium and (**f**) vanadium on YS and UTS. The red and blue dots represent the predicted values of UTS and YS respectively.

It is known that by solution treatment, the second phase carbon nitride could be uniformly dispersed in the matrix and then leads to increased second phase strengthening. However, the continued solution treatment will cause the grain coarsening and the reduction of crystal defects of ASS. So, the effect of second phase strengthening is gradually offset and the strength of ASS decreases ultimately [50–52]. Figure 8 presents the effect of T_s, t_s and T_t on the tensile properties UTS and YS. The results show when the solution treatment conditions are located at [1323~1473 K, 2414 s] or [1378 K, 120–7200 s], both UTS and YS generally show a decreasing trend with increasing T_s and t_s.

Under this situation, the effects of grain coarsening and reduction of defects offset the second phase strengthening. In Figure 8c, both UTS and YS decrease as T_t increases. When the temperature is close to 1300 K, the predicted value of UTS reduces to nearly 0 and even less than YS. This is clearly contrary to previous theories, which is similar to the previous prediction for the influence trend of carbon. The possible reason is a lack of sufficient data. By comparing the predicted results in Figures 7 and 8, the test temperature is the most influential variable for YS and UTS, where the variations of tensile properties are the largest, especially for UTS. This is also consistent with the conclusion of the MIV analysis above.

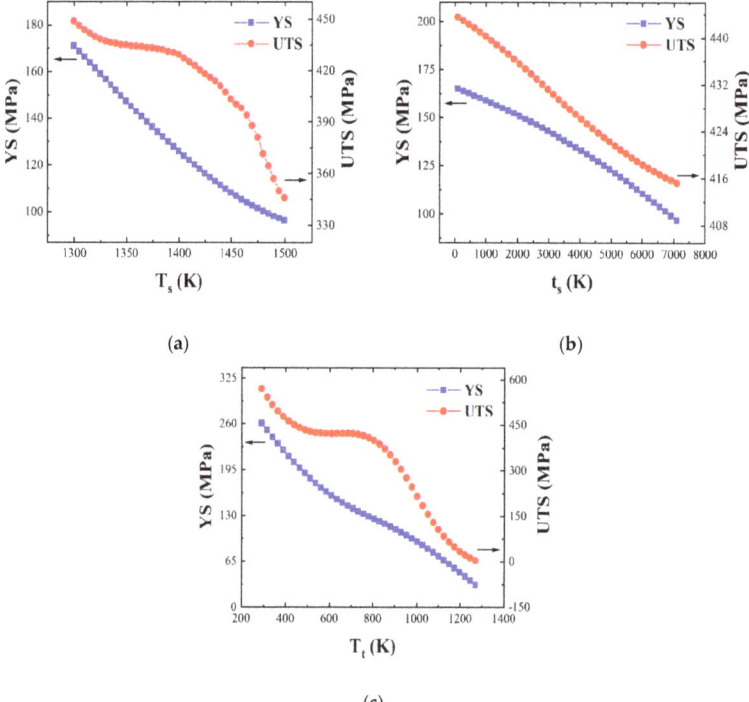

Figure 8. The predicted influence trends of (**a**) temperature, (**b**) time of solution treatment and (**c**) test temperature on YS and UTS. The blue and red dots represent the predicted values of YS and UTS respectively.

4. Conclusions

In this work, we have developed two BPNN models capable of studying and predicting two tensile properties of austenitic stainless steel, YS and UTS, as a function of 20 variables including chemical composition, heat treatment conditions and test temperature. The accuracy of the established models is evaluated based on three statistical parameters, RMSE, R and MAPE. The results indicate that the models have not only highly predictive accuracy on both the training and testing set, but also have good prediction performance for some unknown data. For analyzing the effect of each variable on YS and UTS, we use the MIV method and predict the influence trends of several important variables with the established models. The results correctly reflect the positive and negative correlation between the tensile properties and all features, which are consistent with the previous metallurgical theories.

Compared with experiments and other models, the present models are able to accurately predict the tensile properties of ASS when all features are known. Based on the models, the test temperature and Ni content are found to be two most important factors for both YS and UTS. This work is helpful for the preparation and development of new ASS in the future.

Author Contributions: Conceptualization, Y.W. and X.W.; investigation, Y.W., X.L. and Z.X.; writing-original draft preparation, Y.W. and X.W.; writing-review and editing, X.L., Z.X., R.L., W.L., Y.Z. and Y.X.; supervision, X.W. and C.L. All authors have read and agreed to the published version of the manuscript.

Funding: This work is supported by the National Key Research and Development Program of China (2017YFE0302400 and 2017YFA0402800) the National Natural Science Foundation of China (11735015, 51871207, 51801203, 51671184, 51671185 and U1832206) and Anhui Provincial Natural Science Foundation (1908085J17).

Conflicts of Interest: The authors declare no conflict of interest.

Appendix A

Table A1. The 415 samples with water quenching after solution treatment (label: 1).

Variables	Range	Mean	SD
Cr	16.39%–21.06%	18.31%	1.432%
Ni	8.4%–34.45%	16.4%	9.078%
Mo	0%–2.65%	0.95%	1.114%
Mn	0.81%–1.75%	1.39%	0.3088%
Si	0.39%–0.82%	0.585%	0.1194%
Nb	0%–0.79%	0.0749%	0.2004%
Ti	0%–0.53%	0.188%	0.2098%
V	0%–0.057%	0.0071%	0.0166%
Cu	0%–0.35%	0.139%	0.1101%
N	0%–0.081%	0.0197%	0.01422%
C	0.012%–0.1%	0.0607%	0.01556%
B	0%–0.003%	0.0005%	0.00058%
P	0%–0.038%	0.0214%	0.00858%
S	0%–0.05%	0.011%	0.00829%
Co	0%–0.54%	0.115%	0.152%
Al	0%–0.52%	0.124%	0.188%
T_s	1323–1473 K	1378 K	42.79 K
t_s	120–7200 s	2414 s	2397.2 s
T_t	293–1273 K	731 K	267.42 K

Table A2. 394 samples with air quenching after solution treatment (label: 0).

Variables	Range	Mean	SD
Cr	15.9%–18.3%	17.57%	0.4594%
Ni	9.15%–12.4%	11.1%	1.0906%
Mo	0%–2.72%	0.856%	1.1288%
Mn	0.79%–1.71%	1.386%	0.2784%
Si	0.29%–0.84%	0.482%	0.1335%
Nb	0%–0.9%	0.128%	0.2936%
Ti	0%–0.56%	0.221%	0.2286%
V	0%–0% [2]	0%	0%
Cu	0%–0%	0%	0%
N	0%–0%	0%	0%
C	0.02%–0.1%	0.053%	0.01167%
B	0%–0.004%	0.000081%	0.00056%
P	0%–0.04%	0.0209%	0.00711%
S	0%–0.028%	0.0158%	0.00646%
Co	0%–0%	0%	0%
Al	0%–0%	0%	0%
T_s	1293–1403 K	1348 K	32.7514 K
t_s	120–3600 s	2117 s	1149.61 s
T_t	293–973 K	626 K	221.568 K

[2] The 394 samples with air quenching have the chemical composition except V, Cu, N, Co and Al.

Table A3. The variables of new unseen data of austenitic stainless steel.

Variables	No.15 (SUS 316)	No.42 (SUS 304)
Cr	17.05%	18.53%
Ni	12.6%	9.1%
Mo	2.24%	0.12%
Mn	1.1%	0.88%
Si	0.7%	0.54%
Nb	0.001%	0%
Ti	0.03%	0.02%
V	0%	0%
Cu	0.31%	0.06%
N	0.017%	0.023%
C	0.05%	0.05%
B	0.003%	0%
P	0.033%	0.029%
S	0.003%	0.01%
Co	0%	0.21%
Al	0.02%	0.023%
T_s	1373 K	1373 K
t_s	1800 s	1800 s
Water or air quenching	Water quenching (label: 1)	Water quenching (label: 1)
T_t	298, 673, 773, 873, 973 K	298, 673, 773, 873, 973 K

References

1. Callister, W.D.; Rethwisch, D.G. *Materials Science and Engineering: An Introduction*; John Wiley & Sons: New York, NY, USA, 2007; Volume 7.
2. Krauss, G. *Steels: Processing, Structure, and Performance*; ASM International: Russell, OH, USA, 2015.
3. Reifsnider, K.L.; Tamuzs, V. On nonlinear behavior in brittle heterogeneous materials. *Compos. Sci. Technol.* **2006**, *66*, 2473–2478. [CrossRef]
4. Liu, Y.; Zhao, T. Materials discovery and design using machine learning. *J. Mater.* **2017**, *3*, 159–177. [CrossRef]
5. Murphy, K.P. *Machine Learning: A Probabilistic Perspective*; MIT Press: Cambridge, MA, USA, 2012.
6. Bishop, C.M. *Neural Networks for Pattern Recognition*; Oxford University Press: Oxford, UK, 1995.
7. Zurada, J.M. *Introduction to Artificial Neural Systems*; West Publishing Company: St. Paul, MN, USA, 1992; Volume 8.
8. Cortes, C.; Vapnik, V. Support-vector networks. *Mach. Learn.* **1995**, *20*, 273–297. [CrossRef]
9. Vapnik, V. *The Nature of Statistical Learning Theory*; Springer Science & Business Media: New York, NY, USA, 2013.
10. Rokach, L.; Maimon, O.Z. *Data Mining with Decision Trees: Theory and Applications*; World Scientific: Singapore, 2008; Volume 69.
11. Council, N.R. *Integrated Computational Materials Engineering: A Transformational Discipline for Improved Competitiveness and National Security*; National Academies Press: Washington, DC, USA, 2008.
12. Kalidindi, S.R.; Niezgoda, S.R. Microstructure informatics using higher-order statistics and efficient data-mining protocols. *Jom* **2011**, *63*, 34–41. [CrossRef]
13. Rajan, K. Materials informatics. *Mater. Today* **2005**, *8*, 38–45. [CrossRef]
14. Ward, L.; O'Keeffe, S.C. A machine learning approach for engineering bulk metallic glass alloys. *Acta Mater.* **2018**, *159*, 102–111. [CrossRef]
15. Xue, D.; Balachandran, P.V. Accelerated search for materials with targeted properties by adaptive design. *Nat. Commun.* **2016**, *7*, 11241. [CrossRef]
16. Lin, Y.; Zhang, J. Application of neural networks to predict the elevated temperature flow behavior of a low alloy steel. *Comput. Mater. Sci.* **2008**, *43*, 752–758. [CrossRef]
17. Murugesan, M.; Sajjad, M. Hybrid Machine Learning Optimization Approach to Predict Hot Deformation Behavior of Medium Carbon Steel Material. *Metals* **2019**, *9*, 1315. [CrossRef]
18. Kurt, H.; Oduncuoglu, M. A mathematical formulation to estimate the effect of grain refiners on the ultimate tensile strength of Al-Zn-Mg-Cu alloys. *Metals* **2015**, *5*, 836–849. [CrossRef]

19. Mannodi-Kanakkithodi, A.; Pilania, G. Machine learning strategy for accelerated design of polymer dielectrics. *Sci. Rep.* **2016**, *6*, 20952. [CrossRef] [PubMed]
20. Irani, K.B.; Cheng, J. Applying machine learning to semiconductor manufacturing. *IEEE Expert* **1993**, *8*, 41–47. [CrossRef]
21. Yu, L.; Zunger, A. Identification of potential photovoltaic absorbers based on first-principles spectroscopic screening of materials. *Phys. Rev. Lett.* **2012**, *108*, 068701. [CrossRef] [PubMed]
22. Mayer-Schönberger, V.; Cukier, K. *Big Data: A Revolution that Will Transform How We Live, Work, and Think*; Houghton Mifflin Harcourt: Boston, MA, USA, 2013.
23. NIMS MatNavi. Available online: https://mits.nims.go.jp/index_en.html (accessed on 10 December 2019).
24. MatWeb Material Property Data. Available online: http://www.matweb.com/ (accessed on 15 July 2019).
25. Matmatch. Available online: https://matmatch.com/ (accessed on 22 July 2019).
26. Agrawal, A.; Deshpande, P.D. Exploration of data science techniques to predict fatigue strength of steel from composition and processing parameters. *Integr. Mater. Manuf. Innov.* **2014**, *3*, 90–108. [CrossRef]
27. Sourmail, T.; Bhadeshia, H. Neural network model of creep strength of austenitic stainless steels. *Mater. Sci. Technol.* **2002**, *18*, 655–663. [CrossRef]
28. Guo, Z.; Sha, W. Modelling the correlation between processing parameters and properties of maraging steels using artificial neural network. *Comput. Mater. Sci.* **2004**, *29*, 12–28. [CrossRef]
29. Fragassa, C.; Babic, M. Predicting the tensile behaviour of cast alloys by a pattern recognition analysis on experimental data. *Metals* **2019**, *9*, 557. [CrossRef]
30. Hodgson, P.D.; Kong, L.X. The prediction of the hot strength in steels with an integrated phenomenological and artificial neural network model. *J. Mater. Process. Technol.* **1999**, *87*, 131–138. [CrossRef]
31. Ramana, K.; Anita, T. Effect of different environmental parameters on pitting behavior of AISI type 316L stainless steel: Experimental studies and neural network modeling. *Mater. Design* **2009**, *30*, 3770–3775. [CrossRef]
32. Mandal, S.; Sivaprasad, P. Artificial neural network modeling of composition–process–property correlations in austenitic stainless steels. *Mater. Sci. Eng. A* **2008**, *485*, 571–580. [CrossRef]
33. Sivaprasad, P.; Mandal, S. Artificial neural network modelling of the tensile properties of indigenously developed 15Cr-15Ni-2.2Mo-Ti modified austenitic stainless steel. *Trans. Indian Inst. Met.* **2006**, *59*, 437–445.
34. Desu, R.K.; Krishnamurthy, H.N. Mechanical properties of Austenitic Stainless Steel 304L and 316L at elevated temperatures. *J. Mater. Res. Technol.* **2016**, *5*, 13–20. [CrossRef]
35. The British Steelmakers Creep Committee: *BSCC High Temperature Data*; The Iron and Steel Institute: London, UK, 1973.
36. Materials Algorithms Project. Available online: https://www.phase-trans.msm.cam.ac.uk/map (accessed on 17 June 2019).
37. Shen, C.; Wang, C. Physical metallurgy-guided machine learning and artificial intelligent design of ultrahigh-strength stainless steel. *Acta Mater.* **2019**, *179*, 201–214. [CrossRef]
38. Bhadeshia, H. Neural networks in materials science. *ISIJ Int.* **1999**, *39*, 966–979. [CrossRef]
39. Hornik, K.; Stinchcombe, M. Multilayer feedforward networks are universal approximators. *Neural Netw.* **1989**, *2*, 359–366. [CrossRef]
40. Hecht-Nielsen, R. Theory of the backpropagation neural network. In *Neural Networks for Perception*; Academic Press: New York, USA, 1992; pp. 65–93.
41. Rumelhart, D.E.; Hinton, G.E. Learning representations by back-propagating errors. Cognitive Modeling. *Nature* **1986**, *323*, 533–536. [CrossRef]
42. Burden, F.; Winkler, D. Bayesian regularization of neural networks. In *Artificial Neural Networks*; Humana Press: Totowa, NJ, USA, 2008; pp. 23–42.
43. Statistics and Machine Learning Toolbox. Available online: https://www.mathworks.com/products/statistics.html (accessed on 23 December 2019).
44. Bhadeshia, H. *Bainite in Steels. Cambridge: The Institute of Materials*; The University Press: London, UK, 1992.
45. Dombi, G.W.; Nandi, P. Prediction of rib fracture injury outcome by an artificial neural network. *J. Trauma Acute Care* **1995**, *39*, 915–921. [CrossRef]
46. McGuire, M.F. *Stainless Steels for Design Engineers*; ASM International: Russell, OH, USA, 2008.
47. Martins, L.F.M.; Plaut, R.L. Effect of Carbon on the Cold-worked State and Annealing Behavior of Two 18wt% Cr–8wt% Ni Austenitic Stainless Steels. *ISIJ Int.* **1998**, *38*, 572–579. [CrossRef]

48. Farahat, A.I.Z.; El-Bitar, T. Effect of Nb, Ti and cold deformation on microstructure and mechanical properties of austenitic stainless steels. *Mater. Sci. Eng. A* **2010**, *527*, 3662–3669. [CrossRef]
49. Marshall, P. *Austenitic Stainless Steels: Microstructure and Mechanical Properties*; Springer Science & Business Media: New York, NY, USA, 1984.
50. Totten, G.E. *Steel Heat Treatment: Metallurgy and Technologies*; CRC Press: Boca Raton, FL, USA, 2006.
51. Yin, R. *Metallurgical Process Engineering*; Springer Science & Business Media: New York, NY, USA, 2011.
52. Wu, H.; Yang, B. Effect of Solution Temperature on the Microstructure and Mechanical Properties of Wrought 316LN Stainless Steel. *Adv. Mater. Res.* **2014**, *915*, 576–582. [CrossRef]

© 2020 by the authors. Licensee MDPI, Basel, Switzerland. This article is an open access article distributed under the terms and conditions of the Creative Commons Attribution (CC BY) license (http://creativecommons.org/licenses/by/4.0/).

Article

Computational Modeling and Constructal Design Theory Applied to the Geometric Optimization of Thin Steel Plates with Stiffeners Subjected to Uniform Transverse Load

Grégori Troina [1], Marcelo Cunha [1], Vinícius Pinto [1], Luiz Rocha [2], Elizaldo dos Santos [1], Cristiano Fragassa [3,*] and Liércio Isoldi [1]

1. Graduate Program in Ocean Engineering, Federal University of Rio Grande–FURG, Rio Grande 96203-900, Brazil; gregori.troina@gmail.com (G.T.); marcelolamcunha@hotmail.com (M.C.); viniciustorreseng@gmail.com (V.P.); elizaldosantos@furg.br (E.d.S.); liercioisoldi@furg.br (L.I.)
2. Graduate Program in Mechanical Engineering, University of Vale do Rio dos Sinos–UNISINOS, São Leopoldo 93022-750, Brazil; luizor@unisinos.br
3. Department of Industrial Engineering, University of Bologna–UNIBO, 40165 Bologna, Italy
* Correspondence: cristiano.fragassa@unibo.it; Tel.: +39-347-697-4046

Received: 27 December 2019; Accepted: 1 February 2020; Published: 4 February 2020

Abstract: Stiffened thin steel plates are structures widely employed in aeronautical, civil, naval, and offshore engineering. Considering a practical application where a transverse uniform load acts on a simply supported stiffened steel plate, an approach associating computational modeling, Constructal Design method, and Exhaustive Search technique was employed aiming to minimize the central deflections of these plates. To do so, a non-stiffened plate was adopted as reference from which all studied stiffened plate's geometries were originated by the transformation of a certain amount of steel of its thickness into longitudinal and transverse stiffeners. Different values for the stiffeners volume fraction (φ) were analyzed, representing the ratio between the volume of the stiffeners' material and the total volume of the reference plate. Besides, the number of longitudinal (N_{ls}) and transverse (N_{ts}) stiffeners and the aspect ratio of stiffeners shape (h_s/t_s, being h_s and t_s, respectively, the height and thickness of stiffeners) were considered as degrees of freedom. The optimized plates were determined for all studied φ values and showed a deflection reduction of over 90% in comparison with the reference plate. Lastly, the influence of the φ parameter regarding the optimized plates was evaluated defining a configuration with the best structural performance among all analyzed cases.

Keywords: deflection; plates; stiffeners; numerical simulation; Constructal Design

1. Introduction

According to Timoshenko and Gere [1], thin plates are plane structural components that have one dimension, called thickness, substantially smaller than the other dimensions. Structural elements containing plates are employed in different engineering sectors, such as automotive, aerospace, naval, and civil.

Due to the slenderness of the plates (i.e., these elements are thin, having a low bending stiffness and hence a short resistance against transverse and longitudinal moments) the necessity of incorporating beam structures in order to enhance the bending stiffness has been noted [2]. Different manufacturing processes can be used to obtain stiffened steel plates. Among them, the welding technology plays an important role at the shipbuilding and ship repairing activities [3]. On the other hand, the modern stiffened panels used for the fuselage of aerospace industry have been manufactured by Electromagnetic forming (EMF), as explained in Tan et al. [4,5].

In addition, several researchers have studied the mechanical behavior of stiffened thin steel plates. Rossow and Ibrahimkhail [6], through the internal Constraint Method, analyzed two case studies: a square plate with one central stiffener and a rectangular plate with two orthogonal stiffeners. These problems were also solved computationally in the software NASTRAN® and STRUDL®. Bedair [7] analyzed stiffened plates under transverse loads through the Sequential Quadratic Programming (SQP) method, idealizing the structure as a plate-beam system. Tanaka and Bercin [8] applied the Boundary Element Method (BEM) to analyze the elastic bending of stiffened plates, the examples studied through this methodology were a square plate with a central stiffener and a rectangular plate with two equally spaced parallel stiffeners. Beam-reinforced plates was also the subject of study in the work of Sapountzakis and Katsikadelis [9], where the shear stresses in the bond regions between the plate and stiffeners—which is an important parameter when projecting reinforced prefabricated plates or plates made of composite materials—were estimated. In Salomon [10], several computational models were proposed, based on the Finite Element Method (FEM), for the numerical simulation of stiffened plates with different boundary conditions submitted to bending. The obtained results were compared to each other, indicating that 3D numerical models reproduce the physical problem in a more realistic way; however, 2D numerical models can also be adopted with a good accuracy. The work of Hasan [11] evaluated, through the software NASTRAN®, the maximum stresses and displacements in stiffened plates under static uniform load in order to determine the optimal positioning of rectangular cross-sectional stiffeners. In Silva [12], a numerical study about ribbed slabs was developed with aid of software ANSYS®, by using the beam element BEAM44 to model the ribs and the shell element SHELL63 to model the slab. It was shown that the eccentricity between the slab and the reinforcement ribs leads to a reduction in the deflections. Recently, De Queiroz et al. [13] applied Constructal Design Method (CDM) associated with FEM to investigate the influence of stiffened plate's geometry in its out-of-plane central displacement, inferring that significant reductions of deflection can be reached only by an adequate rearrangement of the plate's geometric configuration. In addition, a geometric optimization by means the Exhaustive Search (ES) technique was also performed.

Regarding the geometric optimization techniques normally adopted for stiffened plates, beyond the ES technique used in Reference [13] one can also highlight: the Genetic Algorithm (GA) adopted in Kallasy and Marcelin [14], Cunha et al. [15], and Putra et al. [16]; as well as the Response Surface Methodology (RSM) employed in Lee et al. [17] and Anyfantis [18].

References [3–18] give an overview about some manufacturing processes for stiffened plates, mechanical behavior of stiffened plates submitted to bending, and geometric optimization techniques applied in stiffened plate problems. Table 1 summarizes this information, showing its relationship with the present work.

There exists a wide variety of theories about plates, which depend on the geometry, loads and boundary conditions. However, in these theories, the differential governing equations are extremely complicated to solve, being possible to solve analytically only for simple geometries, loads and boundary conditions [19]. Thus, numerical simulation is an important tool when analyzing structural elements composed of stiffened plates.

Therefore, since it is possible to quickly and accurately perform simulations of numerical computational models of plates with various geometries, the geometry variation of stiffened plates subjected to uniformly distributed loads was studied in order to evaluate the influence of different degrees of freedom (the number of longitudinal (N_{ls}) and transverse (N_{ts}) stiffeners and the ratio between the stiffener's height and thickness (h_s/t_s)) on the minimization of the central deflection of these structures.

Table 1. Summary of references used for an overview about stiffened plates.

Reference	Type	Year	Scope	Methodology	Relationship
[3]	Book	2017	MP	Theoretical	Exemplify how stiffened plates can be obtained by welding process
[4]	Paper	2016	MP	Experimental and Numerical	Exemplify how stiffened plates can be obtained by electromagnetic forming
[5]	Paper	2017	MP	Experimental and Numerical	Exemplify how stiffened plates can be obtained by electromagnetic forming
[6]	Paper	1978	MB	Numerical	Verification of the computational models
[7]	Paper	1997	MB	Numerical	Exemplify an analysis procedure employed for stiffened plates
[8]	Paper	1997	MB	Numerical	Verification of the computational models
[9]	Paper	2000	MB	Numerical	Exemplify an analysis procedure employed for stiffened plates
[10]	Master's Thesis	2001	MB	Numerical	Verification of the computational models
[11]	Paper	2007	MB	Numerical	Exemplify an analysis procedure employed for stiffened plates
[12]	Master's Thesis	2010	MB	Numerical	Verification of the computational models
[13]	Paper	2019	GO	Numerical	Preliminary study associating CDM, FEM and ES for stiffened plates
[14]	Paper	1997	GO	Numerical	Exemplify the geometric optimization of stiffened plates by means GA
[15]	Paper	2019	GO	Numerical	Exemplify the geometric optimization of stiffened plates by means GA
[16]	Paper	2019	GO	Numerical	Exemplify the geometric optimization of stiffened plates by means GA
[17]	Paper	2015	GO	Numerical	Exemplify the geometric optimization of stiffened plates by means RSM
[18]	Paper	2019	GO	Numerical	Exemplify the geometric optimization of stiffened plates by means RSM

MP-Manufacturing Process; MB-Mechanical Behavior; and GO-Geometric Optimization.

By means the Constructal Design Method (CDM), the present work studied a set of stiffened steel plates derived from a reference plate with length a, width b, and thickness t, which had a fraction of its volume φ transformed, through the reduction of the thickness, into various combinations of longitudinal (N_{ls}) and transverse (N_{ts}) stiffeners, with different heights (h_s) and thicknesses (t_s). Then, these structures were numerically simulated in the software ANSYS®, which is based on the Finite Element Method (FEM). The developed numerical models were discretized with two-dimensional (SHELL93) and three-dimensional (SOLID95) finite elements. Through the Exhaustive Search (ES) technique, the obtained numerical results were compared aiming to determine the optimized geometric configurations that minimize the central out-of-plane displacement of simply supported stiffened plates when subjected to a uniformly distributed load.

It is important to highlight that, based on the finds presented in De Queiroz et al. [13], the present work brings a more comprehensive approach: we attained several values for the stiffeners volume fraction (φ = 0.1, 0.2, 0.3, 0.4, and 0.5) and a larger variation of the number of transverse and longitudinal stiffeners (N_{ls} and N_{ts} from 2 to 6). In addition, here different computational models were adopted, allowing investigation of the accuracy of 2D and 3D models.

Concerning the material of construction adopted for the stiffeners and plates, it is well known that structural steel is a good choice due to its relative low cost, adequate mechanical properties, and ease fabrication (mainly by welding) [3]. Structural steel—also called constructional steel or carpentry steel—is characterized by its carbon content, i.e., the percentage content in terms of weight. The carbon presence increases the yield stress of the material, yet at the same time reduces its ductility and weldability. Because of this, structural steel is normally characterized by a mild carbon content. For example, the steel ASTM A36, used in the present work, has maximum carbon content varying between 0.25% and 0.29%. Its mechanical properties are: yielding stress of 250 MPa, ultimate stress of 400 MPa, modulus of elasticity of 200 GPa, and Poisson's ratio of 0.3 [20,21].

2. Theory of Plates

Yamaguchi [22] defines a plate as a continuous body that is flat before loading and has a specific geometric characteristic: one dimension is much smaller than the other two dimensions.

According to Timoshenko and Woinowsky-Krieger [23], the bending properties of a plate depend greatly on its thickness as compared with its other dimensions. Thus, Szilard [19] classified plates into four types based on the ratio between the thickness t and the smallest planar dimension, i.e., its width b: membranes, when $(t/b) < 0.02$; thin plates when $0.02 < (t/b) < 0.10$; moderately thick plates in the range of $0.10 < (t/b) < 0.20$; and thick plates for $(t/b) > 0.20$.

The load-carrying action of a plate is similar, to a certain extent, to that of beams or cables; thus, plates can be approximated by a gridwork of an infinite number of beams or by a network of an infinite number of cables, depending on the flexural rigidity of the structures. This two-dimensional structural action of plates results in lighter structures, and thus offers numerous economic advantages. Therefore, thin plates combine light weight and form efficiency with high load-carrying capacity, economy and technological effectiveness [24].

The plate-type structures are studied by using the governing equations of the Theory of Elasticity. However, it is extremely complex finding exact solutions for the differential equations inherent to problems involving these structures. The recent trend in the development of the plate theories is characterized by heavy reliance on high-speed computers and by the introduction of more rigorous theories [25].

According to Szilard [19], depending on the nature of the applied loads, the analysis is static or dynamic. Regarding the deflections, the theories of elastic plates are divided into two categories: plates with small and large deflections. For materials in the linear-elastic regime, the theories are based on the Hooke's Law stress-strain relations, while materials in the nonlinear or plastic range have more complex stress-strain relationships. Moreover, there are theories depending on the plate's mechanical properties: isotropic (same material properties in all directions), anisotropic (different material properties in different directions) and composite plates (layers of different materials).

Theory of Stiffened Plates

Salomon [10] presents a division for the analytical approaches for the study of stiffened plates into three broad categories: grillages, orthotropic plate model and plate-beam systems.

According to Salomon [10], the idealization of a plate with stiffeners as a beam grillage requires an effective width of plating varying from 50% to 80% of the spacing between stiffeners. The effective width is the portion of the plate that is used, along with the stiffener cross section, to calculate the moment of inertia as well as the bending and torsional stiffness of the plate. When comparing with the experimental values, the application of this method gives a difference of deflection values between 5% and 10%, and a difference of beam stresses values of generally between 10% and 20%.

In the orthotropic plate approach, the stiffened plate is replaced by an orthotropic non-stiffened plate, which is one the structural properties of which differ along orthogonal axes. This structural anisotropy can be due to manufacturing textures, stiffening beams or even inherent properties of the material.

Similarly to the Kirchhoff plate theory, but now assuming four elastic constants (two modules of elasticity: E_x and E_y; and two Poisson ratios: ν_x and ν_y) to describe the stress-strain relations in the x and y directions, the governing equation of orthotropic plates is given by Reference [19]:

$$D_x \frac{\partial^4 w}{\partial x^4} + 2B \frac{\partial^4 w}{\partial x^2 \partial y^2} + D_y \frac{\partial^4 w}{\partial y^4} = p_z(x, y) \tag{1}$$

where w is the displacement component in z direction; D_x and D_y are the bending stiffness in x and y directions; and B is the effective torsional stiffness of the orthotropic plate.

In order to analyze stiffened plates through the orthotropic plate model, it is necessary to define expressions to determine the sectional bending and torsional properties along the orthotropy direction. Since it is a complicated task to perform, whenever possible, direct tests should be executed in order to determine these properties. However, based on analytical considerations, reasonable approximations can be applied to calculate these stiffnesses, as shown in Szilard [19] and Timoshenko and Woinowsky-Krieger [23] for plates reinforced with rectangular-profile or I shaped beams, corrugated plates and reinforced concrete slabs.

According to Szilard [19], although the real structural behavior of plates reinforced with stiffeners is not exactly replicated by the orthotropic plate model, experimental data indicate a good agreement between results when the stiffeners are small, close and equally spaced.

Lastly, the idealization of stiffened plates as a plate-beam system is the methodology that best reflects the physical problem behavior. In this approach, it is used continuity conditions at the interface between the plate and the reinforcement beams (stiffeners). Due to the mathematical difficulties of analytically solving these problems, this approach was boosted by the advent of digital computers, which enabled the solutions of the models to be obtained through numerical methods. Among the numerical methods, the Finite Element Method (FEM) is the most powerful and effective one to find accurate numerical solutions. Currently, many researchers put effort into developing efficient and accurate FEM models for plates reinforced by stiffeners [10].

3. Computational Modeling

Computational modeling is used to numerically study a wide range of engineering complex problems, whose governing equations are ordinary or partial differential. Finite Difference Method (FDM), Finite Element Method (FEM) and Finite Volume Method (FVM) are the most employed discretization methods when solving numerical models governed by differential equations. These methods are advantageous over other approaches because they transform differential equations into systems of linear equations, give high quality approximations and are highly flexible in representing complex geometries [26].

The FEM, as defined by Burnett [27], is a computer-aided mathematical technique applied in the obtainment of approximate numerical solutions to abstract equations of calculus that predict the behavior of physical systems when subjected to external influences. This method consists in the following steps: division of the domain into sub-regions (sub-domains called finite elements), transforming the governing differential equation into fine-element algebraic equations and numerically solving the elementary equations through a linear equations system. In the FEM, the continuum domain is idealized as an assemblage of interconnected discrete finite-size elements that behave as a binding mechanism in order to hold the discretized system together. More details about FEM can be found in Schäfer [26], Burnett [27], Gallagher [28], Zienkiewicz and Taylor [29], and Bathe [30].

In the present work, the FEM was employed through the ANSYS® Mechanical APDL software to perform a linear static structural analysis of the central displacements of stiffened plates under uniformly distributed transverse loads.

The ANSYS® software has different types of finite elements available, such as beam elements, plate and shell elements (two-dimensional), and solid elements (three-dimensional). In this work, the computational models for the non-stiffened and stiffened plates were developed using the finite elements SHELL93 and SOLID95.

SHELL93 is suitable to model problems involving plane or curved thin-walled structures (Figure 1a). This quadrilateral element has eight nodes with six degrees of freedom per node: translation in x, y, and z directions and rotation around x, y, and z axis. The used interpolation functions are of the quadratic type. Among the capabilities of this element are the analysis of plasticity, large displacement and large strain. Two important structural considerations are made: the stress normal to the element's plane varies linearly through the thickness; and the transverse shear stress is constant through the thickness. Moreover, a simplified version of this element is available, where the meshes are generated by triangular shaped elements [31].

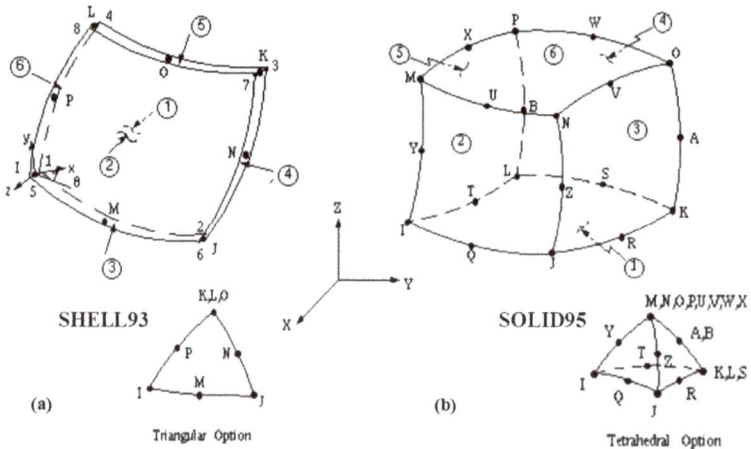

Figure 1. Finite elements adopted: (a) SHELL93; and (b) SOLID95 (Adapted from [31]).

SOLID95 is a high-order hexahedral element that can tolerate irregular shapes without significantly losing accuracy (Figure 1b). The element is also well suited to model curved geometries, being defined by twenty nodes with three degrees of freedom per node: translations in x, y, and z nodal directions, having capability to model stress stiffening, large strain, large deflection and creep. It can also be used in its tetrahedral version, with ten nodes [31].

Furthermore, it can be stated that models employing SOLID95 element give more accurate results compared to the ones with SHELL93. It occurs because solid element considers the complete three-dimensional stress and strain states of a solid body, whereas the two-dimensional element has inherent simplifications of assuming plane stress and plane strain states [31].

In the present study, as already mentioned, the ASTM A36 steel with linear-elastic behavior was considered. However, the ANSYS® software can perform static or dynamic simulations of structural components of any metallic material, considering linear or nonlinear mechanical behavior [31].

4. Constructal Design Method

The Constructal Theory, developed by Adrian Bejan, arose from the observation of the shape complexity of natural systems. It presumes that the generation of flow systems' geometric configurations is a physical phenomenon that is not the result of chance, but rather is based on physical principle called Constructal Law. The cross section of rivers and the way pulmonary veins are interconnected, for instance, and are determined through this principle [32].

Bejan and Lorente [33] stated, through the Constructal Theory, that flow systems are fated to persist imperfectly. Thus, the system evolves so that there is an improvement in the flaws distribution and, hence, the fluid body flows easily. Therefore, the natural phenomenon continually evolves not to eliminate the imperfections but actually to distribute them in order to generate a less imperfect geometrical configuration.

For the application of the Constructal Theory in the analysis of mechanical structures, a similar approach is used to the one considered in flow configurations, where the flow is related with the flow of stresses in the solid component. For instance, the geometric configuration for the structure resulting from the application of this principle is the one that has less stress concentration regions, which is obtained when the stresses are distributed throughout the material, according with the Constructal Law [33].

The Constructal Law is employed through the Constructal Design Method (CDM), which guides the engineer to obtain flow configurations that have the best global performance, under specific conditions [33].

In order to apply the CDM in the problem of stiffened plates subjected to a transverse loading, a flat steel plate without stiffeners having length a = 2000 mm, width b = 1000 mm and thickness t = 20 mm was adopted as reference. From the reference plate a volume fraction φ of its thickness was used to generate different combinations of longitudinal N_{ls} and transverse N_{ts} rectangular stiffeners, considering different values for the h_s/t_s ratio, being h_s and t_s the height and thickness of the stiffeners, respectively. Thus, the stiffened plates have the thickness t_p dependent on the parameter φ and the thickness t_s dependent on the commercial values of adopted steel plates. Figure 2 shows an example of stiffened plate with N_{ls} = 2 and N_{ts} = 3, where these geometric parameters are depicted.

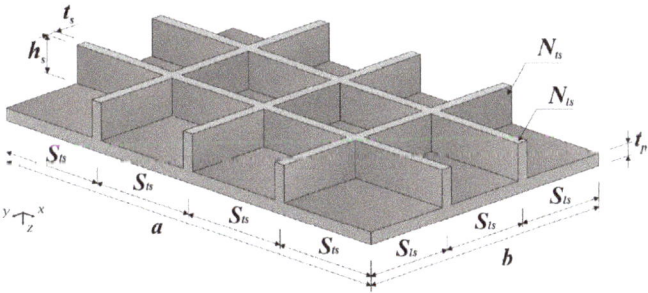

Figure 2. Stiffened plate P(2,3) with 2 longitudinal and 3 transverse stiffeners.

The volume fraction φ of material taken from the reference plate to generate the stiffeners is a constraint parameter of the Constructal Design method, being defined as:

$$\varphi = \frac{V_s}{V_r} = \frac{N_{ls}(ah_s t_s) + N_{ts}[(b - N_{ls}t_s)h_s t_s]}{abt} \qquad (2)$$

where V_s is the volume of the reference plate transformed into stiffeners and V_r is the total volume of the non-stiffened plate used as reference.

The length a and the width b of the reference plate are also adopted as constraints, since they are usually project parameters. Therefore, the volume fraction to be transformed into stiffeners was taken only from the reference plate thickness t. Thus, all plates have the same length and width and all geometric configurations have the same amount of steel, enabling a comparative evaluation of them to be performed.

In addition to the restrictions above mentioned, the application of the Constructal Design needs the definition of degrees of freedom, which in this problem are: h_s/t_s (ratio between height and thickness of the stiffeners); N_{ls} (number of longitudinal stiffeners); and N_{ts} (number of transverse stiffeners). As illustrated in Figure 2, it is worth mentioning that all stiffeners have rectangular cross section, same height, and uniform longitudinal (S_{ls}) and transverse (S_{ts}) spacing, respectively, given by:

$$S_{ls} = \frac{b}{(N_{ls}+1)} \quad (3)$$

$$S_{ts} = \frac{a}{(N_{ts}+1)} \quad (4)$$

Five values for the volume fraction were considered: φ = 0.1, 0.2, 0.3, 0.4, and 0.5. For each φ value, 25 combinations of longitudinal and transverse stiffeners were analyzed, according with the notation P(N_{ls},N_{ts}), through the variation of the following degrees of freedom: N_{ls} = 2, 3, 4, 5, and 6 and N_{ts} = 2, 3, 4, 5, and 6. Moreover, stiffeners thickness according to standard sizes of commercial steel plates were adopted and therefore the stiffeners height h_s and the ratio h_s/t_s derive from these predefined t_s sizes.

Table 2. Applications of CDM in fluid mechanics and/or heat transfer areas.

Reference	Year	Area	Description
[34]	2020	FM and HT	Design and analysis of an array of constructal fork-shaped fins (with two and three branches) adhered to a circular tube and operating under fully wet conditions, aiming the maximization of the net heat transfer rate.
[35]	2020	HT	Design and thermo-economic assessment of a flat plate solar collector has been studied with the multi-objective of improving its thermal efficiency and total annual cost.
[36]	2020	FM and HT	Optimization method applied for designing the layout of grooved evaporator wick structures in vapor chamber heat spreaders, reaching a capillary pressure improvement and temperature gradient homogeneity.
[37]	2020	FM and HT	Overall net heat transfer maximization for a cooler and reheater in the wet flue gas desulfurization equipment of coal-firing thermal power plant was explored, by means the geometric optimization of its flow architectures.
[38]	2020	FM and HT	Geometric optimization and flow parameters modeling for subcooled flow boiling (two-phase flow) were performed, aiming to minimize the thermal resistance of the microchannel heat exchanger.
[39]	2019	HT	Optimization of the geometrical configurations to assemble the ducts of an earth-air heat exchanger with the aim of improving its thermal performance.
[40]	2019	FM	Optimal design of a dual-pressure turbine in an ocean thermal energy conversion system is obtained, being the total power output of the turbine chosen as the optimization objective.

Table 2. *Cont.*

Reference	Year	Area	Description
[41]	2019	FM	New geometries for a comb-like network (single manifold duct ramified to several branches, all subject to a pressure reservoir) were obtained with perform significantly better than those with constant diameter and spacing.
[42]	2019	FM and HT	Design of a shell-and-tube evaporator with ammonia-water working fluid is evaluated, adopting a complex function (composed by heat transfer rate and total pumping power) as optimization objective.
[43]	2019	FM and HT	Geometrical optimization of internal longitudinal fins of a tube (extended inward from the pipe perimeter to a prescribed radius) is carried out ensuring maximum heat transfer and thermal efficiency.
[44]	2018	FM	Study of the geometry influence on the performance of an oscillating water column wave energy converter subject to several real scale waves with different periods, aiming the maximization of its hydrodynamic power.
[45]	2018	FM and HT	Optimization of a converter steelmaking procedure is performed by a complex function considering molten steel yield and useful energy as performance parameters.
[46]	2018	FM and HT	Determination of geometries that maximize the heat transfer and minimize pressure drop for viscoplastic fluids in cross flow around elliptical section tubes.
[47]	2018	FM and HT	Geometry optimization of a phase change material heat storage system is developed with the purpose to find the optimum shape factor for its elemental volume.
[48]	2018	FM	A geometric evaluation of an overtopping wave energy converter in real scale and submitted to incident regular waves was developed, being the goal to promote the maximization of the device available power.
[49]	2017	FM and HT	Study demonstrating how to design pores in building materials so that incoming fresh air can be efficiently tempered with low-grade heat while conduction losses are kept to a minimum.
[50]	2017	FM	As the design of a microdevice manifold should be tapered for uniform flow rate distribution, it is inferred that not only pressure drop but also velocity distribution in the microdevice play an integral role in the flow uniformity.
[51]	2017	FM and HT	Geometrical evaluation of a triangular arrangement of circular cylinders subjected to convective flows, having the multi-objective of maximizing the Nusselt number and minimizing the drag coefficient.
[52]	2017	FM and HT	An iron and steel production whole process is considered, being adopted a complex function composed of steel yield, useful energy, and maximum temperature difference as the optimization objective.
[53]	2017	FM and HT	Evaluation about the influence of geometric parameters of a solar chimney power plant, with the purpose of maximizing its available power.

FM-Fluid Mechanics; and HT-Heat Transfer.

The stiffeners heights h_s of the different geometric configurations were obtained through the Equation (2). However, only those cases that respected the following geometric limitations were simulated: h_s < 0.3 m (avoiding a disproportion between the height of the stiffener and the planar dimensions of the plate) and $h_s/t_s > 1$ (in order to avoid the stiffener thickness from being greater than its height).

Summarizing, the schematic diagram in Figure 3 shows the geometric configurations derived from the application of CDM for each volume fraction φ analyzed, which were simulated in the ANSYS® software Ver 19.3 (ANSYS, Inc., Canonsburg, PA, US).

It is worth highlighting that the CDM application in engineering problems related to fluid mechanics and/or heat transfer areas is a recognized and consecrated procedure for the geometric evaluation as well as geometric optimization. This fact can be proved due the numerous publications about it that can be found in the literature. With the purpose of exemplify the versatility of the CDM in these engineering areas, some recent studies (from 2017 to 2020) are summarized in Table 2.

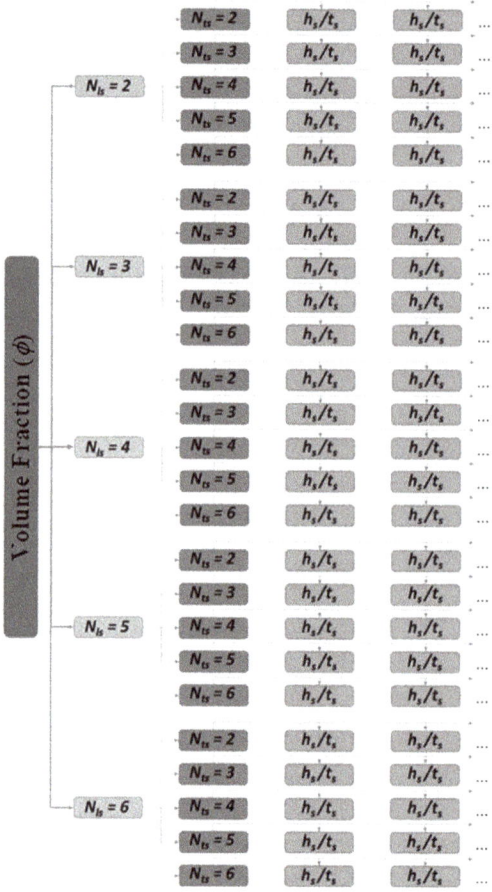

Figure 3. Geometrical configurations proposed by the application of CDM (continuum damage mechanics).

However, the CDM application in structural engineering problems has not been properly explored by the scientific community. Nowadays, there are few publications on this topic, but it is possible to cite Bejan and Lorente [33], Bejan et al. [54], Lorente et al. [55], and Isoldi et al. [56] where, by means of analogies among heat transfer, fluid mechanics, and mechanics of materials, it was conceptually proven that the CDM is also applicable in structural engineering problems. There are also works dedicated to investigating the influence of geometric configurations of plates submitted to elastic or elasto-plastic buckling: Isoldi et al. [57], Rocha et al. [58], Helbig et al. [59], Lorenzini et al. [60], Helbig et al. [61], Helbig et al. [62], Da Silva et al. [63], and Lima et al. [64,65]; while in Cunha et al. [66], De Queiroz et al. [13], Amaral et al. [67], and Pinto et al. [68] the influence of geometry of stiffened plates was analyzed when submitted to bending. Finally, Mardanpour et al. [69] and Izadpanahi et al. [70] applied the CDM in a study about aircraft structures.

5. Results and Discussion

Initially, convergence mesh tests and verifications for the developed computational models were carried out. After that, the verified computational models were used to numerically simulate the geometric configurations of stiffened plates indicated in Figure 3, and the results analyzed and discussed.

5.1. Mesh Convergence Test and Verification of Computational Models

In the present research, the verification procedure of each numerical model developed into ANSYS® was performed in order to evaluate its capability in determining the central deflection of stiffened plates under a uniformly distributed transverse load. The verification was carried out comparing the obtained numerical result in the present study with those from other researches and, when possible, with analytical solution. Before that, a mesh convergence test was performed, aiming to find the numerical solution non-mesh dependent, being the solution used for the computational verification of each model.

To achieve this, different cases of plates with and without stiffeners were analyzed: rectangular plate without stiffeners (reference plate), square plate with a central stiffener, and a square plate with orthogonal stiffeners. The computational models were discretized with two-dimensional (SHELL93) and three-dimensional (SOLID95) finite elements, as earlier mentioned. Moreover, triangular and quadrilateral shapes for SHELL93 were adopted, as well as tetrahedral and hexahedral shapes for SOLID95.

5.1.1. Rectangular Plate without Stiffeners (Reference Plate)

The first mesh convergence/verification was performed analyzing a simply supported (SS) rectangular steel plate without stiffeners, being this the reference plate. A uniform load of 10 kN/m² was applied transversally to the plate (in z direction). As for the plate's material, the ASTM A36 steel has Elastic Modulus E = 200 GPa and Poisson's ratio ν = 0.3. Figure 4 presents the dimensions and the boundary and load conditions of the analyzed case.

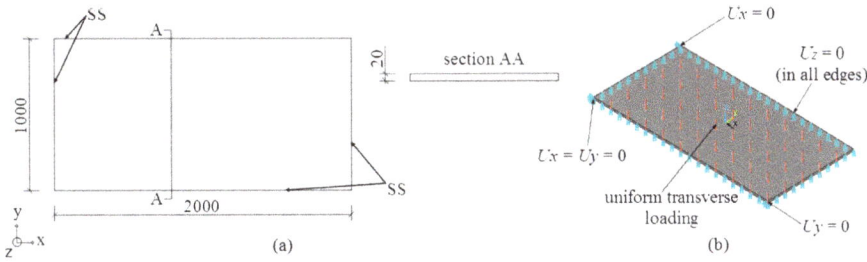

Figure 4. Reference plate: (**a**) physical model (unit: mm) and (**b**) boundary and load conditions.

The analytical solution for this problem can be obtained from the Lévy method presented in Timoshenko and Woinowsky-Krieger [23], providing a central plate deflection of U_z = 0.698 mm.

Regarding the numerical solutions, the occurrence of mesh convergence with the coarser tested meshes can be observed in Figure 5. The exception happened with the tetrahedral SOLID95 model, being necessarily a more refined mesh to obtain a converged solution. Moreover, one can note that the converged obtained results narrowly agree with the analytical solution and, therefore, it can be stated that the computational models for the plate without stiffeners were properly verified.

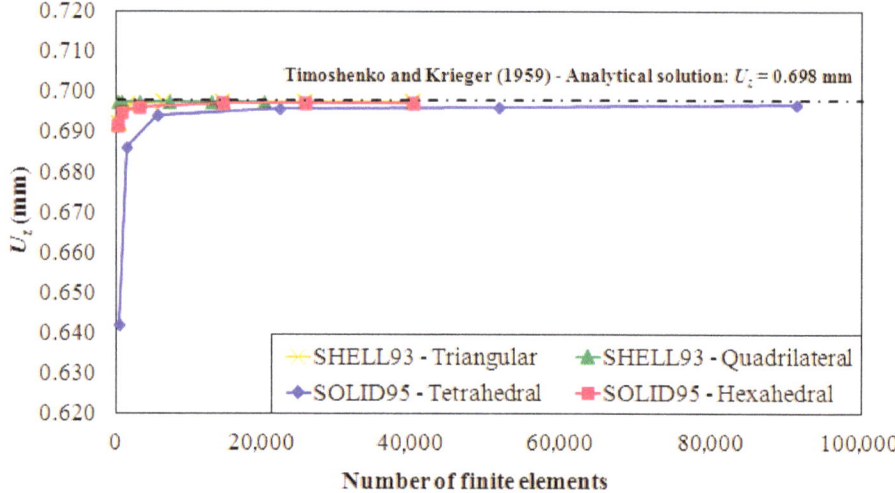

Figure 5. Mesh convergence and verification for the reference plate.

5.1.2. Square Plate with a Central Stiffener

This evaluation of the computational models for stiffened plates was based on the works of Rossow and Ibrahimkhail [6] and Tanaka and Bercin [8]. The case, which was also studied by Silva [12], consists in a thin square plate with a central stiffener. The plate's material is a metallic alloy which has $E = 117.21$ GPa and $v = 0.3$. The plate was simply supported (SS) in the edges (including the stiffener's ends), being submitted to a transverse uniform distributed load of 6.89 kN/m^2, as depicted in Figure 6.

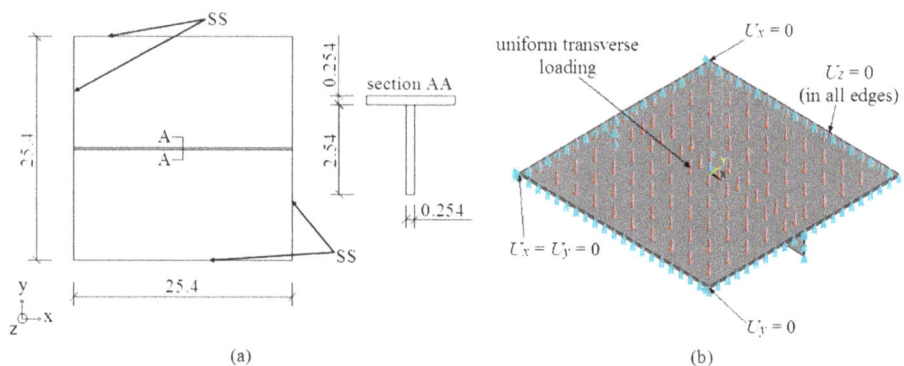

Figure 6. Square plate with a central stiffener: (a) physical model (unit: mm) and (b) boundary and load conditions.

In Figure 7 the numerical results obtained with the four proposed computational models are compared with those presented by Rossow and Ibrahimkhail [6], Tanaka and Bercin [8] and Silva [12].

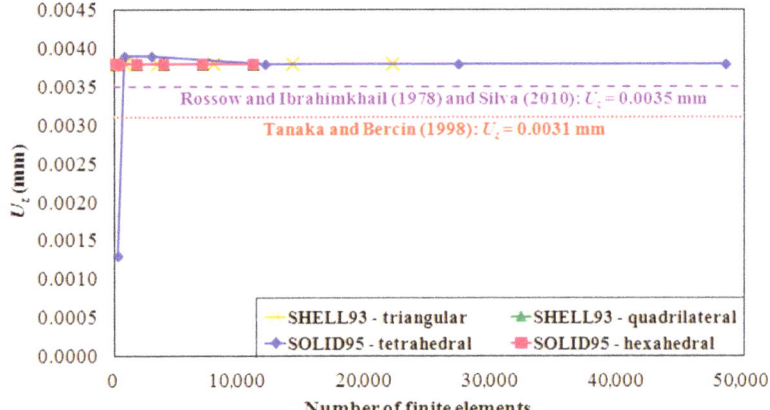

Figure 7. Mesh convergence and verification for the square plate with a central stiffener.

From Figure 7, the mesh convergence test indicates a good agreement among all proposed models in the present work, highlighting that it was not necessary to use more refined meshes to achieve the independent mesh solutions for this case. The only exception again was the tetrahedral SOLID95 model, which requires a more refined mesh to reach convergence. Additionally, the obtained numerical converged results were compared to the solutions found in the aforementioned references. In order to obtain the solutions, each author used a different approach, as previously indicated in the introductory section. Considering these solutions, it is possible to consider the proposed numerical models verified.

5.1.3. Square Plate with Orthogonal Stiffeners

Lastly, we simulated a square thin plate reinforced with orthogonal stiffeners located as shown in Figure 8. The structure consists of a plate and two sets of orthogonal stiffeners, being subjected to a uniform pressure of 9.8 kN/m². The plate edges, including the stiffeners ends, were considered as simply supported (SS). The steel alloy mechanical properties of the plate's material are $E = 210$ GPa and $\nu = 0.3$.

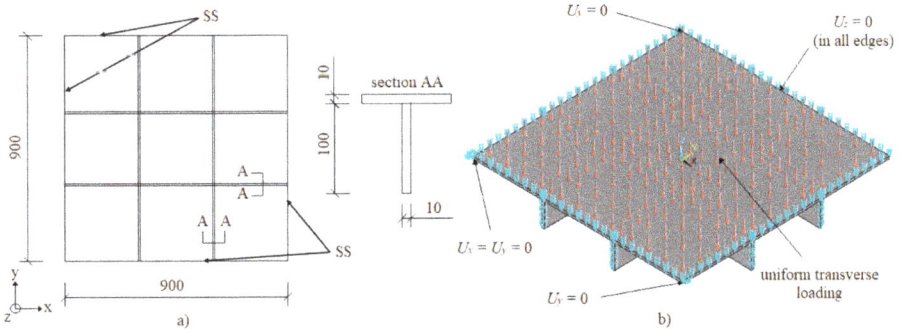

Figure 8. Square plate with orthogonal stiffeners: (**a**) physical model (unit: mm) and (**b**) boundary and load conditions.

This case was also numerically analyzed by Salomon [10] through four types of computational models developed and solved using the finite element software ADINA®. As stated by Salomon [10], the solution of a three-dimensional (3D) model provides the best response prediction of the real structure, since the model does not include pre-defined assumptions about the mechanics of the structure. Therefore, only the 3D numerical solution of Salomon [10], obtained from a 27-node finite element, was taken into account. Figure 9 shows this result confronted with those obtained by the present study.

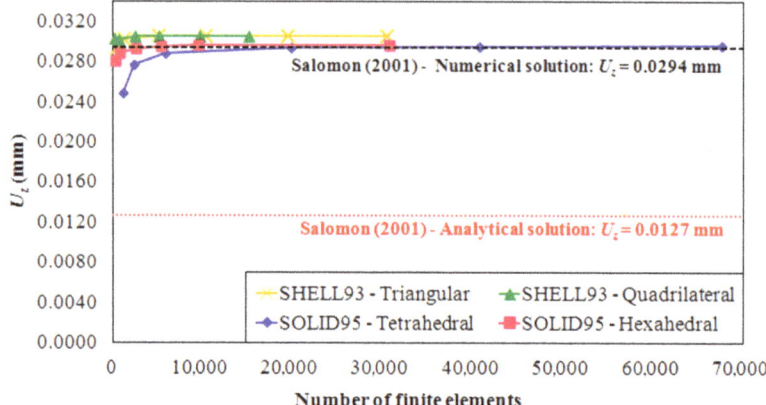

Figure 9. Mesh convergence and verification for the square plate with orthogonal stiffeners.

According to Figure 9, one can infer that the values obtained with the computational models of the present work are convergent, having the same trend observed in previous cases. Furthermore, they are similar to the value found by the reference used to comparison and, as expected, displacements with SHELL93 models were slightly greater than the displacements with the other models, since 2D models present less structural rigidity due to the assumed assumptions of the structure mechanics for thin plates.

The analytical solution presented by Salomon [10] was also included in Figure 9. This solution can be obtained by the application of Equation (1) derived from the orthotropic plate approach (see Section 2), and has, despite the difference, the same order of magnitude of the numerical solutions.

With the purpose to verify, once more, the proposed computational models based on the comparison of results with Salomon [10], it was simulated stiffened plates (see Figure 8) with different stiffeners heights (varying from 10 mm to 100 mm, with increment of 10 mm). These results are presented in Figure 10, showing superimposed results that indicated an excellent agreement among the solutions of the proposed models with the solution of the 3D model used by Salomon [10], allowing us to affirm that the computational models developed in this work were properly verified.

5.2. Case Study

In order to generate the results of this research, only the numerical models SHELL93–quadrilateral and SOLID95–hexahedral were used, because they presented better accuracy and computational efficiency in the simulations developed in the mesh convergence and verification tests (see Section 5.1).

To ensure the adequate mesh refinement to be used in the numerical simulations, convergence tests for each studied φ (volume fraction) were performed. In the executed tests, the most complex geometries were used, i.e., the plates P(6,6) with the thinnest stiffener's thickness. Besides, we applied the same boundary and load conditions as employed in all analyzed plates, namely, simply supported edges and uniformly distributed load of 10 kN/m².

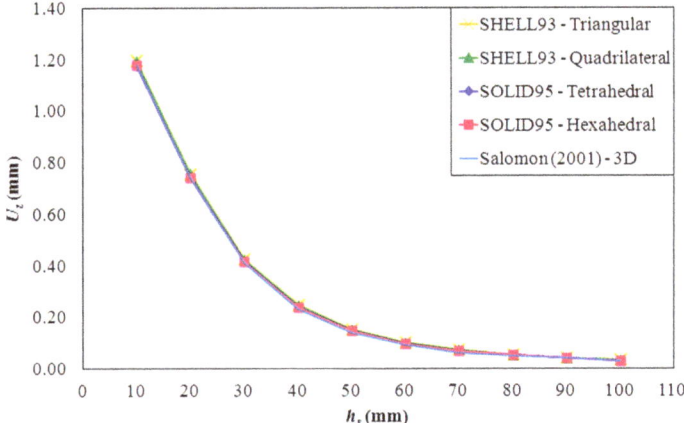

Figure 10. Mesh convergence and verification for the square plate with orthogonal stiffeners for different stiffeners heights.

In the mesh convergence tests, four different mesh refinements were used: M1; M2; M3; and M4, where the element size in each mesh is a fraction of the plate's width (M1 = $b/20$; M2 = $b/40$; M3 = $b/60$; and M4 = $b/80$). To illustrate these mesh convergence tests, Figure 11 shows the results for both models SHELL93–quadrilateral and SOLID95–hexahedral for $\varphi = 0.5$, and it is possible to observe the convergence from the mesh M3 for both element types. Furthermore, the convergence criterion was also reached in the mesh M3 for all φ under analysis. Thus, the meshes used throughout this research have an element size of 1/60 of the plate's width (16.67 mm).

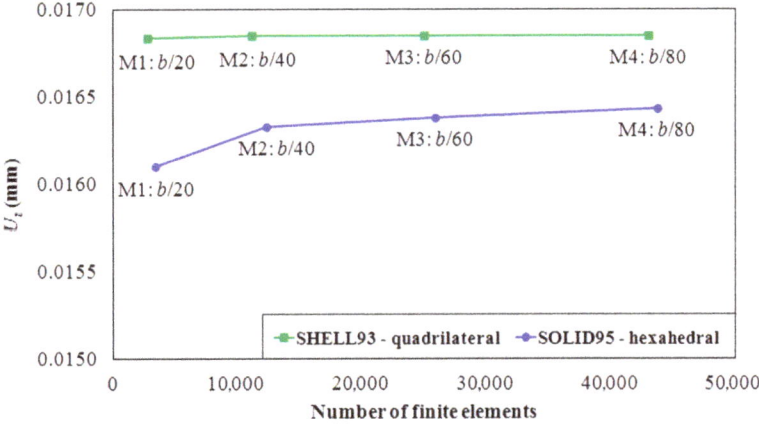

Figure 11. Mesh convergence for $\varphi = 0.5$ for the plate P(6,6) with $h_s/t_s = 49.72$.

Thereafter, the proposed stiffened plates were numerically simulated. As early mentioned, a simply supported non-stiffened steel plate, with $t = 20$ mm, $b = 1$ m, $a = 2$ m, $E = 200$ GPa and $\nu = 0.3$, was adopted as reference. Regarding the load conditions, the plate was subjected to a uniform load of 10 kN/m². The geometrical configurations of the stiffened plates were defined by transforming part of the reference plate into stiffeners. The stiffeners' thicknesses were defined based on commercial values of steel plates, varying from 1/8 in (3.18 mm) to 3 in (76.20 mm).

Then, scatter charts, representing the variation of the plates' central deflection, for each $P(N_{ls}, N_{ts})$, as a function of the degree of freedom h_s/t_s, were plotted. To illustrate the results, Figure 12 shows the aforementioned graphs for the volume fraction $\varphi = 0.5$.

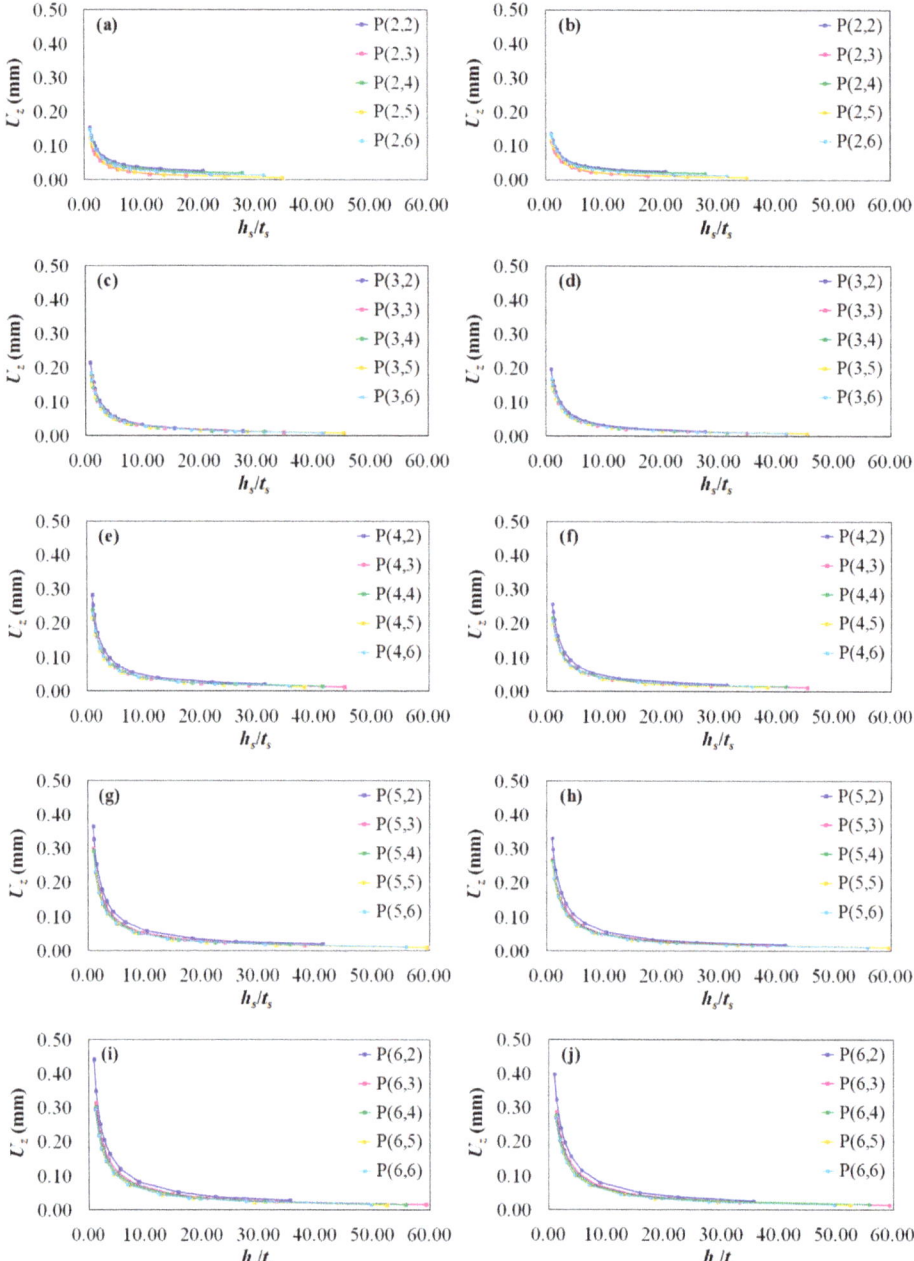

Figure 12. Central deflection for $\varphi = 0.5$ with N_{ls} and N_{ts} varying from 2 to 6: (**a**) SHELL93; (**b**) SOLID95; (**c**) SHELL93; (**d**) SOLID95; (**e**) SHELL93; (**f**) SOLID95; (**g**) SHELL93; (**h**) SOLID95; (**i**) SHELL93; and (**j**) SOLID95.

From Figure 12 one can notice that the mere transformation of a portion of steel from the reference plate into stiffeners enhanced the mechanical behavior (regarding the central deflection) of all stiffened plates, since all analyzed configurations presented a displacement lower than the reference plate's one (see Section 5.1.1).

Through the graphs of Figure 12, it was also noted that an increase in the ratio h_s/t_s entails a reduction in the central out-of-plane displacement U_z of the plates. This occurs because as h_s/t_s increases, the moment of inertia of the new defined structures also increases.

In addition, it was possible to adjust power curves to the numerical results that mathematically described the relation between the central deflection U_z and ratio h_s/t_s with great accuracy. To do so, the following general equation was defined:

$$U_z = C_1 \left(\frac{h_s}{t_s}\right)^{C_2} \quad (5)$$

where C_1 and C_2 are constants that depend on the number of longitudinal N_{ls} and transverse N_{ts} stiffeners.

To exemplify the performed curve fitting, Figure 13 presents the power curves that best fitted the data for the plates P(6,2) and P(6,5) with $\varphi = 0.5$ and SHELL93–quadrilateral and SOLID95–hexahedral elements type. Moreover, the same figure presents the determination coefficient R^2, which indicates how well the curve fits the obtained data. In both cases, the plates P(6,2) and P(6,5) had coefficients R^2 superior to 0.99 (or 99%).

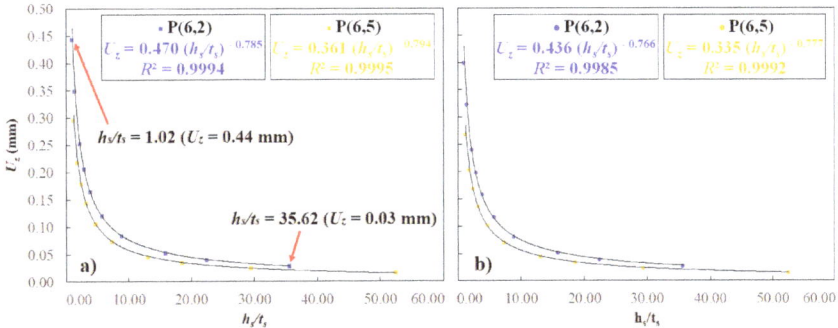

Figure 13. Curve fitting for P(6,2) and P(6,5) with $\varphi = 0.5$: (a) SHELL93 and (b) SOLID95.

In Appendix A we present the coefficients C_1; C_2; and R^2 for each studied geometry and for all volume fractions, as well as the range of degree of freedom h_s/t_s that was numerically simulated. It is important to note that all values of C_2 are negative, which means that the central deflection decreases as the relation h_s/t_s increases, corroborating with the behavior observed in the graphs of Figure 12.

Moreover, an important find which emerges from Figure 13a is that when comparing, for instance, the results of P(6,2) with $h_s/t_s = 1.02$ in relation to P(6,2) with $h_s/t_s = 35.62$, both with SHELL93 model, there is a nearly 94% reduction in transverse displacement in the center of the stiffened plate. Therefore, since the amount of steel used in the manufacture of these stiffened plates is kept constant, one can state that considerable improvements in structural rigidity can be achieved only due to the influence of its geometric configuration variation. It is worth mentioning that this trend occurs for all stiffened plates arrangements considered in this work.

Taking into account that the Constructal Design method is able to evaluate the influence of the geometric parameters (degrees of freedom) on the mechanical behavior regarding the deflection of the stiffened plates under analysis, so the degrees of freedom N_{ls}, N_{ts}, and h_s/t_s were evaluated seeking the lowest central deflection among all studied geometrical configurations, for each adopted volume fraction φ.

Initially, it was determined the optimal relation h_s/t_s, called $(h_s/t_s)_o$, i.e., the ratio h_s/t_s that leads to the minimized central displacements, called $(U_z)_m$, for each combination of $P(N_{ls}, N_{ts})$. The graphs in Figure 14 show the minimized central deflections of the stiffened plates, considering each combination of the number of longitudinal N_{ls} and the number of transverse N_{ts} stiffeners. It is important to highlight that the lines in Figure 14 are adopted only to aid the visualization of how N_{ts} influences the central deflection. Obviously, these lines do not indicate or represent continuity among the discrete values of N_{ts} variation.

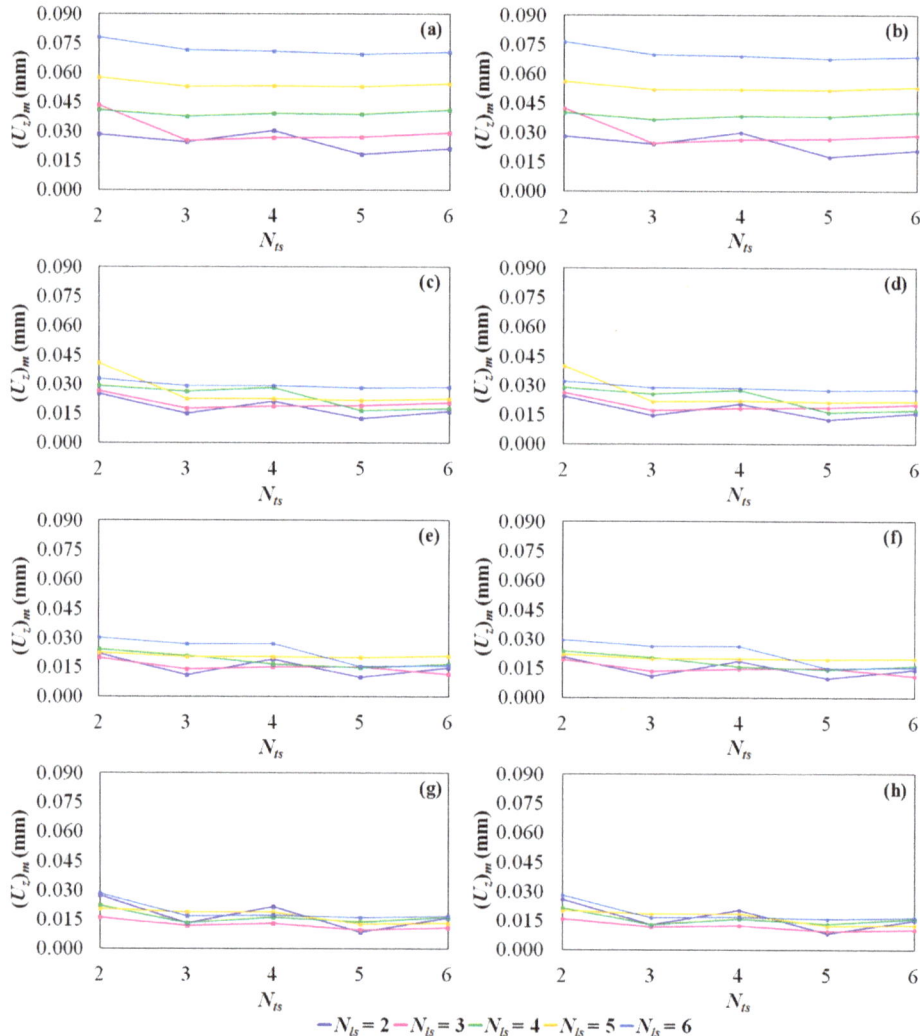

Figure 14. Influence of N_{ts} on the value of $(U_z)_m$ for $\varphi = 0.1$ to 0.5, N_{ls} and $N_{ts} = 2$ to 6: (**a**) SHELL93; (**b**) SOLID95; (**c**) SHELL93; (**d**) SOLID95; (**e**) SHELL93; (**f**) SOLID95; (**g**) SHELL93; and (**h**) SOLID95.

Figure 14 indicates that an increase in the number of stiffeners does not necessarily mean an improvement in the plate's stiffness. By keeping the volume of material constant, the increase in the number of stiffeners decreases the reinforcements' height and its cross-sectional moment of inertia and hence reduces the stiffness of the structure. Another important observation is that the results showed an oscillation trend, where plates with an odd number of transverse stiffeners presented better results than those with an even N_{ts}. This trend is explained by the fact that plates with and odd N_{ts} have a stiffener at its very center, reducing, therefore, the deflection in the structure's geometric center.

For both models (SHELL93 and SOLID95) and each N_{ls}, it was possible determining the optimal number of transverse stiffeners $(N_{ts})_o$, which is the N_{ts} that leads to a twice minimized central deflection of the plates, called $(U_z)_{mm}$. Consequently, the twice optimized ratio h_s/t_s, called $(h_s/t_s)_{oo}$ was also defined. Table 3 shows the aforementioned parameters for each volume fraction φ.

Table 3. Values of $(N_{ts})_o$, $(h_s/t_s)_{oo}$ and $(U_z)_{mm}$ for each N_{ls} and φ.

φ	N_{ls}	$(N_{ts})_o$	t_s (mm)	h_s (mm)	$(h_s/t_s)_{oo}$	$(U_z)_{mm}$ (mm) SHELL93	$(U_z)_{mm}$ (mm) SOLID95
0.1	2	3	3.18	180.19	56.66	0.0652	0.0639
0.1	3	3	3.18	140.21	44.09	0.1115	0.1094
0.1	4	3	3.18	114.75	36.08	0.1679	0.1649
0.1	5	3	3.18	97.11	30.54	0.2310	0.2269
0.1	6	5	3.18	74.41	23.40	0.2902	0.2852
0.2	2	5	3.18	280.52	88.21	0.0180	0.0177
0.2	3	3	3.18	280.42	88.18	0.0251	0.0245
0.2	4	3	3.18	229.50	72.17	0.0373	0.0365
0.2	5	5	3.18	168.61	53.02	0.0528	0.0515
0.2	6	5	3.18	148.82	46.80	0.0696	0.0679
0.3	2	5	4.75	282.20	59.41	0.0126	0.0123
0.3	3	3	4.75	282.04	59.38	0.0174	0.0171
0.3	4	5	3.18	291.70	91.73	0.0163	0.0159
0.3	5	5	3.18	252.91	79.53	0.0216	0.0211
0.3	6	5	3.18	223.23	70.20	0.0279	0.0272
0.4	2	5	6.35	281.95	44.40	0.0100	0.0098
0.4	3	6	4.75	282.72	59.52	0.0113	0.0111
0.4	4	5	4.75	261.02	54.95	0.0146	0.0143
0.4	5	5	4.75	226.35	47.65	0.0197	0.0192
0.4	6	5	3.18	297.64	93.60	0.0153	0.0149
0.5	2	5	8.00	280.27	35.03	0.0086	0.0084
0.5	3	5	6.35	288.83	45.48	0.0096	0.0094
0.5	4	3	6.35	288.33	45.41	0.0132	0.0129
0.5	5	5	4.75	282.94	59.57	0.0124	0.0121
0.5	6	5	4.75	249.77	52.58	0.0160	0.0156

One can note, from Table 3, that the displacements $(U_z)_{mm}$ tend to higher values as the degree of freedom N_{ls} increases. This happens due to the application of the Constructal Design method, the main restriction of which is to keep the total volume of material constant in all analyzed geometric configurations. By keeping the total volume of steel constant, the stiffeners' height must be reduced in order to increase the N_{ls}, directly affecting the moment of inertia.

Based on results of Table 3, it is possible to determine the three-times optimized geometric configuration for each analyzed volume fraction φ. To illustrate this, for $\varphi = 0.1$, the optimized geometry is the plate P(2,3), with three-times optimized ratio $(h_s/t_s)_{ooo} = 56.66$, which presented a three-times minimized $(U_z)_{mmm} = 0.0652$ mm for the simulations with SHELL93 elements and $(U_z)_{mmm} = 0.0639$ mm with SOLID95. For the volume fractions $\varphi = 0.2; 0.3; 0.4;$ and 0.5, the best geometrical configuration was the plate with geometric configuration P(2,5), which caused uniform spacing in longitudinal and transverse directions (see Figure 2) defined as $S_{ls} = S_{ts} = 0.333$ m.

Lastly, we performed an evaluation of the influence of φ on the results of the central displacement of the stiffened plates, with it being possible to identify, among all studied stiffened plates configurations, the one that achieves superior performance. Figure 15 shows the three-times minimized central deflection $(U_z)_{mmm}$ for each analyzed volume fraction.

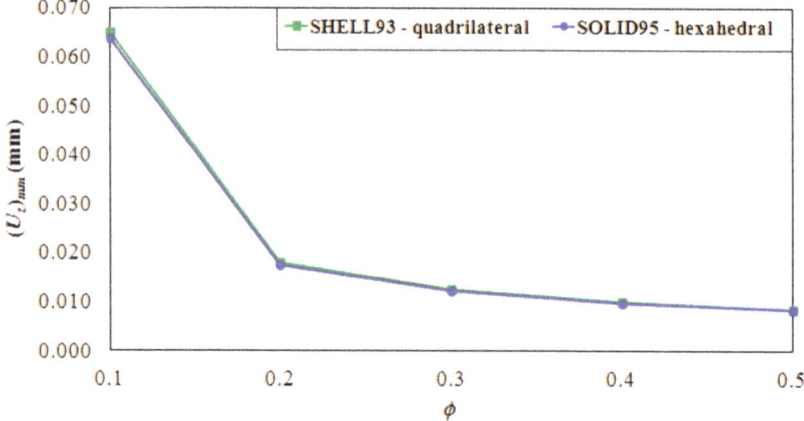

Figure 15. Influence of the volume fraction φ on the transverse central deflection.

From Figure 15, it is possible to observe that there was not a significant difference among the results of the plates' central deflections for volume fractions within the range $0.3 \leq \varphi \leq 0.5$. Moreover, we determined the plate with the best global performance (depicted in Figure 16) among all cases analyzed in the present work, demonstrating that the best configuration is the plate with optimized $\varphi_o = 0.5$, four-times optimized ratio $(h_s/t_s)_{oooo} = 35.03$, three-times optimized $(N_{ts})_{ooo} = 5$ and twice optimized $(N_{ls})_{oo} = 2$, which presented four-times minimized displacement $(U_z)_{mmmm} = 0.0086$ and $(U_z)_{mmmm} = 0.0084$ mm for the models with SHELL93 and SOLID95, respectively.

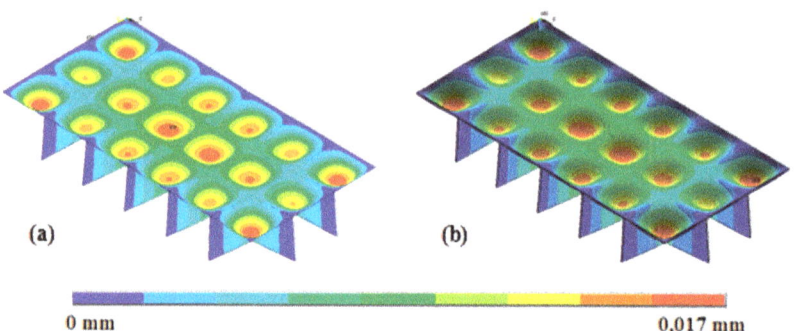

Figure 16. Deformed configuration of the global optimized plate for: (a) SHELL93 and (b) SOLID95.

Therefore, the geometry with the best mechanical performance among all analyzed stiffened plates provided a reduction in the transverse central deflection of 98.77% in comparison with the non-stiffened reference plate. In other words, the optimal geometry presented a displacement 81-times smaller than the one presented by the reference plate. Thus, significant improvements in the structural performance of the stiffened plates were reached when changing the geometric configuration of these structures.

6. Further Considerations

This paper allied the computational modeling, Constructal Design Method and Exhaustive Search technique in the geometric evaluation of stiffened steel plates. From the generated results, the following considerations emerge:

- Despite keeping the same steel volume, same dimensions (except thickness), same load and same support conditions, in a general way, all proposed stiffened plates presented lower central deflection when compared with non-stiffened reference plate. Those apparently obvious results demonstrate how the proposed method is in the right direction toward the correct prediction of physical effects. Furthermore, they show how, for a fixed amount of weight, stiffened plates are more efficient. Adequate stiffening is therefore necessary. Welding techniques are also required.
- The number of transverse stiffeners (N_{ts}), number of longitudinal stiffeners (N_{ls}), and ratio between height and thickness of stiffeners (h_s/t_s) have a deep influence on plates' stiffness and hence in its deflections. These results also permit us to give additional trust to the method, confirming an obvious physical reality regarding the fact that more stiffeners influence the response of the plate.
- For stiffened plates fabricated with the same amount of steel volume, the increase in the number of stiffeners does not necessarily imply a reduction of its central deflection, highlighting the importance of performing studies involving geometric evaluation in order to efficiently use the material when projecting and constructing these kind of structures.
- In a general way, the central deflection values for the stiffened plates tend to stabilize as the ratio h_s/t_s increases. This asymptotic trend indicates that for values of $h_s/t_s \geq 20$ there is no significant reduction of central transverse displacement. It is an important finding because over-incrementing h_s/t_s can be harmful to the structure as it increases the reinforcements' slenderness and, consequently, intensifies the mechanical element's propensity to instability problems (local buckling).
- For each combination of N_{ls} and N_{ts} a power curve was fitted to mathematically describe the relation between the central displacement and the ratio h_s/t_s. The coefficients of determination R^2 presented values from 92% up to 99.99%, evidencing the great accuracy of the performed curve fitting. The equations derived from these curves are highly useful in determining the central displacements of the plates for values of h_s/t_s within the simulated range and even to extrapolate these results for different values of h_s/t_s.
- The global optimized stiffened plate, i.e., the best performance among all analyzed geometric configurations, was the one with optimized $\varphi_o = 0.5$, four-times optimized $(h_s/t_s)_{oooo} = 35.03$, three times optimized $(N_{ts})_{ooo} = 5$ and twice optimized $(N_{ls})_{oo} = 2$, which presented a four-times minimized deflection of $(U_z)_{mmmm} = 0.0086$ mm for the simulation performed using SHELL93 elements and $(U_z)_{mmmm} = 0.0084$ mm for the simulation with SOLID95 elements. This geometric configuration reached a reduction of 98.77% in the transverse central displacement, if compared with the reference plate.
- Concerning the excellent convergence between the obtained results for both numerical models (SHELL93 and SOLID95), one can indicate the employment of shell finite element for stiffened plate simulations, since it has accuracy and needs a somewhat lower amount of processing time.
- Through the application of the Constructal Design Method, recommendations were obtained about the best geometric configurations of stiffened plates with the aim of minimizing the central out-of-plane displacement of these structures. In addition, it was also possible to draw conclusions which can serve as a support for researches related to this topic, about the mechanical behavior of structures composed of plates and stiffeners.
- The present study specifically considered the mechanical behavior related to the transverse displacements of plates. To do so, an ideal structure (with no imperfections) having a linear (geometric and material) behavior was considered. This simple approach was adopted, aiming to show the applicability of the CDM in this kind of engineering problem. Therefore, in future works

a stress analysis, as well as a Geometrically and Materially Nonlinear Analysis with Imperfections Included (GMNIA), can be performed. Moreover, other geometric parameters can be varied, other types of stiffeners can be investigated, and other types of metallic alloys can be tested.

7. Conclusions

Here we presented and discussed in detail a new computational method employed for modeling of thin steel plates with stiffeners subjected to uniform transverse loads. Stiffened thin steel plates are commonly used in a large range on engineering sectors (such as aeronautical, civil, naval or offshore engineering). Specifically, an approach associating Constructal Design Method and Exhaustive Search technique was here developed, merged in a computational model, validated and then applied with the scope to minimize the central deflections of these plates. A non-stiffened plate was adopted as reference for validation, but also for a plain comprehension of outcomes. Various values for the stiffeners volume fraction (φ) were investigated searching for their influence, together with the configuration providing the best structural performance. In these terms, the number of longitudinal and transverse stiffeners, and the aspect ratio of stiffener shape were considered as degrees of freedom. For several combinations of these parameters, the optimized plates were determined showing, in some cases, a deflection reduction of over 90%.

Author Contributions: Conceptualization, G.T., E.d.S., L.I., L.R.; methodology, G.T., M.C., V.P.; software, G.T., M.C.; validation, G.T., M.C., L.I.; formal analysis, L.R., E.d.S., C.F., L.I.; investigation, G.T., M.C., E.d.S. and L.I.; resources, E.d.S., C.F. and L.I.; data curation, L.I.; writing—original draft preparation, G.T., M.C. and V.P.; writing—review and editing, V.P., L.I. and C.F.; visualization, L.R., C.F. and E.d.S; supervision, L.R. and L.I.; project administration, E.d.S., C.F. and L.I. All authors have read and agreed to the published version of the manuscript.

Funding: The authors thank FAPERGS (Research Support Foundation of Rio Grande do Sul), CNPq (Brazilian National Council for Scientific and Technological Development), CAPES (Brazilian Coordination for Improvement of Higher Education Personnel) and MAECI (Italian Ministry of Foreign Affairs and International Cooperation) for their institutional and financial supports. In particular, E. dos Santos, L. Isoldi, and L. Rocha thank CNPq for research grants (Processes: 306024/2017-9, 306012/2017-0, and 307847/2015-2, respectively).

Conflicts of Interest: The authors declare no conflict of interest.

Appendix A

Table A1. Values of C_1; C_2; and R^2 for $\varphi = 0.1$.

$P(N_{ls}, N_{ts})$	h_s/t_s Range	SHELL93			SOLID95		
		C_1	C_2	R^2	C_1	C_2	R^2
P(2,2)	$1.05 \leq h_s/t_s \leq 66.07$	0.663	−0.519	0.9854	0.650	−0.518	0.9840
P(2,3)	$1.18 \leq h_s/t_s \leq 56.66$	0.666	−0.542	0.9831	0.653	−0.541	0.9818
P(2,4)	$1.04 \leq h_s/t_s \leq 49.60$	0.643	−0.511	0.9792	0.631	−0.511	0.9777
P(2,5)	$1.23 \leq h_s/t_s \leq 44.11$	0.672	−0.524	0.9790	0.661	−0.524	0.9778
P(2,6)	$1.11 \leq h_s/t_s \leq 39.71$	0.657	−0.499	0.9750	0.647	−0.498	0.9736
P(3,2)	$1.03 \leq h_s/t_s \leq 49.56$	0.767	−0.448	0.9664	0.749	−0.445	0.9630
P(3,3)	$1.23 \leq h_s/t_s \leq 44.09$	0.769	−0.471	0.9704	0.752	−0.469	0.9680
P(3,4)	$1.11 \leq h_s/t_s \leq 39.71$	0.736	−0.454	0.9671	0.721	−0.452	0.9646
P(3,5)	$1.01 \leq h_s/t_s \leq 36.12$	0.710	−0.440	0.9631	0.695	−0.438	0.9602
P(3,6)	$1.35 \leq h_s/t_s \leq 33.12$	0.749	−0.457	0.9703	0.736	−0.456	0.9684
P(4,2)	$1.10 \leq h_s/t_s \leq 39.66$	0.839	−0.384	0.9522	0.820	−0.381	0.9474
P(4,3)	$1.01 \leq h_s/t_s \leq 36.08$	0.790	−0.386	0.9512	0.770	−0.383	0.9463
P(4,4)	$1.35 \leq h_s/t_s \leq 33.10$	0.820	−0.410	0.9623	0.804	−0.408	0.9595
P(4,5)	$1.25 \leq h_s/t_s \leq 30.58$	0.789	−0.401	0.9600	0.773	−0.399	0.9569
P(4,6)	$1.26 \leq h_s/t_s \leq 28.41$	0.765	−0.388	0.9578	0.750	−0.386	0.9546
P(5,2)	$1.34 \leq h_s/t_s \leq 33.05$	0.904	−0.341	0.9474	0.884	−0.337	0.9417
P(5,3)	$1.24 \leq h_s/t_s \leq 30.54$	0.857	−0.347	0.9484	0.836	−0.343	0.9430
P(5,4)	$1.16 \leq h_s/t_s \leq 28.38$	0.824	−0.342	0.9480	0.805	−0.339	0.9427
P(5,5)	$1.08 \leq h_s/t_s \leq 26.51$	0.796	−0.337	0.9469	0.777	−0.334	0.9413
P(5,6)	$1.02 \leq h_s/t_s \leq 24.87$	0.774	−0.329	0.9455	0.756	−0.326	0.9395
P(6,2)	$1.15 \leq h_s/t_s \leq 28.33$	0.892	−0.277	0.9314	0.869	−0.272	0.9217
P(6,3)	$1.07 \leq h_s/t_s \leq 26.47$	0.853	−0.286	0.9341	0.829	−0.280	0.9248
P(6,4)	$1.01 \leq h_s/t_s \leq 24.84$	0.825	−0.285	0.9348	0.803	−0.281	0.9255
P(6,5)	$1.49 \leq h_s/t_s \leq 23.40$	0.864	−0.320	0.9558	0.845	−0.317	0.9519
P(6,6)	$1.41 \leq h_s/t_s \leq 22.12$	0.840	−0.315	0.9553	0.823	−0.313	0.9516

Table A2. Values of C_1; C_2; and R^2 for $\varphi = 0.2$.

$P(N_{ls}, N_{ts})$	h_s/t_s Range	SHELL93			SOLID95		
		C_1	C_2	R^2	C_1	C_2	R^2
P(2,2)	$1.35 \leq h_s/t_s \leq 59.28$	0.400	−0.660	0.9987	0.390	−0.658	0.9986
P(2,3)	$1.16 \leq h_s/t_s \leq 50.86$	0.393	−0.706	0.9985	0.379	−0.701	0.9981
P(2,4)	$1.02 \leq h_s/t_s \leq 44.53$	0.389	−0.669	0.9982	0.376	−0.665	0.9977
P(2,5)	$1.42 \leq h_s/t_s \leq 88.21$	0.410	−0.701	0.9989	0.398	−0.698	0.9987
P(2,6)	$1.28 \leq h_s/t_s \leq 79.41$	0.413	−0.681	0.9986	0.402	−0.678	0.9983
P(3,2)	$1.01 \leq h_s/t_s \leq 44.48$	0.550	−0.657	0.9956	0.528	−0.649	0.9942
P(3,3)	$1.41 \leq h_s/t_s \leq 88.18$	0.534	−0.682	0.9983	0.514	−0.677	0.9978
P(3,4)	$1.28 \leq h_s/t_s \leq 79.41$	0.516	−0.672	0.9977	0.498	−0.667	0.9971
P(3,5)	$1.17 \leq h_s/t_s \leq 72.23$	0.507	−0.673	0.9965	0.487	−0.667	0.9956
P(3,6)	$1.07 \leq h_s/t_s \leq 66.24$	0.501	−0.663	0.9954	0.482	−0.657	0.9944
P(4,2)	$1.27 \leq h_s/t_s \leq 79.31$	0.701	−0.637	0.9955	0.675	−0.632	0.9943
P(4,3)	$1.16 \leq h_s/t_s \leq 72.17$	0.641	−0.647	0.9942	0.612	−0.640	0.9929
P(4,4)	$1.07 \leq h_s/t_s \leq 66.21$	0.606	−0.635	0.9936	0.581	−0.628	0.9921
P(4,5)	$1.29 \leq h_s/t_s \leq 61.15$	0.614	−0.653	0.9942	0.589	−0.648	0.9931
P(4,6)	$1.20 \leq h_s/t_s \leq 56.82$	0.597	−0.642	0.9933	0.574	−0.637	0.9920
P(5,2)	$1.06 \leq h_s/t_s \leq 66.10$	0.791	−0.597	0.9882	0.753	−0.588	0.9854
P(5,3)	$1.28 \leq h_s/t_s \leq 61.08$	0.760	−0.623	0.9909	0.725	−0.616	0.9891
P(5,4)	$1.20 \leq h_s/t_s \leq 56.77$	0.715	−0.618	0.9903	0.683	−0.610	0.9884
P(5,5)	$1.12 \leq h_s/t_s \leq 53.02$	0.682	−0.615	0.9889	0.649	−0.607	0.9868
P(5,6)	$1.06 \leq h_s/t_s \leq 49.74$	0.657	−0.607	0.9878	0.627	−0.599	0.9855
P(6,2)	$1.18 \leq h_s/t_s \leq 56.66$	0.903	−0.572	0.9839	0.861	−0.563	0.9805
P(6,3)	$1.11 \leq h_s/t_s \leq 52.94$	0.823	−0.580	0.9839	0.780	−0.570	0.9805
P(6,4)	$1.05 \leq h_s/t_s \leq 49.68$	0.772	−0.576	0.9836	0.734	−0.567	0.9804
P(6,5)	$1.32 \leq h_s/t_s \leq 46.80$	0.785	−0.601	0.9871	0.748	−0.593	0.9849
P(6,6)	$1.25 \leq h_s/t_s \leq 44.23$	0.753	−0.594	0.9863	0.719	−0.587	0.9840

Table A3. Values of C_1; C_2; and R^2 for $\varphi = 0.3$.

$P(N_{ls}, N_{ts})$	h_s/t_s Range	SHELL93			SOLID95		
		C_1	C_2	R^2	C_1	C_2	R^2
P(2,2)	$1.04 \leq h_s/t_s \leq 31.42$	0.255	−0.688	0.9989	0.244	−0.681	0.9991
P(2,3)	$1.04 \leq h_s/t_s \leq 42.75$	0.245	−0.753	0.9995	0.234	−0.744	0.9993
P(2,4)	$1.07 \leq h_s/t_s \leq 37.44$	0.247	−0.695	0.9989	0.236	−0.687	0.9990
P(2,5)	$1.37 \leq h_s/t_s \leq 59.41$	0.259	−0.754	0.9992	0.250	−0.748	0.9992
P(2,6)	$1.23 \leq h_s/t_s \leq 53.49$	0.264	−0.719	0.9989	0.255	−0.714	0.9989
P(3,2)	$1.06 \leq h_s/t_s \leq 37.38$	0.378	−0.734	0.9992	0.360	−0.724	0.9986
P(3,3)	$1.36 \leq h_s/t_s \leq 59.38$	0.352	−0.745	0.9993	0.336	−0.736	0.9992
P(3,4)	$1.23 \leq h_s/t_s \leq 53.49$	0.344	−0.738	0.9993	0.328	0.730	0.9990
P(3,5)	$1.13 \leq h_s/t_s \leq 48.67$	0.341	−0.745	0.9991	0.325	−0.735	0.9986
P(3,6)	$1.04 \leq h_s/t_s \leq 44.64$	0.341	−0.738	0.9988	0.325	−0.729	0.9982
P(4,2)	$1.22 \leq h_s/t_s \leq 53.39$	0.503	−0.718	0.9989	0.480	−0.710	0.9984
P(4,3)	$1.12 \leq h_s/t_s \leq 48.60$	0.455	−0.732	0.9986	0.430	−0.721	0.9978
P(4,4)	$1.03 \leq h_s/t_s \leq 44.60$	0.430	−0.717	0.9984	0.407	−0.707	0.9976
P(4,5)	$1.49 \leq h_s/t_s \leq 91.73$	0.428	−0.732	0.9991	0.408	−0.725	0.9990
P(4,6)	$1.39 \leq h_s/t_s \leq 85.23$	0.419	−0.720	0.9991	0.401	−0.714	0.9989
P(5,2)	$1.02 \leq h_s/t_s \leq 44.50$	0.614	−0.704	0.9967	0.579	−0.691	0.9950
P(5,3)	$1.47 \leq h_s/t_s \leq 91.62$	0.569	−0.723	0.9990	0.539	−0.714	0.9986
P(5,4)	$1.38 \leq h_s/t_s \leq 85.15$	0.533	−0.718	0.9989	0.506	−0.709	0.9984
P(5,5)	$1.29 \leq h_s/t_s \leq 79.53$	0.511	−0.721	0.9986	0.483	−0.712	0.9980
P(5,6)	$1.22 \leq h_s/t_s \leq 74.61$	0.496	−0.717	0.9983	0.469	−0.707	0.9976
P(6,2)	$1.36 \leq h_s/t_s \leq 84.99$	0.751	−0.700	0.9979	0.712	−0.691	0.9971
P(6,3)	$1.28 \leq h_s/t_s \leq 79.41$	0.667	−0.708	0.9977	0.627	−0.697	0.9967
P(6,4)	$1.21 \leq h_s/t_s \leq 74.52$	0.618	−0.702	0.9976	0.583	−0.692	0.9966
P(6,5)	$1.14 \leq h_s/t_s \leq 70.20$	0.586	−0.705	0.9970	0.550	−0.693	0.9958
P(6,6)	$1.09 \leq h_s/t_s \leq 66.35$	0.562	−0.699	0.9967	0.529	−0.688	0.9954

Table A4. Values of C_1; C_2; and R^2 for $\varphi = 0.4$.

$P(N_{ls},N_{ts})$	h_s/t_s Range	SHELL93			SOLID95		
		C_1	C_2	R^2	C_1	C_2	R^2
P(2,2)	$1.07 \leq h_s/t_s \leq 29.55$	0.184	−0.662	0.9952	0.172	−0.653	0.9956
P(2,3)	$1.20 \leq h_s/t_s \leq 35.96$	0.175	−0.784	0.9994	0.165	−0.772	0.9995
P(2,4)	$1.06 \leq h_s/t_s \leq 31.50$	0.176	−0.672	0.9962	0.165	−0.660	0.9966
P(2,5)	$1.09 \leq h_s/t_s \leq 44.40$	0.183	−0.781	0.9994	0.174	−0.771	0.9995
P(2,6)	$1.15 \leq h_s/t_s \leq 39.98$	0.187	−0.714	0.9979	0.178	−0.706	0.9981
P(3,2)	$1.04 \leq h_s/t_s \leq 31.44$	0.275	−0.771	0.9997	0.260	−0.756	0.9995
P(3,3)	$1.09 \leq h_s/t_s \leq 44.37$	0.250	−0.775	0.9996	0.235	−0.760	0.9994
P(3,4)	$1.15 \leq h_s/t_s \leq 39.98$	0.247	−0.771	0.9995	0.233	−0.760	0.9994
P(3,5)	$1.06 \leq h_s/t_s \leq 36.39$	0.246	−0.781	0.9996	0.232	−0.767	0.9993
P(3,6)	$1.38 \leq h_s/t_s \leq 59.52$	0.248	−0.770	0.9991	0.237	−0.762	0.9991
P(4,2)	$1.14 \leq h_s/t_s \leq 39.88$	0.371	−0.751	0.9995	0.351	−0.739	0.9992
P(4,3)	$1.04 \leq h_s/t_s \leq 36.32$	0.334	−0.775	0.9994	0.312	−0.759	0.9989
P(4,4)	$1.38 \leq h_s/t_s \leq 59.47$	0.314	−0.742	0.9987	0.298	−0.733	0.9988
P(4,5)	$1.28 \leq h_s/t_s \leq 54.95$	0.313	−0.773	0.9995	0.295	−0.762	0.9993
P(4,6)	$1.19 \leq h_s/t_s \leq 51.07$	0.307	−0.754	0.9993	0.290	−0.744	0.9992
P(5,2)	$1.35 \leq h_s/t_s \leq 59.33$	0.482	−0.762	0.9993	0.456	−0.751	0.9991
P(5,3)	$1.26 \leq h_s/t_s \leq 54.85$	0.425	−0.769	0.9994	0.398	−0.755	0.9990
P(5,4)	$1.18 \leq h_s/t_s \leq 51.00$	0.397	−0.764	0.9993	0.373	−0.751	0.9989
P(5,5)	$1.11 \leq h_s/t_s \leq 47.65$	0.381	−0.767	0.9992	0.355	−0.753	0.9986
P(5,6)	$1.05 \leq h_s/t_s \leq 44.72$	0.371	−0.763	0.9990	0.347	−0.749	0.9983
P(6,2)	$1.16 \leq h_s/t_s \leq 50.86$	0.585	−0.754	0.9989	0.549	−0.741	0.9981
P(6,3)	$1.10 \leq h_s/t_s \leq 47.55$	0.512	−0.761	0.9988	0.475	−0.744	0.9978
P(6,4)	$1.04 \leq h_s/t_s \leq 44.64$	0.472	−0.752	0.9987	0.439	−0.737	0.9978
P(6,5)	$1.53 \leq h_s/t_s \leq 93.60$	0.455	−0.759	0.9992	0.427	−0.748	0.9991
P(6,6)	$1.45 \leq h_s/t_s \leq 88.46$	0.437	−0.751	0.9992	0.411	−0.741	0.9990

Table A5. Values of C_1; C_2; and R^2 for $\varphi = 0.5$.

$P(N_{ls},N_{ts})$	h_s/t_s Range	SHELL93			SOLID95		
		C_1	C_2	R^2	C_1	C_2	R^2
P(2,2)	$1.06 \leq h_s/t_s \leq 20.84$	0.150	−0.592	0.9921	0.134	−0.578	0.9917
P(2,3)	$1.16 \leq h_s/t_s \leq 17.91$	0.138	−0.823	0.9998	0.129	−0.804	0.9999
P(2,4)	$1.02 \leq h_s/t_s \leq 27.79$	0.140	−0.608	0.9890	0.127	−0.591	0.9891
P(2,5)	$1.18 \leq h_s/t_s \leq 35.03$	0.143	−0.807	0.9994	0.135	−0.793	0.9996
P(2,6)	$1.07 \leq h_s/t_s \leq 31.55$	0.148	−0.683	0.9949	0.137	−0.670	0.9955
P(3,2)	$1.01 \leq h_s/t_s \leq 27.72$	0.218	−0.800	0.9996	0.204	−0.781	0.9996
P(3,3)	$1.17 \leq h_s/t_s \leq 35.00$	0.196	−0.802	0.9995	0.182	−0.783	0.9997
P(3,4)	$1.07 \leq h_s/t_s \leq 31.55$	0.192	−0.797	0.9997	0.179	−0.778	0.9997
P(3,5)	$1.13 \leq h_s/t_s \leq 45.48$	0.191	−0.797	0.9995	0.178	−0.781	0.9995
P(3,6)	$1.04 \leq h_s/t_s \leq 41.73$	0.194	−0.796	0.9996	0.181	−0.780	0.9995
P(4,2)	$1.05 \leq h_s/t_s \leq 31.45$	0.293	−0.764	0.9994	0.273	−0.746	0.9995
P(4,3)	$1.12 \leq h_s/t_s \leq 45.40$	0.261	−0.796	0.9995	0.242	−0.778	0.9995
P(4,4)	$1.03 \leq h_s/t_s \leq 41.69$	0.245	−0.751	0.9988	0.227	−0.735	0.9990
P(4,5)	$1.13 \leq h_s/t_s \leq 38.53$	0.245	−0.801	0.9997	0.228	−0.784	0.9995
P(4,6)	$1.05 \leq h_s/t_s \leq 35.82$	0.240	−0.766	0.9995	0.223	−0.751	0.9994
P(5,2)	$1.01 \leq h_s/t_s \leq 41.55$	0.380	−0.789	0.9996	0.351	−0.770	0.9991
P(5,3)	$1.11 \leq h_s/t_s \leq 38.43$	0.335	−0.799	0.9996	0.309	−0.779	0.9992
P(5,4)	$1.04 \leq h_s/t_s \leq 35.75$	0.313	−0.793	0.9996	0.289	−0.774	0.9991
P(5,5)	$1.39 \leq h_s/t_s \leq 59.57$	0.301	−0.793	0.9993	0.280	−0.779	0.9993
P(5,6)	$1.31 \leq h_s/t_s \leq 55.90$	0.295	−0.791	0.9994	0.276	−0.777	0.9993
P(6,2)	$1.02 \leq h_s/t_s \leq 35.62$	0.470	−0.785	0.9994	0.436	−0.766	0.9985
P(6,3)	$1.37 \leq h_s/t_s \leq 59.43$	0.412	−0.794	0.9994	0.383	−0.778	0.9993
P(6,4)	$1.30 \leq h_s/t_s \leq 55.80$	0.378	−0.780	0.9993	0.352	−0.766	0.9992
P(6,5)	$1.23 \leq h_s/t_s \leq 52.58$	0.361	−0.794	0.9995	0.335	−0.777	0.9992
P(6,6)	$1.17 \leq h_s/t_s \leq 49.72$	0.348	−0.782	0.9995	0.322	−0.767	0.9992

References

1. Timoshenko, S.; Gere, J. *Theory of Elastic Stability*; McGraw-Hill: New York, NY, USA, 1961.
2. Yasuhisa, Y.; Yu, T.; Masaki, M.; Tetsuo, O. *Design of Ship Hull Structures: A Practical Guide for Engineers*; Springer: Berlin, Germany, 2009.
3. Mandal, N.R. *Ship Construction and Welding*; Springer: Singapore, 2017.
4. Tan, J.; Zhan, M.; Liu, S. Guideline for Forming Stiffened Panels by Using the Electromagnetic Forces. *Metals* **2016**, *6*, 267. [CrossRef]
5. Tan, J.; Zhan, M.; Gao, P.; Li, H. Electromagnetic Forming Rules of a Stiffened Panel with Grid Ribs. *Metals* **2017**, *7*, 559. [CrossRef]
6. Rossow, M.P.; Ibrahimkhail, A.K. A Constraint Method Analysis of Stiffened Plates. *Comput. Struct.* **1978**, *8*, 51–60. [CrossRef]
7. Bedair, O. Analysis of stiffened plates under lateral loading using sequential quadratic programming (SQP). *Comput. Struct.* **1997**, *6*, 63–80. [CrossRef]
8. Tanaka, M.; Bercin, A.N. Static bending analysis of stiffened plates using the boundary element method. *Eng. Anal. Bound. Elem.* **1997**, *21*, 147–154. [CrossRef]
9. Sapountzakis, E.; Katsikadelis, J. Analysis of Plates Reinforced with Beams. *Comput. Mech.* **2000**, *26*, 66–74. [CrossRef]
10. Salomon, A. An Evaluation of Finite Element Models of Stiffened Plates. Master's Thesis, Massachusetts Institute of Technology, Cambridge, MA, USA, 2001.
11. Hasan, M. Optimum design of stiffened square plates for longitudinal and square ribs. *ALKEJ J.* **2007**, *3*, 13–30.
12. Silva, H.B.S. Análise Numérica da Influência da Excentricidade na Ligação Placa-viga em Pavimentos Usuais de Edifícios. Master's Thesis, Universidade de São Paulo, São Carlos, Brazil, 10 January 2010.
13. De Queiroz, J.; Cunha, M.L.; Pavlovic, A.; Rocha, L.A.O.; Dos Santos, E.D.; Troina, G.S.; Isoldi, L.A. Geometric Evaluation of Stiffened Steel Plates Subjected to Transverse Loading for Naval and Offshore Applications. *J. Mar. Sci. Eng.* **2019**, *7*. [CrossRef]
14. Kallassy, A.; Marcelin, J.L. Optimization of stiffened plates by genetic search. *Struct. Optim.* **1997**, *13*, 134–141. [CrossRef]
15. Cunha, M.L.; Estrada, E.D.S.D.; da Silva Troina, S.D.; dos Santos, E.D.; Rocha, L.A.O.; Isoldi, L.A. Verification of a genetic algorithm for the optimization of stiffened plates through the constructal design method. *Res. Eng. Struct. Mater.* **2019**, *5*, 437–446.
16. Putra, G.L.; Kitamura, M.; Takezawa, A. Structural optimization of stiffener layout for stiffened plate using hybrid GA. *Int. J. Nav. Arch. Ocean Eng.* **2019**, *11*, 809–818. [CrossRef]
17. Lee, J.-C.; Shin, S.-C.; Kim, S.-Y. An optimal design of wind turbine and ship structure based on neuro-response surface method. *Int. J. Nav. Arch. Ocean Eng.* **2015**, *7*, 750–769 [CrossRef]
18. Anyfantis, K.N. Evaluating the influence of geometric distortions to the buckling capacity of stiffened panels. *Thin Wall. Struct.* **2019**, *140*, 450–465. [CrossRef]
19. Szilard, R. *Theories and Applications of Plate Analysis: Classical Numerical and Engineering Methods*; John Wiley & Sons: Hoboken, NJ, USA, 2004.
20. Salmon, C.G.; Johnson, J.E. *Steel Structures: Design and Behavior, Emphasizing Load and Resistance Factor Design*; HarperCollins College Publishers: New York, NY, USA, 1996.
21. Bernuzzi, C.; Cordova, B. *Structural Steel Design to Eurocode 3 and AISC Specifications*; John Wiley & Sons, Ltd: Oxford, UK, 2016.
22. Yamaguchi, E. *Basic Theory of Plates and Elastic Stability. Structural Engineering Handbook*; CRC Press LLC: Boca Raton, FL, USA, 1999.
23. Timoshenko, S.; Woinowsky-Krieger, S. *Theory of Plates and Shells*, 2nd ed.; McGraw-Hill: New York, NY, USA, 1959.
24. Ventsel, E.; Krauthammer, T. *Thin Plates and Shells: Theory, Analysis and Applications*; CRC Press: New York, NY, USA, 2001.
25. Ugural, A.C. *Plates and Shells: Theory and Analysis*; CRC Press: New York, NY, USA, 2018.

26. Schäfer, M. *Computational Engineering—Introduction to Numerical Methods*; Springer: Berlin, Germany, 2006.
27. Burnett, D. *Finite Element Analysis—From Concepts to Applications*; Addison–Wesley: Boston, MA, USA, 1989.
28. Gallagher, R. *Finite Element Analysis: Fundamentals*; Prentice-Hall: Englewood Cliffs, NJ, USA, 1975.
29. Zienkiewicz, C.; Taylor, R. *The Finite Element Method*; McGraw-Hill: London, UK, 1989.
30. Bathe, K.J. *Finite Element Procedures*; Prentice-Hall: Upper Saddle River, NJ, USA, 1996.
31. ANSYS Inc. ANSYS User's Manual: Analysis Systems. 2009. Available online: http://research.me.udel.edu/~||lwang/teaching/MEx81/ansys56manual.pdf (accessed on 31 January 2020).
32. Bejan, A. *Shape and Structure, from Engineering to Nature*; Cambridge University Press: Cambridge, MA, USA, 2000.
33. Bejan, A.; Lorente, S. *Design with Constructal Theory*; Wiley: Hoboken, NJ, USA, 2008.
34. Hazarika, S.A.; Deshmukhya, T.; Bhanja, D.; Nath, S. A novel optimum constructal fork-shaped fin array design for simultaneous heat and mass transfer application in a space-constrained situation. *Int. J. Therm. Sci.* **2020**, *150*, 106225. [CrossRef]
35. Ganjehkaviri, A.; Mohd Jaafar, M.N. Multi-objective particle swarm optimization of flat plate solar collector using constructal theory. *Energy* **2020**, *194*, 116846. [CrossRef]
36. Li, B.; Yin, X.; Tang, W.; Zhang, J. Optimization design of grooved evaporator wick structures in vapor chamber heat spreaders. *Appl. Therm. Eng.* **2020**, *166*, 114657. [CrossRef]
37. Lee, J. Maximal Heat Transfer Density in Heat Exchangers of Wet Flue Gas Desulfurization Equipment. *Heat Transfer Eng.* **2020**, 41. [CrossRef]
38. Ariyo, D.O.; Bello-Ochende, T. Constructal design of subcooled microchannel heat exchangers. *Int. J. Heat Mass Trans.* **2020**, *146*, 118835. [CrossRef]
39. Brum, R.S.; Ramalho, J.V.A.; Rodrigues, M.K.; Rocha, L.A.O.; Isoldi, L.A.; dos Santos, E.D. Design evaluation of Earth-Air Heat Exchangers with multiple ducts. *Renew. Energy* **2019**, *135*, 1371–1385. [CrossRef]
40. Wu, Z.; Feng, H.; Chen, L.; Xie, Z.; Cai, C.; Xia, S. Optimal design of dual-pressure turbine in OTEC system based on constructal theory. *Energ. Convers. Manag.* **2019**, *201*, 112179. [CrossRef]
41. Lugarini, A.; Franco, A.T.; Errera, M.R. Flow distribution uniformity in a comb-like microchannel network. *Microfluid. Nanofluid.* **2019**, *23*, 44. [CrossRef]
42. Cai, C.; Feng, H.; Chen, L.; Wu, Z.; Xie, Z. Constructal design of a shell-and-tube evaporator with ammonia-water working fluid. *Int. J. Heat Mass Trans.* **2019**, *135*, 541–547. [CrossRef]
43. Salimpour, M.R.; Darabi, J.; Mahjoub, S. Design and Optimization of Internal Longitudinal Fins of a Tube Using Constructal Theory. *J. Eng. Thermophys.* **2019**, *28*, 239–254. [CrossRef]
44. Gomes, M.; das, N.; Lorenzini, G.; Rocha, L.A.O.; dos Santos, E.D.; Isoldi, L.A. Constructal Design Applied to the Geometric Evaluation of an Oscillating Water Column Wave Energy Converter Considering Different Real Scale Wave Periods. *J. Eng. Thermophys.* **2018**, *27*, 173–190. [CrossRef]
45. Liu, X.; Feng, H.; Chen, L. Constructal Design of a Converter Steelmaking Procedure Based on Multi-objective Optimization. *Arab. J. Sci. Eng.* **2018**, *43*, 5003. [CrossRef]
46. Hermany, L.; Lorenzini, G.; Klein, R.J.; Zinani, F.F.; dos Santos, E.D.; Isoldi, L.A.; Rocha, L.A.O. Constructal design applied to elliptic tubes in convective heat transfer cross-flow of viscoplastic fluids. *Int. J. Heat Mass Trans.* **2018**, *116*, 1054–1063. [CrossRef]
47. Solé, A.; Falcoz, Q.; Cabeza, L.F.; Neveu, P. Geometry optimization of a heat storage system for concentrated solar power plants (CSP). *Renew. Energy* **2018**, *123*, 227–235. [CrossRef]
48. Martins, J.C.; Goulart, M.M.; Gomes, M.; das, N.; Souza, J.A.; Rocha, L.A.O.; Isoldia, L.A.; dos Santos, E.D. Geometric evaluation of the main operational principle of an overtopping wave energy converter by means of Constructal Design. *Renew. Energy* **2018**, *118*, 727–741. [CrossRef]
49. Craig, S.; Grinham, J. Breathing walls: The design of porous materials for heat exchange and decentralized ventilation. *Energy Build.* **2017**, *149*, 246–259. [CrossRef]
50. Cetkin, E. Constructal Microdevice Manifold Design With Uniform Flow Rate Distribution by Consideration of the Tree-Branching Rule of Leonardo da Vinci and Hess–Murray Rule. *J. Heat Transf.* **2017**, *139*, 082401. [CrossRef]
51. Barros, G.M.; Lorenzini, G.; Isoldi, L.A.; Rocha, L.A.O.; dos Santos, E.D. Influence of mixed convection laminar flows on the geometrical evaluation of a triangular arrangement of circular cylinders. *Int. J. Heat Mass Trans.* **2017**, *114*, 1188–1200. [CrossRef]

52. Feng, H.; Chen, L.; Liu, X.; Xie, Z. Constructal design for an iron and steel production process based on the objectives of steel yield and useful energy. *Int. J. Heat Mass Trans.* **2017**, *111*, 1192–1205. [CrossRef]
53. Vieira, R.S.; Petry, A.P.; Rocha, L.A.O.; Isoldi, L.A.; dos Santos, E.D. Numerical evaluation of a solar chimney geometry for different ground temperatures by means of constructal design. *Renew. Energy* **2017**, *109*, 222–234. [CrossRef]
54. Bejan, A.; Lorente, S.; Lee, J. Unifying constructal theory of tree roots, canopies and forests. *J. Theor. Biol.* **2008**, *254*, 529–540. [CrossRef]
55. Lorente, S.; Lee, J.; Bejan, A. The "flow of stresses" concept: the analogy between mechanical strength and heat convection. *Int. J. Heat Mass Trans.* **2010**, *53*, 2963–2968. [CrossRef]
56. Isoldi, L.A.; Real, M.V.; Correia, A.L.G.; Vaz, J.; Dos Santos, E.D.; Rocha, L.A.O. Flow of Stresses: Constructal Design of Perforated Plates Subjected to Tension or Buckling. In *Constructal Law and the Unifying Principle of Design—Understanding Complex Systems*, 1st ed.; Rocha, L.A.O., Lorente, S., Bejan, A., Eds.; Springer: New York, NY, USA, 2013; pp. 195–217.
57. Isoldi, L.A.; Real, M.V.; Vaz, J.; Correia, A.L.G.; Dos Santos, E.D.; Rocha, L.A.O. Numerical Analysis and Geometric Optimization of Perforated Thin Plates Subjected to Tension or Buckling. *Mar. Syst. Ocean Tech.* **2013**, *8*, 99–107. [CrossRef]
58. Rocha, L.A.O.; Isoldi, L.A.; Real, M.V.; Dos Santos, E.D.; Correia, A.L.G.; Biserni, C.; Lorenzini, G. Constructal design applied to the elastic buckling of thin plates with holes. *Cent. Eur. J. Eng.* **2013**, *3*, 475–483. [CrossRef]
59. Helbig, D.; Real, M.V.; Dos Santos, E.D.; Isoldi, L.A.; Rocha, L.A.O. Computational modeling and constructal design method applied to the mechanical behavior improvement of thin perforated steel plates subject to buckling. *J. Eng. Therm.* **2016**, *25*, 197–215. [CrossRef]
60. Lorenzini, G.; Helbig, D.; Da Silva, C.C.C.; Real, M.V.; Dos Santos, E.D.; Isoldi, L.A.; Rocha, L.A.O. Numerical Evaluation of the Effect of Type and Shape of Perforations on the Buckling of Thin Steel Plates by means of the Constructal Design Method. *Int. J. Heat Technol.* **2016**, *34*, S9–S20. [CrossRef]
61. Helbig, D.; Da Silva, C.C.C.; Real, M.V.; Dos Santos, E.D.; Isoldi, L.A.; Rocha, L.A.O. Study About Buckling Phenomenon in Perforated Thin Steel Plates Employing Computational Modeling and Constructal Design Method. *Lat. Am. J. Solids Struct.* **2016**, *13*, 1912–1936. [CrossRef]
62. Helbig, D.; Cunha, M.L.; Da Silva, C.C.C.; Dos Santos, E.D.; Iturrioz, I.; Real, M.V.; Isoldi, L.A.; Rocha, L.A.O. Numerical study of the elasto-plastic buckling in perforated thin steel plates using the constructal design method. *Res. Eng. Struct. Mat.* **2018**, *4*, 169–187. [CrossRef]
63. Da Silva, C.C.C.; Helbig, D.; Cunha, M.L.; Dos Santos, E.D.; Rocha, L.A.O.; Real, M.V.; Isoldi, L.A. Numerical Buckling Analysis of Thin Steel Plates with Centered Hexagonal Perforation Through Constructal Design Method. *J. Braz. Soc. Mech. Sci. Eng.* **2019**, *41*, 309. [CrossRef]
64. Lima, J.P.S.; Rocha, L.A.O.; Santos, E.D.; Real, M.V.; Isoldi, L.A. Constructal design and numerical modeling applied to stiffened steel plates submitted to elasto-plastic buckling. *Proc. Roman. Acad. Ser. A* **2018**, *19*, 195–200.
65. Lima, J.P.S.; Cunha, M.L.; dos Santos, E.D.; Rocha, L.A.O.; Real, M.V.; Isoldi, L.A. Constructal Design for the ultimate buckling stress improvement of stiffened plates submitted to uniaxial compressive load. *Eng. Struct.* **2020**, *203*, 109883. [CrossRef]
66. Cunha, M.L.; Troina, G.S.; Rocha, L.A.O.; Dos Santos, E.D.; Isoldi, L.A. Computational modeling and Constructal Design method applied to the geometric optimization of stiffened steel plates subjected to uniform transverse load. *Res. Eng. Struct. Mater.* **2018**, *4*, 139–149. [CrossRef]
67. Amaral, R.R.; Troina, G.S.; Nogueira, C.M.; Cunha, M.L.; Rocha, L.A.O.; Santos, E.D.; Isoldi, L.A. Computational modeling and constructal design method applied to the geometric evaluation of stiffened thin steel plates considering symmetry boundary condition. *Res. Eng. Struct. Mater.* **2019**, *5*, 393–402. [CrossRef]
68. Pinto, V.T.; Cunha, M.L.; Troina, G.S.; Martins, K.L.; dos Santos, E.D.; Isoldi, L.A.; Rocha, L.A.O. Constructal design applied to geometrical evaluation of rectangular plates with inclined stiffeners subjected to uniform transverse load. *Res. Eng. Struct. Mater.* **2019**, *5*, 379–392.

69. Mardanpour, P.; Izadpanahi, E.; Rastkar, S.; Lorente, S.; Bejan, A. Constructal design of aircraft: flow of stresses and aeroelastic stability. *AIAA J.* **2019**. [CrossRef]
70. Izadpanahi, E.; Moshtaghzadeh, M.; Radnezhad, H.R.; Mardanpour, P. Constructal approach to design of wing cross-section for better flow of stresses. *AIAA J.* **2020**. [CrossRef]

© 2020 by the authors. Licensee MDPI, Basel, Switzerland. This article is an open access article distributed under the terms and conditions of the Creative Commons Attribution (CC BY) license (http://creativecommons.org/licenses/by/4.0/).

Article

Dry Sliding Wear Performance of ZA27/SiC/GraphiteComposites

Nenad Miloradović *, Rodoljub Vujanac, Slobodan Mitrović and Danijela Miloradović

Faculty of Engineering, University of Kragujevac, Sestre Janjić 6, 34000 Kragujevac, Serbia
* Correspondence: mnenad@kg.ac.rs; Tel.: +381-34-335990

Received: 25 May 2019; Accepted: 24 June 2019; Published: 26 June 2019

Abstract: The paper describes the wear performance of zinc-aluminium ZA27 alloy, reinforced with silicon-carbide (SiC) and graphite (Gr) particles. The compo-casting technique produced the composite samples. The tested samples were: ZA27 alloy, ZA27/5%SiC composite, and ZA27/5%SiC/3%Gr hybrid composite. A block-on-disc tribometer was used during wear tests under the dry sliding conditions by varying the normal loads and sliding speeds. The sliding distance was constant during tests. The microstructure of the worn surfaces of the tested materials was analysed using the scanning electronic microscope (SEM) and the energy dispersive spectrometry (EDS).

Keywords: hybrid composite; wear performance; ZA27 alloy

1. Introduction

Zinc–aluminium alloys are used for various industrial applications. They have favourable properties, such as: A low melting point, high tensile strength and hardness, easy machinability, high castability, good corrosion resistance, and low manufacturing costs.

The effects of different reinforcement materials, like SiC, graphite, Al_2O_3 and ZrO_2 on tribological behaviour of composites based on ZA27 alloy have been the research topic of many authors. Their research results clearly show that reinforcement particles improve the tribological behaviour of the base Zn-Al alloy.

To improve the wear performance of zinc-based alloys, SiC is added as reinforcement to form composites. Thus, the dimensional stability and wear resistance of the composites were improved [1,2]. The sliding wear tests were conducted using a pin-on-disc method. The composite melt was prepared with 10 wt.% SiC particles of 50 μm–100 μm size. The author concludes that the wear-rate of the zinc-based alloy and its composite containing SiC particles increases with load in dry and lubricated conditions.

Effect of the SiC particles reinforcement on the dry sliding wear behaviour of ZA27 alloy composites was investigated in Reference [3]. The sliding wear tests were conducted at loads of 3 kg, 4 kg, 5 kg and 6 kg (29.4 N, 39.2 N, 49.1 N and 58.9 N, respectively) and rotational speeds of 200 rpm, 250 rpm and 300 rpm. It has been noticed that the observed composites exhibit abrasion wear at low loads and dominant delamination wear at high loads.

Authors of several papers have investigated and reported a positive tribological effect of graphite particles on ZA27 alloys in the dry sliding tests [4–6]. In Reference [4], the graphite particulate size was 50 μm–100 μm and its contents in the composites were 0%, 1%, 3% and 5%. It was found that the increase in graphite content within the ZA27 matrix resulted in a significant increase of ductility and strength, but a decrease of hardness. In Reference [5], ZA27 based composites were prepared with 0%, 4%, 6% and 8% of graphite particulate addition using the compo casting method. Experimental results show that the damping capacity increases with addition of graphite.

Unlike more conventional materials such as cast iron [7], this effect was obtained with the tribo-induced graphite film on the contact surface of the composite. The composite material based on

ZA27 alloy with 2 wt.% of graphite particles was tested under dry and lubricated sliding conditions [8]. The authors have shown that the difference between the wear resistance of composite and the matrix alloy increases with the increase of the applied load and the sliding speed. This was attributed to solid lubrication provided by the formation of the graphite-rich film on the tribo-surface.

The metal-matrix composites (MMCs) have better mechanical properties, such as: Higher tensile strength, temperature stability and elastic modulus compared to unreinforced alloys [9,10]. The results from Reference [9] show that ZA alloys wear-rate strongly depends on the applied load. This dependence was found to be nonlinear. In addition, the transfer of Fe from the disc to the sample was noticed in dry sliding conditions. The research confirmed that the wear properties of the matrix alloys improve with the addition of SiC particles. Based on the experimental results from Reference [10], the authors have modelled dry sliding wear behaviour using a statistical approach method. They observed the effect of interactions between the sliding speed, the applied load and the sliding distance on wear behaviour of metal matrix composites. The increase of the sliding speed during tests generated the decrease of the wear volume loss. The increase in applied load and sliding distance caused an increase in the wear volume loss. The developed model proved to be an effective tool for evaluation of composites dry sliding wear performance.

The Taguchi design of the experiments as means for optimisation of the wear test parameters was used to describe the tribological behaviour of ZA27/10%SiC composites and ZA27/10%SiC/3%Gr hybrid composites in Reference [11]. It has been found that the contact load has the greatest influence on the wear-rate (85.85%), while the sliding speed (7.09%) and graphite content (5.24%) have a smaller influence. The interactions between main parameters (graphite, load, sliding speed) have significantly less influence than the parameters themselves.

The papers [12,13] deal with the wear characteristics of ZA27/5%SiC/3%Gr and ZA27/10%SiC/1%Gr hybrid composites, respectively. Both composites were produced by using the compo-casting method and tested for different applied loads, sliding speeds and sliding distances in order to prove their better resistance to wear. Based on experimental tests conducted with the "block-on-disc" tribometer, variations in wear volume loss in dry sliding conditions were presented. By monitoring the wear process, the advantages of the tested hybrid composites as advanced tribo-materials was observed.

The hard reinforcements (such as SiC) enhance the hardness and abrasive wear resistance of Zn-Al alloy, but reduce its machinability and conductivity [14]. It has been concluded that graphite, as a solid lubricant with good conductivity, can be dispersed in Zn-Al alloy together with SiC. The increase in SiC content induces the increase of the ultimate tensile strength of the composite material.

Study in Reference [15] was focused on processing and physical and mechanical characterisation of the ZA-27 metal matrix composites reinforced with SiC particles, with different weight percentage (0 to 9 wt.%). It was concluded that the observed composites have low porosity and improved micro-hardness and tensile strength compared to the pure ZA27 alloy. The authors conclude that the erosion characteristics of these composites can also be successfully analysed using Taguchi experimental design scheme.

In Reference [16], experiments were conducted on Al–Gr composites and Al–SiC–Gr hybrid composites using the pin-on-disc equipment. Investigations have shown that the optimally combined % reinforcement is around 7.5% for any value of the sliding distance, sliding speed, and load within the considered range. Application of aluminium based composite materials in automotive industry enables the production of the lighter, safer and more energy efficient vehicles.

Tribological tests of aluminium based composite materials were described in Reference [17]. A matrix alloy was reinforced with 10% Gr and 3% SiC. It was shown that an increase in the SiC fraction causes an increase of the wear resistance of hybrid composites, while decreasing the wear.

Results from References [18–20] indicate that the alumina particle content can have a positive influence on the tribological behaviour of the ZA27 alloy-based hybrid composite. The tribological tests were conducted using the pin-on-disc wear test equipment under the dry sliding condition [18]. Wear process was studied using analysis of variance technique and regression equations. In Reference [19], the alumina percentage in the composites was varied from 0 to 9%. The hybrid composite with 9% Al_2O_3 and 3% graphite was found to have the highest wear resistance. The paper [20] investigated the composites with 3, 5, and 10 wt.% of Al_2O_3 particles that were produced by the compo-casting procedure. The difference in the wear resistance between the composite and the matrix alloy was increased with the increase of Al_2O_3 particle content.

Composite, which is improved by reinforcements with alumina and graphite nano-particles, has preferable microstructure stability and tribological behaviour than conventional MMCs [21].

In References [22,23], the authors state that nano-particles can improve the base material, and contribute to wear resistance, damping properties, and mechanical strength. However, thus obtained nanocomposites are seldom used in commercial applications, because of their recent development. Due to their good thermal conductivity, wear resistance and high specific strength, they are suitable as materials for aircraft brakes, bicycle frames, heat sinks and solders, but also in aerospace and automotive industries for making the cylinder liners, aircraft fins, disk brakes and calipers.

Increase of mass or volumetric share of SiC, Al_2O_3 and graphite changes the tribological characteristics of MMCs. By combining the appropriate share of reinforcement materials, the optimal values of tribological characteristics of materials are achieved [24,25].

Good characteristics of the zinc-aluminium alloys have inspired researchers to reinforce them with already mentioned reinforcement materials. In this way, much more enhanced mechanical and tribological properties may be obtained. Based on the available research, some authors have separately studied the effects of reinforcing ZA27 with SiC and Gr. However, the majority of the research was dedicated to improving tribological properties of the ZA27 alloy-based composites. Not many of them have been focused on hybrid composites with the same substrate. Thus, this paper aims to investigate the tribological behaviour of hybrid composites compared to the base alloy. Acquired knowledge may be used to improve the existing solutions and lead to new industrial and engineering applications.

Zinc aluminium alloy ZA27 was used as a base material for the production of composites in this research. The following composites were analysed: ZA27/5%SiC/3%Gr hybrid composite and ZA27/5%SiC composite. Other researchers have rarely tested composites with a given percentage of reinforcement particles, therefore, the test results presented in this paper contribute to a better understanding of their wear performance.

2. Materials and Methods

The tested composite materials were successfully prepared using the compo-casting procedure, which is economical and enables a good distribution of the reinforcements in the composite structure. Previous preparation of reinforcement particles is not necessary, so the production becomes simpler and cheaper.

In order to obtain the composite materials, equipment consisting of the process part (melting furnace, mixer and crucible) and the part for temperature measurement, control and regulation (thermocouple and instrumentation). A two- phase compo-casting procedure was used. First, the infiltration of the particles into the semi-solid melt of the basic alloy was obtained by mixing. The graphite particles of size 15 µm were added, while the average size of the SIC particles was 26 µm. After the operating temperature of 465 °C was reached, mixing of melt with the rotational speed of the active part of the mixer of 500 min^{-1}, lasting 2.5 min has been conducted in order to facilitate the infiltration of particles and to prevent thickening of melt. The second phase included the hot pressing of the obtained composite casts.

2.1. Structure of the Tested Materials

Table 1 shows the chemical composition of the ZA27 alloy used for obtaining the composite materials.

Table 1. The chemical composition of ZA27 alloy.

Element	Al	Cu	Mg	Fe	Sn	Cd	Pb	Zn
wt.%	26.3	2.54	0.018	0.062	0.002	0.005	0.004	balance

The hardness of the tested samples was: ZA27—124 Hv, ZA27/5%SiC composite—147 Hv, ZA27/5%SiC/3%Gr hybrid composite—145 Hv.

The metallographic analysis is a very powerful research tool that enables the review of the grain structure and the size of the reinforcement particles. Thus, the microstructure of the tested materials was observed by computer-aided optical microscope Meiji Techno's MT8500 (Meiji Techno Co., Ltd., Saitama, Japan). The specimens were polished and etched according to standard metallographic practice.

Figure 1a presents the microstructure of ZA27 alloy. The visible uniformity of the structure caused favourable mechanical properties of the tested materials. It can be seen that the structure of the sample of ZA27 alloy is typically dendrite. The microstructure of the matrix alloy revealed primary α dendrites, eutectoid $\alpha + \eta$ and meta-stable ε phase. Similar observations have been made in References [1,8,9,14]. The structure is found to be homogeneous, consisting of light-coloured cores rich with aluminium, with a grey coloured extracted eutecticum made of α- and η-phase rich with zinc.

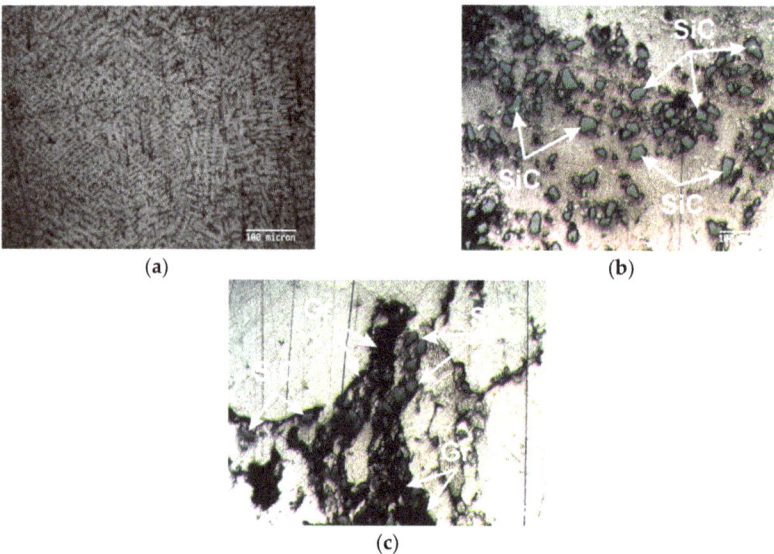

Figure 1. The microstructure of the tested materials: (**a**) ZA27 alloy; (**b**) ZA27/5%SiC composite; (**c**) ZA27/5% SiC/3%Gr hybrid composite.

Microstructure of ZA27/5%SiC composite is given in Figure 1b and microstructure of ZA27/5%SiC/3%Gr hybrid composites displayed in Figure 1c. The particles of SiC and Gr are indicated in the figures.

Reasonably homogeneous dispersion of the reinforcing particles is visible. The silicon carbide particles can be clearly identified when compared with graphite. The soft particles of graphite did not maintain their original size, because their erosion had occurred during the mixing procedure.

2.2. Testing Methods

The measuring equipment consisted of five major component groups: Tribology testing device (tribometer), sensors, measuring amplifier, data acquisition system, and computer.

The tests of tribological characteristics of the ZA27 alloy and both composites were done using the block-on-disc tribometer with computer support. Tribometer configuration contains the power system (electric motor and drive), the system for setting the normal load and sliding speed, the system for self-aligning of the rotational disc and the stationary block, and the measuring system. The design of the special block carrier enables the transfer of the normal load in the direction of the disc axis. In this way, a contact along the whole length of the block on disk is obtained.

The dry sliding tests were conducted for sliding speeds of 0.25 m/s, 0.5 m/s and 1 m/s, normal loads of 10 N, 20 N and 30 N and sliding distance of 600 m. The contact pairs were prepared according to the demands of ASTM G77-05 standard. All discs, made of 90MnCrV8 steel with a hardness of 62–64 HRC, had the diameter of 35 mm and surface roughness of $R_a = 0.3$ µm. Dimensions of the blocks were 6.35 mm × 15.75 mm × 10.16 mm. The blocks were made of the tested ZA27 alloy, ZA27/5%SiC composite and ZA27/5%SiC/3%Gr hybrid composite. The roughness of the contact surface of the test blocks was $R_a = 0.3$ µm.

All the experiments were repeated five times per sample/condition. The wear track width at the block contact surface was selected as the basic parameter of wear. The measurement of the wear track width was conducted using the universal measuring microscope UIM-21 (GOMZ, Saint Petersburg, Russia). Its final values were calculated as the arithmetic mean of the five conducted measurements. Wear track width for different normal loads and sliding speeds were used for calculation of material wear-rate. Wear volume was calculated in accordance with ASTM G77-83 (based on the measured wear track width and the geometry of the contact pair, shown in References [17,20]) and used to calculate the wear-rate expressed in mm^3/m.

In order to understand the microstructure of the worn surfaces better, the SEM analysis of the tested alloy and MMCs was performed. The scanning electronic microscope JEOL JSM-6610LV (JEOL Ltd., Tokyo, Japan) was used. It has high resolution, and it is equipped with modern EDS module. Before the SEM analysis, the samples were cleaned thoroughly. The samples were immersed in ethanol, and then kept in an ultrasonic bath in order to achieve reliable results. The position of the samples in the sample chamber was tracked by a camera, which enabled a direct selection of position for signal recording.

3. Results and Discussion

The variations of the wear-rate in the dry sliding conditions are presented in corresponding diagrams for adopted sliding distance, sliding speeds and contact loads. Table 2 contains the measurement results of wear track width and calculated wear-rate for the three types of materials, and all combinations of sliding speeds and applied loads.

In order to observe the evolution of the wear process and to make comparisons, the results of the wear-rate for ZA27 alloy and composites were given in the same diagrams. In order to clearly state the difference between wear-rate levels, diagrams are given for one force level and one sliding speed level (Figures 2–4). Diagrams of the wear-rate obtained for the normal contact load of 10 N and three different sliding speeds are presented in Figure 2a–c.

Diagrams of wear-rate obtained for the normal contact load of 20 N and three different sliding speeds are presented in Figure 3a–c.

Diagrams of wear-rate obtained for normal contact load of 30 N and three different sliding speeds are presented in Figure 4a–c.

The research results are in accordance with similar research related to ZA27 based alloys and composites published in the literature [2,3,11,13,20]. It has been reported that the composite possesses better wear resistance than a matrix alloy under dry sliding conditions. The wear-rate of composites, as well as of the matrix alloy increased with the increase of applied load and sliding speed.

Based on previous diagrams, it may be seen that the wear-rate curves have the same character both for ZA27 alloy and for the considered composites, but the composite materials exhibit smaller wear-rates compared to that of ZA27 alloy in all tests. In the initial phase of the tests, very intensive wear was observed, followed by a period of stabilisation. The intensive initial wear of the tested materials causes the larger slope of the curves at the beginning of the tests.

Table 2. Values of wear track width and wear-rate.

Tested Materials		ZA27		ZA27/5%SiC		ZA27/5%SiC/3%Gr	
v/m/s	F/N	Wear Track Width/mm	Wear-Rate/mm^3/m× 10^{-3}	Wear Track Width/mm	Wear-Rate/mm^3/m× 10^{-3}	Wear Track Width/mm	Wear-Rate/mm^3/m× 10^{-3}
0.25	10	2.974	1.329	2.867	1.096	2.682	0.897
	20	3.768	2.706	3.681	2.323	3.551	2.085
	30	4.025	3.299	3.869	2.669	3.795	2.547
0.5	10	3.166	1.603	2.997	1.253	2.751	0.968
	20	3.859	2.907	3.712	2.383	3.561	2.103
	30	4.223	3.812	3.917	2.801	3.807	2.575
1	10	3.424	2.029	3.111	1.401	2.814	1.037
	20	3.985	3.202	3.795	2.547	3.564	2.108
	30	4.451	4.466	3.948	2.868	3.864	2.688

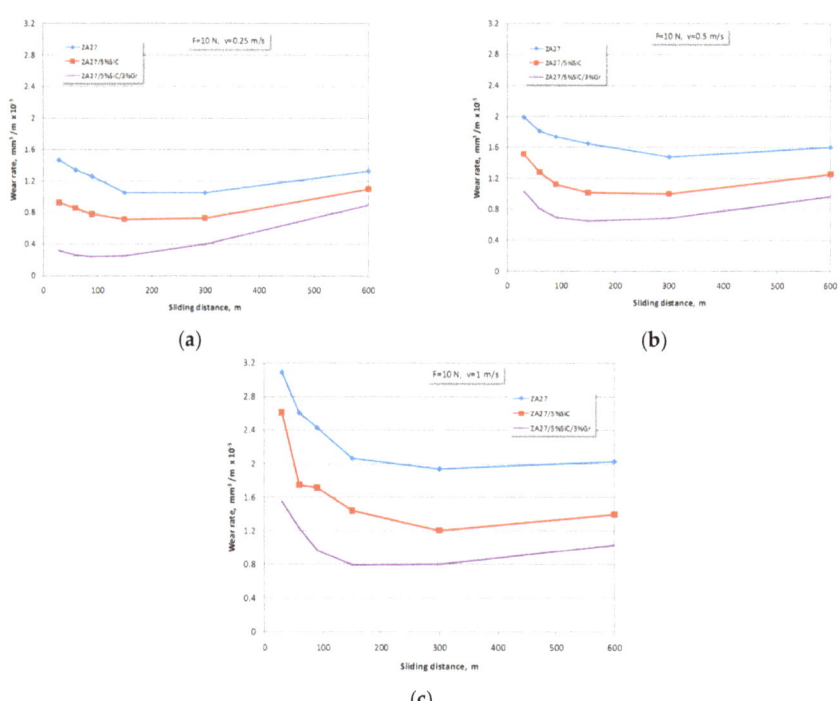

Figure 2. Wear-rate of the tested materials for normal load of 10 N and sliding speed of: (a) $v = 0.25$ m/s; (b) $v = 0.5$ m/s; (c) $v = 1$ m/s.

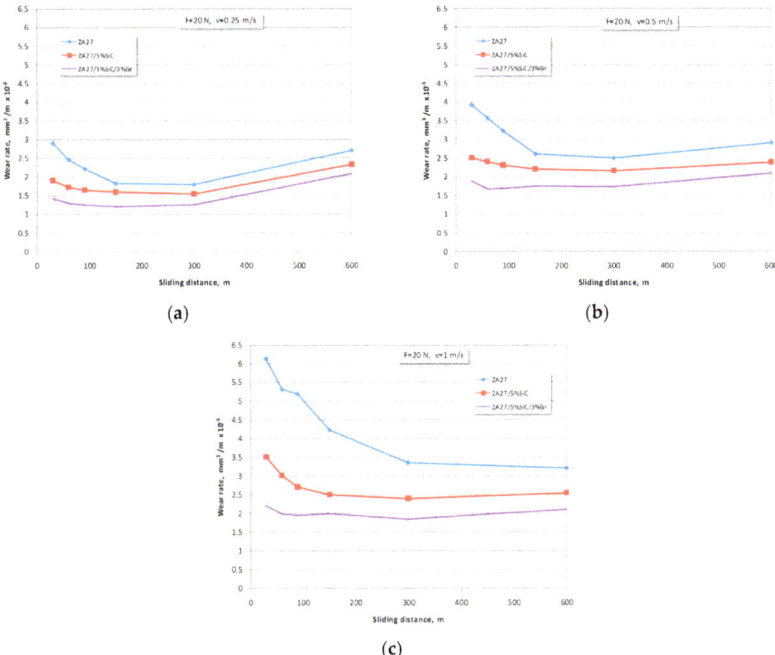

Figure 3. Wear-rate of the tested materials for normal load of 20 N and sliding speed of: (**a**) $v = 0.25$ m/s; (**b**) $v = 0.5$ m/s; (**c**) $v = 1$ m/s.

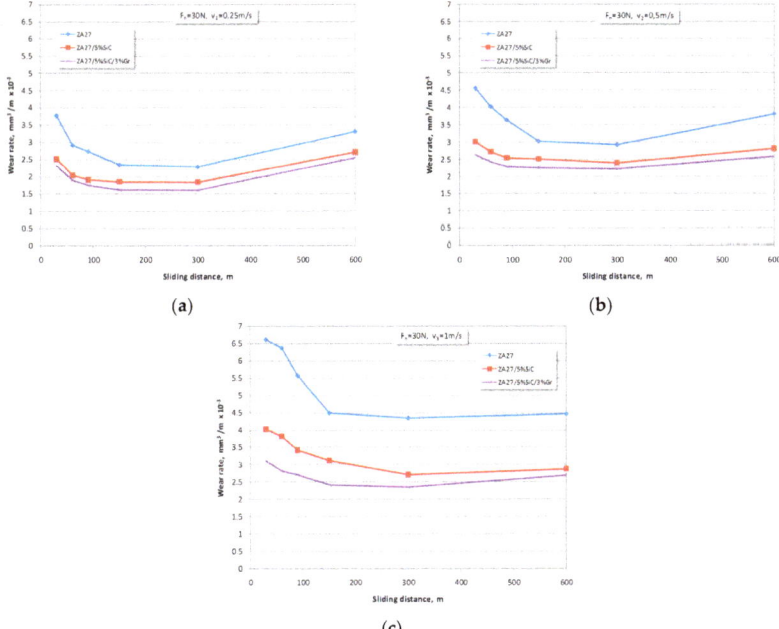

Figure 4. Wear-rate of the tested materials for normal load of 30 N and sliding speed of: (**a**) $v = 0.25$ m/s; (**b**) $v = 0.5$ m/s; (**c**) $v = 1$ m/s.

Analytical and graphical dependences between the wear-rate and the adopted parameters are presented in Figure 5a–c. The power regression functions in the form given by Equation (1) were used:

$$wear\ rate = C \cdot F_n^a \cdot v^b, \qquad (1)$$

where C, a, b are constants, F_n is the normal contact load, and v is the sliding speed.

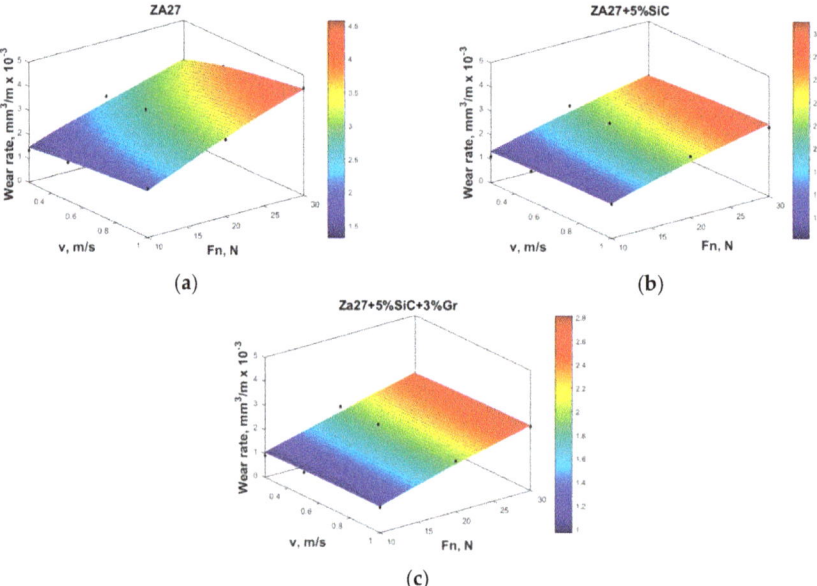

Figure 5. The wear-rate regression surface for: (a) ZA27 alloy ($wear-rate = 0.3422 \cdot F_n^{0.7533} \cdot v^{0.1979}$), $R^2 = 0.9878$, adj $R^2 = 0.9837$; (b) ZA27/5%SiC composite ($wear-rate = 0.2403 \cdot F_n^{0.7589} \cdot v^{0.0956}$), $R^2 = 0.9379$, adj $R^2 = 0.9172$; (c) ZA27/5%SiC/3%Gr hybrid composite ($wear-rate = 0.1670 \cdot F_n^{0.8222} \cdot v^{0.03125}$), $R^2 = 0.9606$, adj $R^2 = 0.9475$.

By using the experimental results and curve fitting tool of the software package Matlab (R2015a, MathWorks, Inc., Natick, MA, USA), the regression coefficients were calculated with 95% confidence bounds. The goodness of fit was represented by R^2 and adjusted R^2 values. The very good correlation between the experimental data and empirical distributions was evident.

The obtained values of the wear-rates and their respective standard deviations for the ZA27 alloy and tested composites in the dry sliding conditions were given in comparative histograms in Figure 6 for the sliding distance of 600 m. From the presented histograms, it can be clearly seen that the highest values of the wear-rate belong to the ZA27 alloy for all the testing regimes. Histograms also show that wear-rate increases by increasing the normal load and sliding speed. The largest value of the wear-rate corresponds to the highest sliding speed (1 m/s) and the highest normal load (30 N). On the other hand, the lowest sliding speed (0.25 m/s) and the lowest load (10 N) give the smallest wear values.

Figure 7a–c shows corresponding SEM micrographs for a normal load of 10 N, sliding speed of 0.25 m/s and sliding distance of 600 m. A set of parallel grooves and scratches can be seen on the worn surface of all the tested materials. These grooves and scratches were made by sliding of the significantly harder rotation disc of the tribometer along the worn surfaces. In addition, these parallel and continuous scratches point to abrasive wear induced by the penetration of the hard SiC particles into a softer surface (mark A in Figure 7b,c).

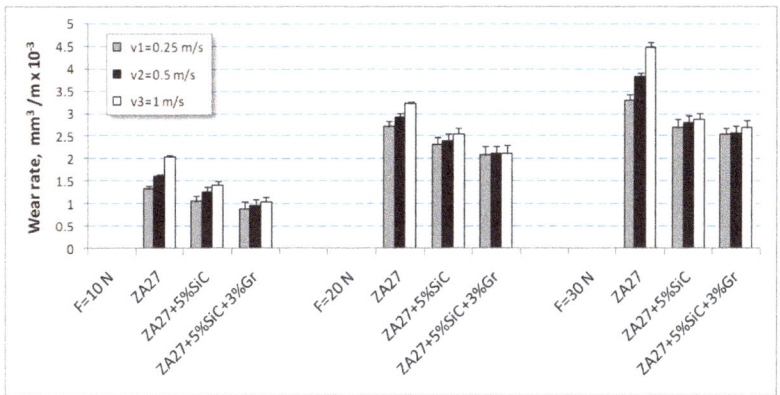

Figure 6. The wear-rate of the tested materials in the dry sliding conditions.

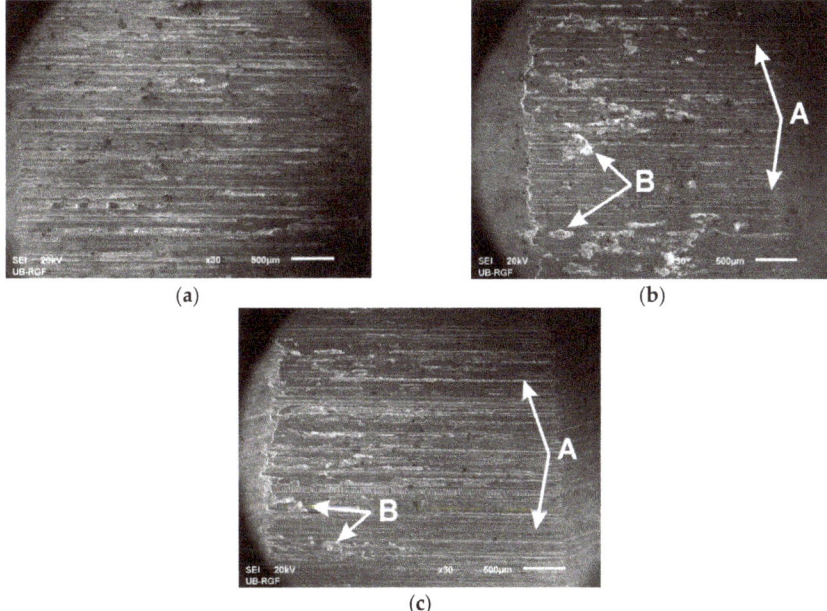

Figure 7. The appearance of the worn surfaces of the tested materials taken by SEM: (**a**) ZA27 alloy; (**b**) ZA27/5%SiC composite; (**c**) ZA27/5%SiC/3%Gr hybrid composite.

Morphology of the worn surfaces and the obtained wear-rates point to the fact that the generation of the tribo-layers on the contact surface of the observed composites influences their overall wear behaviour.

The existence of the transfer of material between the contact surfaces has indicated that there was adhesive wear present. The adhesion wear occurs as a result of the creation and destruction of alternating friction connections. A part of worn material is ejected from the contact in the form of wear debris. The other part is stuck on the wear track. The pits irregular form and depth are also visible on the surface of the composite (mark B in Figure 7b,c).

Research from Reference [26] also states that wear tracks show typical abrasive wear for ZA27 alloy composite reinforced with 6% garnet at low loads. The authors point to the existence of the delamination process at higher loads. Removal of the surface layers by delamination was also detected in Reference [2].

In Reference [1], deep grooves were observed on the surfaces of the zinc-based matrix alloy, while they were not present on the smoother surface of the composite. Sticking of fine debris particles was observed in both cases. Similar observations have been made in References [27,28]. Rougher worn surface with deep grooves, damages and transfer material was attributed to intensive abrasive and adhesive wear.

The presence of the oxide layer formed on ZA27 was detected in Reference [29]. This layer protects the surface from deformation and damage by reducing the wear-rate.

The results obtained in this paper are in accordance with [30]. Deep grooves were observed on the worn surface of ZA27 alloy. Addition of graphite in hybrid composite makes the grooves shallower.

Figure 8a shows the results of the chemical analysis of the micro-constituents of all tested materials obtained by the energy dispersive spectrometry.

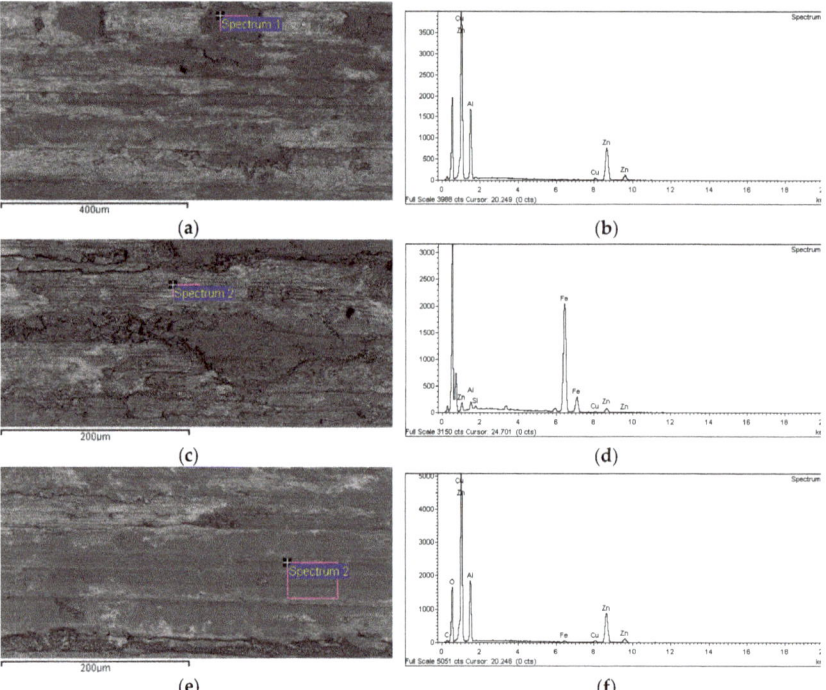

Figure 8. SEM micrographs and EDS analysis of the worn surfaces: (**a**) ZA27 alloy SEM micrograph; (**b**) ZA27 alloy EDS analysis; (**c**) ZA27/5%SiC composite SEM micrograph; (**d**) ZA27/5%SiC composite EDS analysis; (**e**) ZA27/5%SiC/3%Gr hybrid composite SEM micrograph; (**f**) ZA27/5%SiC/3%Gr hybrid composite EDS analysis.

Spectrum analysis confirms the presence of Zn, Al, Si, Gr (C). The presence of Fe is detected due to material transfer from the rotational disc to the composite block (the same as in Reference [9]).

Generally, the wear performance of the tested materials is influenced by the formation of a mechanically mixed layer (MML) between the contact surfaces. Stability and thickness of the MML depend on the intensity of formation and destruction of the MML on the worn surface. The larger extent of material transfer is caused by severe wear conditions and higher frictional heating [20].

Creation of the MML is more prominent in alloys reinforced with ceramic particles. During the process of the dry sliding wear, hard ceramic particles scratch the counter surface materials. This phenomenon may explain the presence of Fe in the EDS spectrum analysis of the worn surface in Figure 8b. It can be concluded that tribo-layers contain SiC fractures, graphite, iron oxide and other particles contained by the observed hybrid composite, which was confirmed by EDS analysis.

The wear debris, fragmentation of oxide layer, transfer of materials between the contact surfaces and mixing of constituents contributes to the formation of the MML under an applied load. This stable layer prevents further direct contact between the contact surfaces. The decomposition of the unstable transfer layer may lead to restated contact and increase of wear.

4. Conclusions

Based on the existing theoretical knowledge and extensive experimental investigations, tribological knowledge of ZA27/SiC/Gr composites may be completed. Obtained excellent wear properties of the tested composite materials confirm that they can be used as advanced tribo-materials.

Based on the extensive experimental research presented in the paper, followed by corresponding analyses, the following conclusions can be made:

- The obtained experimental results indicate that reinforcement with the SiC particles and graphite can significantly improve tribological properties of the ZA27 alloy. The wear-rates of the tested composites are smaller than the wear-rate of theZA27 alloy for all applied sliding speeds and normal loads.
- The wear-rate behaviour has the same character for all the tested materials.
- Increase of the normal load causes the increase in wear-rate for the tested materials.
- The wear-rate also increases with the increase of the sliding speed.
- The microstructure analysis using an optical microscope and SEM shows that the particles are well dispersed in the ZA27 alloy matrix, as well as in the tested composites.
- The hybrid ZA27/SiC/Gr composites are tribologically superior to the ZA27/SiC composite in all the test conditions, due to lubrication obtained by the existence of the graphite film on the contact surface of the hybrid composite.

Author Contributions: Conceptualisation, N.M. and S.M.; methodology, N.M.; software, R.V. and D.M.; formal analysis, N.M., R.V. and D.M.; investigation, N.M.; resources, S.M.; writing—original draft preparation, N.M. and D.M.; writing—review and editing, N.M.; visualisation, N.M. and R.V.; supervision, S.M.

Funding: This research was funded by the Ministry of Education, Science and Technological Development of the Republic of Serbia, grant number TR35021.

Conflicts of Interest: The authors declare no conflict of interest.

References

1. Prasad, B.K. Investigation into sliding wear performance of zinc-based alloy reinforced with SiC particles in dry and lubricated conditions. *Wear* **2007**, *262*, 262–273. [CrossRef]
2. Prasad, B.K. Abrasive wear characteristics of a zinc-based alloy and zinc-alloy/SiC composite. *Wear* **2002**, *252*, 250–263. [CrossRef]
3. Sharma, S.C.; Girish, B.M.; Kramath, R.; Satish, B.M. Effect of SiC particle reinforcement on the unlubricated sliding wear behavior of ZA-27 alloy composites. *Wear* **1997**, *213*, 33–40. [CrossRef]
4. Seah, K.H.W.; Sharma, S.C.; Girish, B.M. Mechanical properties of cast ZA-27 graphite particulate composites. *Mater. Des.* **1996**, *16*, 271–275. [CrossRef]
5. Girish, B.M.; Prakash, K.R.; Satish, B.M.; Jain, P.K.; Prabhakar, P. An investigation into the effects of graphite particles on the damping behavior of ZA-27 alloy composite material. *Mater. Des.* **2011**, *32*, 1050–1056. [CrossRef]
6. Fragassa, C.; Minak, G.; Pavlovic, A. Tribological aspects of cast iron investigated via fracture toughness. *Tribol. Ind.* **2016**, *38*, 1–10.

7. Fragassa, C.; Babic, M.; Minak, G. Predicting the tensile behaviour of cast alloys by a pattern recognition analysis on experimental data. *Metals* **2019**, *9*, 557. [CrossRef]
8. Babic, M.; Mitrovic, S.; Džunic, D.; Jeremic, B.; Bobic, I. Tribological Behavior of Composites Based on ZA-27 Alloy Reinforced with Graphite Particles. *Tribol. Lett.* **2010**, *37*, 401–410. [CrossRef]
9. Auras, R.; Schvezov, C. Wear Behaviour, Microstructure, and Dimensional Stability of As-Cast Zinc-Aluminium/SIC (Metal Matrix Composites) Alloys. *Metall. Mater. Trans. A* **2004**, *35A*, 1579–1590. [CrossRef]
10. Basavarajappa, S.; Chandramohan, G. Dry Sliding Wear Behaviour of Metal Matrix Composites: A Statistical Approach. *J. Mater. Eng. Perform.* **2006**, *15*, 656–660. [CrossRef]
11. Miloradović, N.; Stojanović, B.; Nikolić, R.; Gubeljak, N. Analysis of wear properties of Zn-based composites using the Taguchi method. *Mater. Test.* **2018**, *60*, 265–272. [CrossRef]
12. Mitrović, S.; Babić, M.; Miloradović, N.; Bobić, I.; Stojanović, B.; Dzunić, D.; Pantić, M. Wear Characteristics of Hybrid Composites Based on ZA27 Alloy Reinforced with Silicon Carbide and Graphite Particles. *Tribol. Ind.* **2014**, *36*, 204–210.
13. Miloradovic, N.; Stojanovic, B. Tribological behaviour of ZA27/10SiC/1Gr hybrid composite. *J. Balkan Tribological Assoc.* **2013**, *19*, 97–105.
14. Kiran, T.S.; Prasanna Kumar, M.; Basavarajappa, S.; Vishwanatha, B.M. Mechanical properties of as-cast ZA-27/Gr/SiCp hybrid composite for the application of journal bearing. *J. Eng. Sci. Technol.* **2013**, *8*, 557–565.
15. Mishra, S.K.; Biswas, S.; Satapathy, A. A study on processing, characterization and erosion wear behavior of silicon carbide particle filled ZA-27 metal matrix composites. *Mater. Des.* **2014**, *55*, 958–965. [CrossRef]
16. Suresha, S.; Sridhara, B.K. Effect of silicon carbide particulates on wear resistance of graphitic aluminium matrix composites. *Mater. Des.* **2010**, *31*, 4470–4477. [CrossRef]
17. Stojanović, B.; Babić, M.; Miloradović, N.; Mitrović, S. Tribological behaviour of A356/10SiC/3Gr hybrid composite in dry-sliding conditions. *Mater. Technol.* **2015**, *49*, 117–121.
18. Radhika, N.; Subramaniam, R. Wear behavior of aluminium/alumina/graphite hybrid metal matrix composites using Taguchi's techniques. *Ind. Lubr. Tribol.* **2013**, *65*, 166–174. [CrossRef]
19. Joshi, A.G.; Desai, R.S.; Prashanth, M.V.; Sandeep, S. Study on Tribological Behaviour of ZA-27/Al_2O_3/Gr MMC. *Int. J. Emerging Technol.* **2016**, *7*, 117–122.
20. Babic, M.; Mitrovic, S.; Zivic, F.; Bobic, I. Wear Behavior of Composites Based on ZA-27 Alloy Reinforced by Al_2O_3 Particles Under Dry Sliding Condition. *Tribol. Lett.* **2010**, *38*, 337–346. [CrossRef]
21. Güler, O.; Çuvalci, H.; Gökdağ, M.; Çanakçi, A.; Çelebi, M. Tribological Behaviour of ZA27/Al_2O_3/Graphite Hybrid Nanocomposites. *Part. Sci. Technol.* **2017**, *36*, 1–9. [CrossRef]
22. Casati, R.; Vedani, M. Metal Matrix Composites Reinforced by Nano-Particles—A Review. *Metals* **2014**, *4*, 65–83. [CrossRef]
23. Guaglianoni, W.C.; Cunha, M.A.; Bergmann, C.P.; Fragassa, C.; Pavlovic, A. Synthesis, Characterization and Application by HVOF of a WCCoCr/NiCr Nanocomposite as Protective Coating Against Erosive Wear. *Tribol. Ind.* **2018**, *40*, 477–487. [CrossRef]
24. Stojanovic, B.; Babic, M.; Mitrovic, S.; Vencl, A.; Miloradovic, N.; Pantic, M. Tribological characteristics of aluminium hybrid composites reinforced with silicon carbide and graphite. A review. *J. Balkan Tribological Assoc.* **2013**, *19*, 83–96.
25. Christopher, C.M.L.; Sasikumar, T.; Santulli, C.; Fragassa, C. Neural network prediction of aluminum–silicon carbide tensile strength from acoustic emission rise angle data. *FME Trans.* **2018**, *46*, 253–258. [CrossRef]
26. Ranganath, G.; Sharma, S.C.; Krishna, M. Dry sliding wear of garnet reinforced zinc/aluminium metal matrix composites. *Wear* **2001**, *251*, 1408–1413. [CrossRef]
27. Babic, M.; Mitrovic, S.; Jeremic, B. The influence of heat treatment on the sliding wear behavior of a ZA-27 alloy. *Tribol. Int.* **2010**, *43*, 16–21. [CrossRef]
28. Pola, A.; Montesano, L.; Gelfi, M.; La Vecchia, G.M. Comparison of the sliding wear of a novel Zn alloy with that of two commercial Zn alloys against bearing steel and leaded brass. *Wear* **2016**, *368–369*, 445–452. [CrossRef]

29. Savaşkan, T.; Maleki, R.A.; Tan, H.O. Tribological properties of Zn-25Al-3Cu-1Si alloy. *Tribol. Int.* **2015**, *81*, 105–111. [CrossRef]
30. Kumar, N.S. Mechanical and Wear Behavior of ZA-27/SiC/Gr Hybrid Metal Matrix Composites. *Mater. Today Proc.* **2018**, *5*, 19969–19975. [CrossRef]

© 2019 by the authors. Licensee MDPI, Basel, Switzerland. This article is an open access article distributed under the terms and conditions of the Creative Commons Attribution (CC BY) license (http://creativecommons.org/licenses/by/4.0/).

Article

Process and High-Temperature Oxidation Resistance of Pack-Aluminized Layers on Cast Iron

Xing Wang, Yongzhe Fan, Xue Zhao, An Du *, Ruina Ma * and Xiaoming Cao

Key Lab for New Type of Functional Materials in Hebei Province, Tianjin Key Lab Material Laminating Fabrication and Interface, School of Material Science and Engineering, Hebei University of Technology, Tianjin 300132, China; wangxinggr@163.com (X.W.); fyz@hebut.edu.cn (Y.F.); zhaoxue@hebut.edu.cn (X.Z.); caoxiaoming@hebut.edu.cn (X.C.)
* Correspondence: duan@hebut.edu.cn (A.D.); maryna@126.com (R.M.);
  Tel.: +86-159-0226-6022 (A.D.); +86-136-4207-4657 (R.M.)

Received: 15 May 2019; Accepted: 1 June 2019; Published: 4 June 2019

Abstract: Pack aluminizing of spheroidal graphite cast iron with different aluminizing temperature and time was studied. Results showed that the thickness of aluminized layer increased with the increasing temperature and time. The optimized process parameters are as follow: the aluminizing packed temperature is 830 °C and the time is 3 h. The aluminized layer consisted of the inner FeAl and the outer Fe_2Al_5. Some graphite nodules were observed in the aluminide layer after aluminizing. The mass gain of the aluminized cast iron was 0.405 mg/cm^2, being 1/12 of the untreated substrate after oxidation. The high temperature oxidation resistance can be improved effectively by pack aluminizing, even though there were graphite nodules in the aluminide layer.

Keywords: spheroidal graphite cast iron; pack aluminizing; microstructure; high-temperature oxidation resistance

---

1. Introduction

Spheroidal graphite cast irons have good thermal conductivity, excellent fatigue resistance and high strength. Therefore, components such as engine parts, brake pads and heat exchangers with the requests of high temperature and excellent fatigue resistance are made of cast irons [1,2]. In recent years, the maximum combustion temperature increased with the raising of car speed, resulting in the use of coatings to protect spheroidal graphite cast iron [3].

Previous researches have shown that FeAl intermetallic phases such as Fe_2Al_5, $FeAl_3$ and FeAl could effectively increase the high-temperature oxidation resistance [4–6]. This is because intermetallic phases such as Fe_2Al_5, $FeAl_3$, and FeAl can protect the metal surface by continuously forming Al_2O_3 film. Plasma spraying [7], slurry aluminizing [8], hot dipping and pack aluminizing [9,10] are the major technologies to form FeAl coatings. Packing aluminizing is a method that the workpiece and the powder aluminizing agent are loaded into a sealed aluminizing box, and then the surface is aluminized by heating, heat preservation, and diffusion annealing. Among these methods, pack aluminizing offers the advantage of forming coating over different shapes and sizes of components, such as engine parts [11].

The majority of reports of pack aluminizing were about carbon steel, stainless steel and alloys [12–14]. Above studies have shown that pack aluminizing could be successfully applied to carbon steel, stainless steel and alloys at low temperature. However, it was under the help of other surface process, such as magnetron sputtering, deep rolling process and plasma nitriding [13,15–17]. Without these treatments, the aluminizing layer obtained at low temperature was very thin with one phase [18]. Thick aluminizing layer and more phases can be obtained at high temperature, but pores appeared in the FeAl layers [19]. For Ni-based alloys and Ti-based alloys, the aluminizing temperature should be higher than 1050 °C and 1000 °C respectively to form the pack aluminizing coatings [20,21].

Cast irons are consisted of a large number of C and Si elements which hinder the Al diffusion. In addition, spheroidal graphites located in substrate are obstacle [21]. Many efforts have been made on hot dip aluminizing to produce iron aluminum intermetallic layer on cast iron [6,22]. However, the adhesion of alumina scale over the aluminide coating has been a major concern. The coating formed by pack aluminizing has good adhesion. Currently, there is no information available concerning pack aluminizing on spheroidal graphite cast irons. As a result, basic researches on pack aluminizing technology were conducted in our study.

This paper reported the effects of aluminizing temperature and time on the microstructure and thickness of aluminide layer during pack aluminizing on spheroidal graphite cast irons. Moreover, the high temperature oxidation resistance was also evaluated.

2. Experiment

Spheroidal graphite cast irons (QT420, Ningbo Zhongbo metal material Co. Ltd., 4 grade, fully ferritic, Zhejiang, China) were employed as substrate material. The chemical composition provided by supplier of the cast iron is shown in Table 1. The dimension of the specimens approximately 10 mm × 10 mm × 4 mm. The specimens were ground by using silicon carbide waterproof abrasive paper to a 1200-grade finish and ultrasonically cleaned in ethanol. The pack powder mixture was consisted of 25Al-2NH$_4$Cl-73Al$_2$O$_3$ (wt. %). The samples were placed in cylinders which were sealed with quartz grains and refractory clay. The sealed cylinders were placed inside a box-type furnace and pack aluminized at the temperature range from 800 °C to 880 °C for 1 h to 5 h.

Table 1. Composition of the cast iron (wt %).

C	Si	Mn	P	S	Fe
3.0	1.6	0.7	0.5	0.12	allowance

The packing aluminized samples were polished. The cross-sectional microstructure of aluminide layer was observed by a Hitachi S-4800 microscope (SEM, Suzhou Seins Instrument Co. Ltd., Suzhou, China). The average value of aluminized coating thickness was obtained from 6 lengths, which include the maximum and the minimum. The XRD test of the permeable layer cross section adopts the method of layer by layer separation: using silicon carbide sandpaper to polish the permeable layer, and determine the grinding thickness through the metallographic microscope. X-ray diffractometer (XRD, Brook AXS Co. Ltd., Los Angeles, CA, USA) was used to determine the phase composition of the infiltration layer and the oxide layer. CuK$_\alpha$ was adopted, and the sweeping speed was 4 °/min. The test Angle was: the infiltration layer was 20–83°, and the oxidation layer was 10–90°. The phases of the aluminide layer were identified with a D8-Discover X-ray Diffraction.

The aluminized samples were polished with 400# sandpaper until the surface appears bright white to remove the oxide scale on the surface. Then they were put into a ceramic crucible, and then heated by a box-type furnace. High temperature oxidation experiments were carried out at 750 °C in static air for 72 h to estimate high temperature oxidation resistance. The masses of specimens were measured with a precision electronic balance (0.1 mg accuracy) after oxidation experiments.

3. Results and Discussion

3.1. Effect of Pack Aluminizing Time on the Aluminide Layers

The cross-sectional microstructures of the aluminizing samples with different pack aluminizing times are shown in Figure 1. It can be noticed that the spheroidal graphites are observed in the aluminide layer where the coating is not formed successfully as shown in Figure 1a–e. The irregular coatings result from the spheroidal graphites in the coatings which inhibit the diffusion of Al and Fe.

In addition, the growth direction of the pack aluminide layers formed inside the matrix is confirmed. Because the diffusion rate of Al atoms is faster than that of Fe atoms [23], the spheroidal graphites are trapped gradually in the aluminide layers.

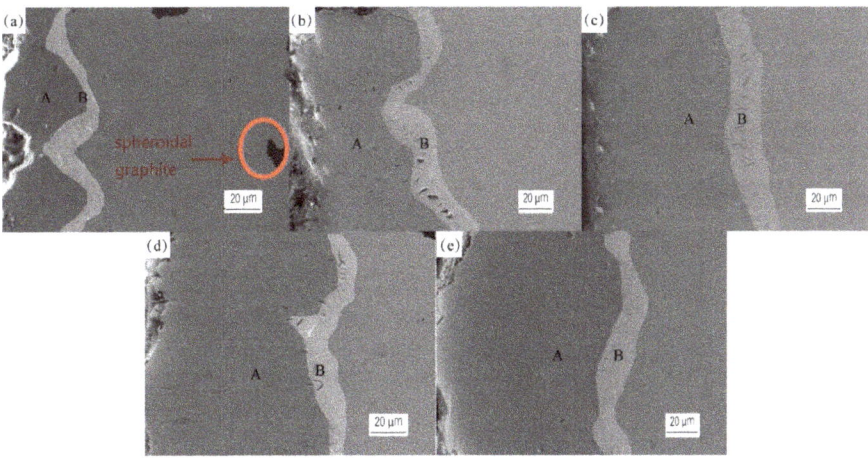

Figure 1. The cross-sectional microstructures of aluminizing samples at 830 °C (**a**) for 1 h; (**b**) for 2 h; (**c**) for 3 h; (**d**) for 4 h; (**e**) for 5 h.

There are two distinct regions in aluminized coating which include the outer layer A and the inner layer B. The layer A obtained at different time is dense and with no pores while some needles can be observed in the layer B, and the growth direction of needles are perpendicular to the aluminide layer (Figure 1a–e). The appearance of the needle can be explained from two aspects: on the one hand, because of the presence of dense oxide film, only Al atom migrated outward during the oxidation process. When Al atom enters the oxide film from the metal/oxide film section, it will leave a vacancy in the metal lattice, and the vacancy migrates inward to the metal/oxide film section in the opposite direction, and the vacancy diffuses to the appropriate position and precipitates to form needle-like holes. On the other hand, the difference of Fe and Al concentration between the high-temperature infiltration layer and the matrix finally leads to the formation of the needle [3]. The gamma region is strongly shrunk because of the infiltration of Al atoms. Recrystallization of gamma-alpha (Al_2O_3) occurs after austenie saturation [24,25]. After aluminizing for 1 h, the interface between the aluminide layer and substrate is fairly irregular owing to the preferential and fast aluminide layer growth in the early stage of aluminizing (Figure 1a). In addition, the diffusion depth of Al atoms is relatively small, so the aluminide layer is very thin. With prolonging the aluminizing time, the interface of aluminide layer and substrate becomes smooth. Besides, the thickness of the layer also increases owing to the further diffusion of the active Al, as shown in Figure 2. About 95 µm thick aluminide layer is produced on cast iron at 3 h. Aluminate coating was prepared through the method of slurry aluminizing with a activator of $AlCl_3$, of which the thickness is only about 60 µm [26]. And the interface between the intermetallic layer and cast iron is smooth and regular (Figure 1d). The aluminide layer thickness increases rapidly in the first 3 h then slows down afterward, The Al atoms are not integrated into the matrix lattice, because the solid solubility of Al atoms in Fe based solid solution reaches the limit with the long diffusion time. Thus, the longer aluminizing time has no obvious influence on the aluminizing layer thickness. In addition, the edges of samples will rise with the prolonging aluminizing time owing to bond of pack powder mixture [18].

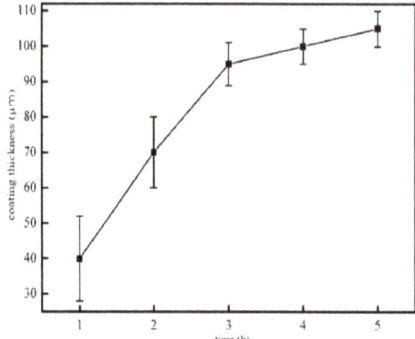

Figure 2. Aluminide layer thickness changing with aluminizing time.

3.2. Effect of Aluminizing Temperature on the Aluminizing Layer

Figure 3 shows the cross section micrograph of the sample with the aluminizing time of 3 h at four different temperatures and Figure 4 illustrates the relationship between the aluminide layer thickness and temperature. The relationship is the same as the Fick's law of diffusion which describes that the thickness of the diffusion layers increases with elevated temperature [27]. The low kinetic energy of the active Al results in the slow diffusion rate at 800 °C. With the rise of aluminizing temperature, the diffusion rate increases, so the thickness of the aluminide layer increases. However, the aluminide layer thickness increases slowly when the aluminide layer is up to a suitable thickness. It can be seen from Figure 3 that only 20 μm thick aluminide layer is produced on cast iron at 800 °C. With the increase of aluminizing temperature, the aluminide layer thickness increases quickly, which is 140 μm at 880 °C. However, the interface between the aluminide layer and the cast iron is irregular (Figure 3d), and the edges of samples rise because of the accumulation of the active Al.

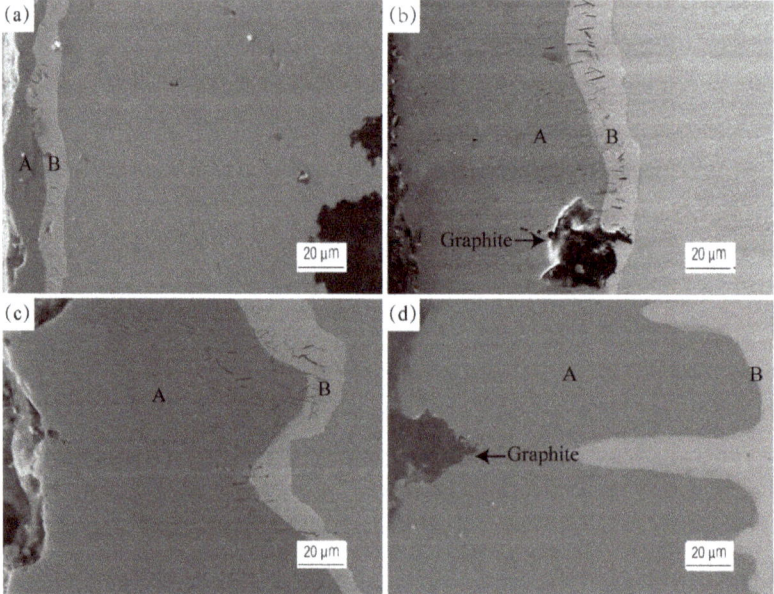

Figure 3. The cross-sectional microstructures of aluminizing samples at (**a**) 800 °C; (**b**) 830 °C; (**c**) 850 °C; (**d**) 880 °C.

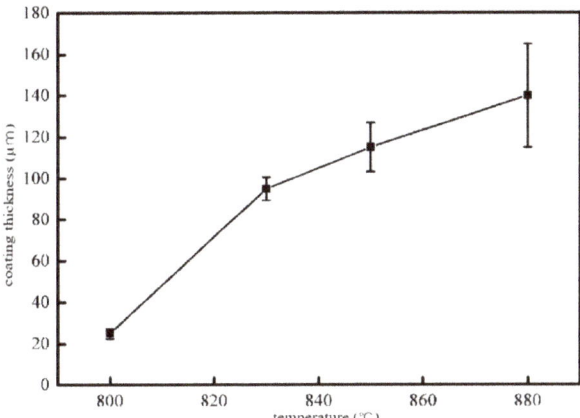

Figure 4. Aluminide layer thickness changing with aluminizing temperature.

3.3. Microstructure of Aluminizing Layer

Above all, in order to obtain a better quality coating, the technological parameters of packing aluminizing adopted in this report are packed at 830 °C for 3 h. The aluminide layer thickness is about 95 µm with the adopted optimized process.

Figure 5 shows the energy-dispersive spectrometry (EDS)test location of the aluminized layer. Figure 5a shows the surface of the aluminized layer, and Figure 5b shows the section of the aluminized layer. In Figure 5a point I was selected on the permeability layer surface. Point II represents the permeability layer between the outer area and the inner layer, point III is located in the outer and inner regions near the permeability layer, point IV carbonitriding layer between the inner, V in matrix and close to the carbonitriding layer inner regions, VI for the inside of a matrix. The analysis results are shown in Table 2. The percentage of Al atoms in the outer layer of the infiltration layer is between 69.55% and 73.88%, while the percentage of Fe atoms is less than 30%. In the inner layer of the permeable layer, the percentage of Al atoms is about 50%, and the percentage of Fe atoms is also 50%. It can be inferred from FeAl binary phase diagram that the outer layer of infiltration layer is Fe_2Al_5 phase and inner layer of the infiltration layer is FeAl phase. In the matrix and near the inner layer of the permeable layer, the content of Al atom is 13.94%, and the content of Fe atom is 86.07%. Therefore, it can be speculated that Al atoms in this region dissolve in Fe atoms to form a solid solution. With the increasing distance to the surface of the permeable layer, the content of Al atom in the matrix decreased to 5%.

Table 2. The EDS analysis of aluminide layer under optimal technology.

Region	Al (at. %)	Fe (at. %)
I	73.88	26.22
II	71.89	28.11
III	69.55	30.45
IV	49.63	50.37
V	13.93	86.07
VI	05.21	94.79

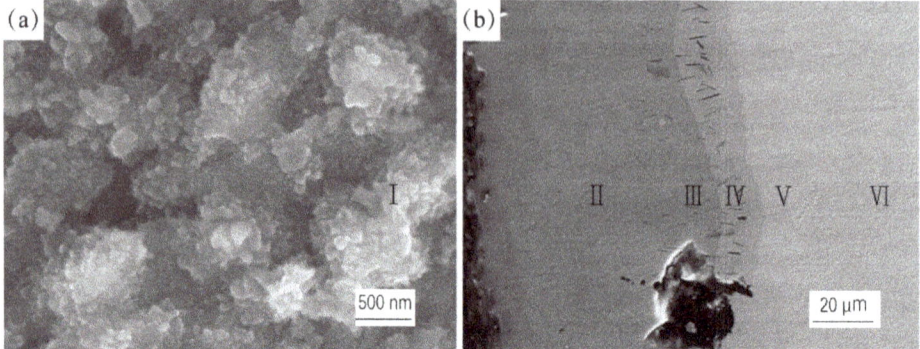

Figure 5. The test region of energy-dispersive spectrometry (EDS) under optimal technology (**a**) the surface of aluminide layer; (**b**) the cross-sectional of aluminide layer.

Figure 6 are the X-ray diffraction patterns of pack aluminizing cast iron at 830 °C for 3 h. Combined with EDS, the Figure 6 revealed that the layer A is Fe_2Al_5 (PDF#47-1435) and the layer B is FeAl (PDF#33-0020) (Figures 1 and 3). Therefore, the intermetallic layer consists of the outer Fe_2Al_5 and the inner FeAl.

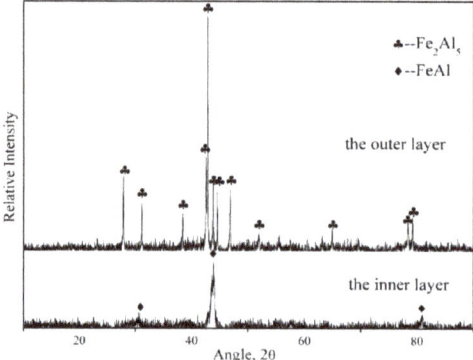

Figure 6. X-ray diffraction patterns of pack aluminizing spheroidal graphite cast iron.

3.4. High-Temperature Oxidation Kientics

Samples (830 °C/3 h) were selected for oxidation experiments. Figure 7 shows the mass gain curves of substrate and the aluminide layer oxidized in static air at 750 °C. It is obvious that the weight gain of the aluminizing cast iron is less than that of substrate after oxidation at 750 °C for 72 h. When the oxidation time is up to 72 h, the mass gain of the coated cast iron is 0.435 mg/cm^2. The high-temperature oxidation test indicated that the aluminide layer could protect the substrate from oxidation permeating effectively. In addition, when the oxidation time reaches 72 h, the mass gain of the coating exceeded 2.0 mg/cm^2 by packing aluminizing on Ti-45Al-8Nb-0.5 (B, C) alloy [21]. The comparison of mass gain indicates that the aluminide layer after pack aluminizing has better high-temperature oxidation resistance than the hot-dipping layer. Using hot dip plating method in Ti-6Al-4V alloy aluminum plating, the coating weight increase is 0.582 mg·cm$^{-1/2}$·h$^{-1/2}$ [28].

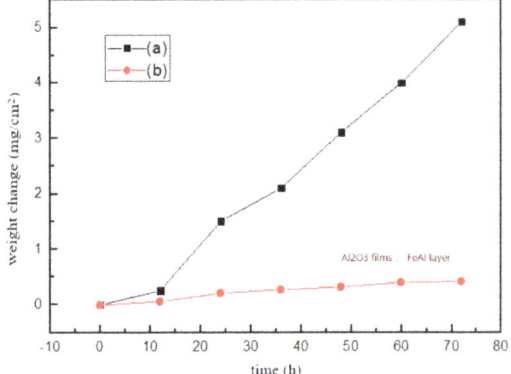

Figure 7. Oxidation kinetics at 750 °C in air (**a**) spheroidal graphite cast iron (**b**) pack aluminizing.

3.5. Microstructures and Phases of Oxidized Specimens

3.5.1. Cast Iron Substrate in Original Condition

It can be seen that the oxide layer consists of two parts and the aggregate thickness of oxide layer is 180 μm. The X-ray diffraction and EDS analysis proved that the oxide layer consists of the outside Fe_3O_4 (PDF#26-1136) and the inside Fe_2O_3 (PDF#54-0489) (Figure 8a). Visual inspections of bare substrate oxidized at 750 °C for 72 h indicated that the appearance of samples is blank and loose. The cross-sectional micrograph of cast iron is plotted in Figure 9a. Furthermore, the majority of spheroidal graphites included in oxide layer are oxidized as shown in Figure 9a. The oxidation of spherical graphite is attributed to the continuous infiltration of O atoms to form carbides [29]. In this study, the weight gain curve of bare substrate raises slowly in the first 12 h of oxidation. The reason is that the weight reduction caused by decarbonization is less than weight gains caused by oxygen entering.

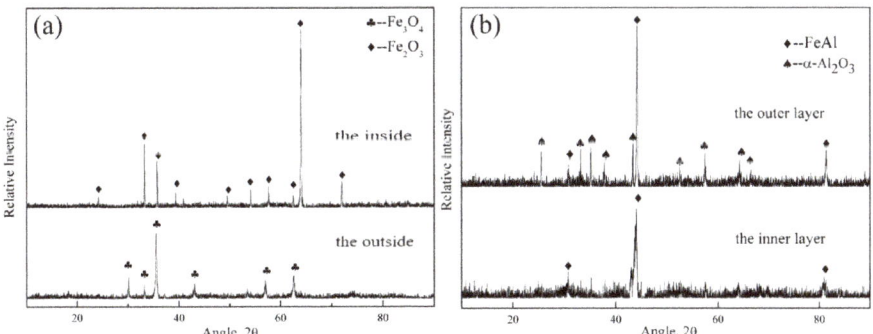

Figure 8. X-ray diffraction patterns of substrate oxidized at 750 °C for 72 h (**a**) spheroidal graphite cast iron (**b**) pack aluminizing.

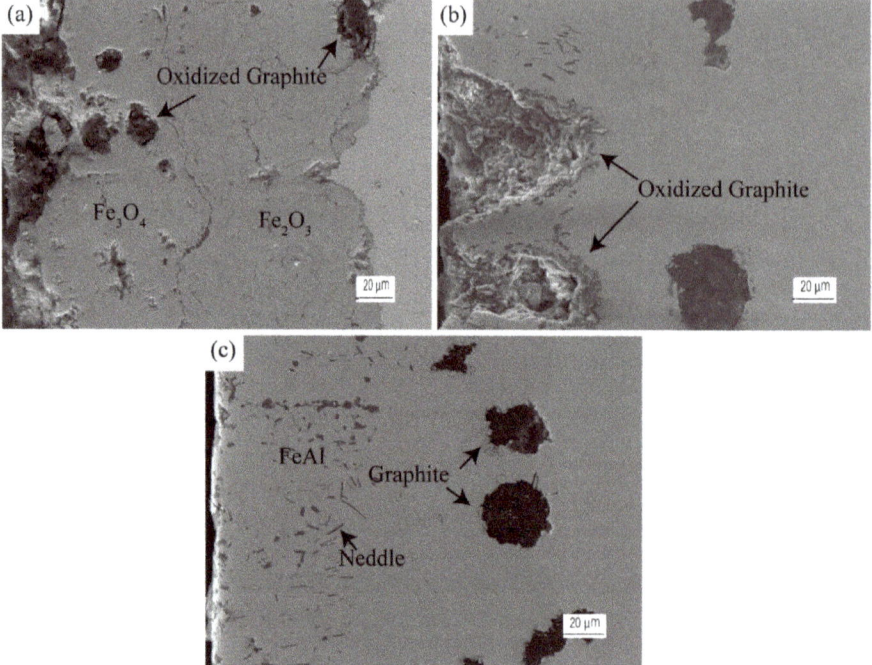

Figure 9. The cross-sectional micrographs of samples after the 750 °C oxidation for 72 h (**a**) spheroidal graphite cast iron (**b**) and (**c**) spheroidal graphite cast iron after pack aluminizing.

3.5.2. Cast Iron after Pack Aluminizing

Visual inspections of aluminide cast iron oxidized at 750 °C for 72 h indicated that the appearance of samples is gray and has a metallic luster on the surface. Figure 9b,c shows the cross-sectional micrograph of aluminide cast iron after oxidation at 750 °C for 72 h. It can be seen that spheroidal graphites contained in aluminide layer are also oxidized. Oxygen penetrated along the cracks and pores in the aluminide layer left caused by oxidized spheroidal graphites. However, the spheroidal graphites in substrate are not oxidized. In addition, there is only one FeAl layer after oxidation and there are many needles in FeAl layer. During high temperature oxidation, the FeAl layer begins to expand.

Figure 8b shows the phases of aluminide coating on cast iron after high temperature oxidation at 750 °C for 72 h. It shows that the aluminide layer consists of Al_2O_3 films and FeAl layer after oxidation. The Fe_2Al_5 which formed in the pack aluminizing process, was not found [26]. The reason of disappearance of Fe_2Al_5 are that the inward diffusion of Al from aluminide layer and the outward diffusion of Fe. Therefore, the Fe_2Al_5 and transforms to FeAl phase. In the early stage of oxidation, Al_2O_3 films which form on the surface of aluminide layer can prevent the oxidation from permeating the substrate effectively. Combining with Figure 7b, the reason for the slow growth of weight gain curve of aluminized substrate in early oxidation is the joint effect of decarbonization and oxygen permeating.

In this study, the pores and cracks produced by oxidized spheroidal graphites lead to oxidation of the certain region of substrate. However, the aluminide layer still plays a role. The high temperature tests indicated that the weight gain of the aluminizing specimens is less than that of substrate after oxidation at 750 °C for 72 h. The coating obtained by pack aluminizing can protect the substrate form oxidation permeating effectively at 750 °C.

4. Conclusions

Spheroidal graphite cast irons were aluminized by pack aluminizing. The thickness of the aluminide layer increased with increasing aluminizing time up to 4 h, after that, it increased very slowly with increasing time. The rise of temperature also caused the increase of thickness. However, if the temperature exceeded 880 °C, the interface between the aluminide layer and the cast iron was irregular. Furthermore, the protrude edges of sample resulted from accumulation of active Al. The phases of aluminide layer were made up the inner FeAl and outer Fe_2Al_5.

The spheroidal graphite in the substrate was observed in the aluminide layer after pack aluminizing. The mass gain of the coated cast iron was 0.435 mg/cm^2, being 1/12 of spheroidal graphite cast iron after oxidation resulting from the formation of Al_2O_3 films on the cast iron substrate. The aluminide layer could effectively prevent the substrate from oxidation at 750 °C, even though there was spheroidal graphite in the aluminide layer.

Author Contributions: Writing-Original Draft Preparation, X.W.; Conceptualization, Y.F.; Data Curation, X.Z.; Methodology, A.D.; Resources, R.M.; Formal Analysis, X.C.

Funding: This research received no external funding.

Acknowledgments: This work was supported by (The National Natural Science Foundation of China); Under Grand (number 51501055 and 51601056).

Conflicts of Interest: The authors declare no conflict of interest.

References

1. Di Cocco, V.; Iacaviello, F.; Rossi, A.; Cavallini, M.; Natali, S. Graphite nodules and fatigue crack propagation micromechanisms in a ferritic ductile cast iron. *Fatigue Fract. Eng. Mater. Struct.* **2013**, *36*, 893–902. [CrossRef]
2. Li, Y.; Dong, S.; Yan, S.; Liu, X.; He, P.; Xu, B. Surface remanufacturing of ductile cast iron by laser cladding Ni-Cu alloy coatings. *Surf. Coat. Technol.* **2018**, *347*, 20–28. [CrossRef]
3. Zhang, Y.; Fan, Y.; Zhao, X.; Du, A.; Ma, R.; Wu, J.; Cao, X. Influence of Graphite Morphology on Phase, Microstructure, and Properties of Hot Dipping and Diffusion Aluminizing Coating on Flake/Spheroidal Graphite Cast Iron. *Metals* **2019**, *9*, 450. [CrossRef]
4. Bates, B.L.; Wang, Y.Q.; Zhang, Y.; Pint, B.A. Formation and oxidation performance of low-temperature pack aluminide coatings on ferritic-martensitic steels. *Surf. Coat. Technol.* **2009**, *204*, 766–770. [CrossRef]
5. Wang, C.J.; Badaruddin, M. The dependence of high temperature resistance of aluminized steel exposed to water-vapour oxidation. *Surf. Coat. Technol.* **2010**, *205*, 1200–1205. [CrossRef]
6. Jiang, W.; Fan, Z.; Li, G.; Liu, X.; Liu, F. Effects of hot-dip galvanizing and aluminizing on interfacial microstructures and mechanical properties of aluminum/iron bimetallic composites. *J. Alloys Compd.* **2016**, *688*, 742–751. [CrossRef]
7. Dhakar, B.; Chatterjee, S.; Sabiruddin, K. Phase stabilization of spraying mechanically blended alumina-chromia powders. *Mater. Manuf. Processes* **2017**, *32*, 355–364. [CrossRef]
8. Pedraza, F.; Boulesteix, C.; Proy, M.; Lasanta, I.; de Miguel, T.; Illana, A.; Pérez, F.J. Behavior of Slurry Aluminized Austenitic Stainless Steels under Steam at 650 and 700 °C. *Oxid. Met.* **2017**, *87*, 443–454. [CrossRef]
9. Abro, M.A.; Lee, D.B. High temperature corrosion of hot-dip aluminized steel in Ar/1% SO_2 gas. *Met. Mater. Int.* **2017**, *23*, 92–97. [CrossRef]
10. Wu, L.-K.; Wu, J.-J.; Wu, W.-Y.; Hou, G.-Y.; Cao, H.-Z.; Tang, Y.-P.; Zhang, H.-B.; Zheng, G.-Q. High temperature oxidation resistance of gamma-TiAl alloy with pack aluminizing and electrodeposited SiO_2 composite coating. *Corros. Sci.* **2019**, *146*, 18–27. [CrossRef]
11. Kochmanska, A.E. Microstructure of Al-Si Slurry Coatings on Austenitic High-Temperature Creep Resisting Cast Steel. *Adv. Mater. Sci. Eng.* **2018**, *2018*, 5473079. [CrossRef]
12. Huang, M.; Wang, Y.; Zhang, M.-X.; Huo, Y.-Q.; Gao, P.-J. Effect of magnetron-sputtered Ai film on low-temperature pack-aluminizing coating for oil casing steel N80. *Surf. Rev. Lett.* **2014**, *21*, 1450053. [CrossRef]

13. Yutanorm, W.; Juijerm, P. Diffusion enhancement of low-temperature pack aluminizing on austenitic stainless steel AISI 304 by deep rolling process. *Kovove Mater.* **2016**, *54*, 227–232. [CrossRef]
14. Ma, F.; Gao, Y.; Zeng, Z.; Liu, E. Improving the Tribocorrosion Resistance of Monel 400 Alloy by Aluminizing Surface Modification. *J. Mater. Eng. Perform.* **2018**, *27*, 3439–3448. [CrossRef]
15. Zhan, Z.; Liu, Z.; Liu, J.; Li, L.; Li, Z.; Liao, P. Microstructure and high-temperature corrosion behaviors of aluminide coatings by low-temperature pack aluminizing process. *Appl. Surf. Sci.* **2010**, *256*, 3874–3879. [CrossRef]
16. Son, Y.I.; Chung, C.H.; Gowkanapalli, R.R.; Moon, C.H.; Park, J.S. Kinetics of Fe_2Al_5 phase formation on 4130 steel by Al pack cementation and its oxidation resistance. *Met. Mater. Int.* **2015**, *21*, 1–6. [CrossRef]
17. Wang, J.; Wu, D.J.; Zhu, C.Y.; Xiang, Z.D. Low temperature pack aluminizing kinetics of nickel electroplated on creep resistant ferritic steel. *Surf. Coat. Technol.* **2013**, *236*, 135–141. [CrossRef]
18. Majumdar, S.; Paul, B.; Kain, V.; Dey, G.K. Formation of Al_2O_3/Fe-Al layers on SS 316 surface by pack aluminizing and heat treatment. *Mater. Chen. Phys.* **2017**, *190*, 31–37. [CrossRef]
19. Lee, J.-W.; Kuo, Y.-C. A study on the microstructure and cyclic oxidation behavior of the pack aluminized Hastelloy X at 1100 °C. *Surf. Coat. Technol.* **2006**, *201*, 3867–3871. [CrossRef]
20. Priest, M.S.; Zhang, Y. Synthesis of clean aluminide coatings on Ni-based superalloys via a modified pack cementation process. *Mater. Corros.* **2015**, *66*, 1111–1119. [CrossRef]
21. Szkliniarz, W.; Moskal, G.; Szkliniarz, A.; Swadźba, R. The Influence of Aluminizing Process on the Surface Condition and Oxidation Resistance of Ti–45Al–8Nb–0.5(B, C) Alloy. *Coatings* **2018**, *8*, 113. [CrossRef]
22. Lin, M.-B.; Wang, C.-J.; Volinsky, A.A. Isothermal and thermal cycling oxidation of hot-dip aluminide coating on flake/spheriodal graphite cast iron. *Surf. Coat. Technol.* **2015**, *206*, 1595–1599. [CrossRef]
23. Lin, M.-B.; Wang, C.-J. Microstructure and high temperature oxidation behavior of hot-dip aluminized coating on high silicon cast iron. *Surf. Coat. Technol.* **2010**, *205*, 1220–1224. [CrossRef]
24. Wang, S.; Zhou, L.; Li, C.; Li, Z.; Li, H. Morphology and Wear Resistance of Composite Coatings Formed on a TA2 Substrate Using Hot-Dip Aluminising and Micro-Arc Oxidation Technologies. *Materials* **2019**, *12*, 799. [CrossRef] [PubMed]
25. Boulesteix, C.; Pedraza, F. Suitable sealants for cracked aluminized austenitic steels and their oxidation behavior. *Surf. Coat. Technol.* **2017**, *327*, 9–17. [CrossRef]
26. Sun, Y.; Dong, J.; Zhao, P.; Dou, B. Formation and phase transformation of aluminide coating prepared by low-temperature aluminizing process. *Surf. Coat. Technol.* **2017**, *330*, 234–240. [CrossRef]
27. Cheng, W.-J.; Wang, C.-J. Growth of intermtallic layer in the aluminide mild steel during hot-dipping. *Surf. Coat. Technol.* **2009**, *204*, 824–828. [CrossRef]
28. Jeng, S.-C. Oxidation behavior and microstructural evolution of hot-dipped aluminum coating on Ti-6Al-4V alloy at 800 °C. *Surf. Coat. Technol.* **2013**, *235*, 867–874. [CrossRef]
29. Tholence, F.; Norell, M. Nitride precipitation during high temperature of ductile cast irons in synthetic exhaust gases. *J. Phys. Chem.* **2005**, *66*, 530–534. [CrossRef]

© 2019 by the authors. Licensee MDPI, Basel, Switzerland. This article is an open access article distributed under the terms and conditions of the Creative Commons Attribution (CC BY) license (http://creativecommons.org/licenses/by/4.0/).

Article

Comparative Study of Jet Slurry Erosion of Martensitic Stainless Steel with Tungsten Carbide HVOF Coating

Galileo Santacruz [1,*], Antonio Shigueaki Takimi [1], Felipe Vannucchi de Camargo [1,2], Carlos Pérez Bergmann [1] and Cristiano Fragassa [2]

[1] Post-Graduation Program in Mining, Metallurgical and Materials Engineering, Federal University of Rio Grande do Sul, Av. Osvaldo Aranha 99, Porto Alegre 90035-190, Brazil; antonio.takimi@gmail.com (A.S.T.); felipe.vannucchi@unibo.it (F.V.d.C.); bergmann@ufrgs.br (C.P.B.)
[2] Department of Industrial Engineering, University of Bologna, Viale del Risorgimento 2, 40136 Bologna, Italy; cristiano.fragassa@unibo.it
* Correspondence: galileo.santacruz@ufrgs.br; Tel.: +55-(51)-3308-3637

Received: 30 April 2019; Accepted: 17 May 2019; Published: 24 May 2019

Abstract: This work evaluates the behavior of a martensitic stainless steel (AISI 410) thermally treated by quenching and tempering with a tungsten carbide (86WC-10Co-4Cr) coating obtained by high-velocity oxygen fuel (HVOF) thermal spray deposition, analyzing the volume loss under erosive attacks at 30° and 90° incidence angles by using jet slurry erosion equipment with electrofused alumina erodent particles. Firstly, the characterization of the samples was carried out in terms of the microstructure (SEM), thickness, roughness, porosity, and microhardness. Then, samples were structurally characterized in the identification of the phases (XRD and EDS) present in the coating, as well as the particle size distribution (LG) and morphology of the erodent. It was determined that the tungsten carbide coating presented better resistance to jet slurry erosion wear when compared to the martensitic stainless steel analyzed, which is approximately two times higher for the 30° angle. The more ductile and brittle natures of the substrate and the coating, respectively, were evidenced by their higher volumetric erosion at 30° for the first and 90° for the latter, as well as their particular material removal mechanisms. The enhanced resistance of the coating is mainly attributed to its low porosity and high WC-Co content, resulting in elevated mechanical resistance.

Keywords: tribology; wear; slurry erosion; coating; cermet

1. Introduction

The erosion of metallic materials due to slurry erosion is a recurrent and common problem in various industrial applications, such as hydraulic turbines for hydroelectric plants and mud pumps, as well as the processing of minerals and pipes for the petrochemical industry. The effect caused by the particulate material in the liquid represents a major industrial problem, affecting the life of components, reducing their performance, and demanding continuous preventive and corrective maintenance, bearing elevated financial and environmental costs to the supply chain of products as a whole. Thus, when designing parts that will be subjected to erosive wear, the proper choice of engineering materials plays a key role in decreasing this wear rate, and assures adequate tribological behavior [1].

Supported by recent research, the technology of surface treatments and coatings of thin films have been demonstrated to be a valid and efficient way to improve the resistance of metallic-coated materials [2], whereas understanding and studying the surface wear mechanisms is vital to predicting failures by erosion and corrosion, providing adequate dimensioning of parts and relying on optimized properties aggregated by the coatings.

Among the coated materials, some of the most relevant ones are martensitic stainless steels, given their wide range of industrial applications. These are commonly adopted for regular ends, such as in the automotive industry [3], but may also be used in the cutting-edge design of complex structures, such as nuclear reactors [4] and power plants [5]. Martensitic stainless steels are high-carbon-content alloys (0.1–1.2%) with chromium (12–18%), presenting magnetic properties, elevated strength, low-temperature toughness, and good corrosion resistance [6], as well as better mechanical resistance to erosive particles than austenitic steel [7].

Cermet materials have shown excellent resistance to erosive wear, increasing the resistance, for instance, of hydro-turbine parts [8], being able to noticeably improve the resistance of originally uncoated metallic pieces to erosion, corrosion, or both, especially at high temperatures [9]. Kumar et al. [8] demonstrated the enhanced abrasive wear resistance properties provided by WC-Co, which can have their corrosion resistance increased with the addition of a small amount of Cr. For this reason, the composition of WC-10% Co-4% Cr is widely considered in coatings for applications in components of hydric companies to prolong their operational life [8]. Coatings of WCCoCr are usually deposited with a High-Velocity Oxy-Fuel (HVOF) thermal spray process [10].

HVOF has demonstrated to be highly effective in the obtention of low-porosity adherent coatings [11–13], whereas the mechanical properties of the coated material are shielded by coatings with high compressive and tensile strengths and elevated hardness. Furthermore, as demonstrated by Liu et al. [14] and Hou et al. [15], HVOF coatings can enhance the resistance of parts against cavitation erosion–corrosion due to its features of high flame velocity, low flame temperature, and the possibility of producing dense coatings with less decarburization and low oxide content.

These coatings are mainly prepared with WC-based cermets, once the alliance of the hard WC phase with a ductile metal in different proportions can produce materials with a significantly wide range of properties [16]. This metallic phase is usually cobalt or nickel, the latter contributing to the enhancement of the material's corrosion resistance. In general, the high hardness and fracture toughness of WC cermets make them suitable materials to improve abrasion resistance in many industrial applications, such as hydraulic machinery and coastal installations [17].

The present investigation evaluates the erosive wear of the coating 86WC-10% Co-4% Cr when deposited by HVOF using feedstock materials on quenched and tempered martensitic stainless steel (AISI 410), and how it upgrades its original erosion resistance. The volume loss under jet slurry erosion for incidence angles of 30° and 90° between the axis of symmetry of the fluid flow and the surface of the samples was evaluated, including controlling parameters such as the angle and speed of impact, test temperature, and concentration of erosive particles in the suspension. Electrofused alumina was used as erodent material. The materials were characterized with regard to their microstructure by scanning electron microscopy (SEM), light microscopy (LM), thickness, roughness, porosity, microhardness, and phase structure by X-ray diffraction (XRD) and energy dispersive spectroscopy (EDS). Also, the particle size distribution was analyzed by laser granulometry (LG), and the morphology of the erodent was identified.

2. Materials and Methods

2.1. Substrate Material

Martensitic stainless steel (AISI 410) cylinders which were 30 mm in diameter and 10 mm in height were used [18]. The steel samples were austenitized at 1263 K for 35 min, then oil-quenched and tempered at 793 K for 35 min afterward. The microstructure obtained after this procedure was composed of martensite with some precipitated carbides on the grain contours, a typical outcome of quenching and tempering thermal treatments [6] (Figure 1).

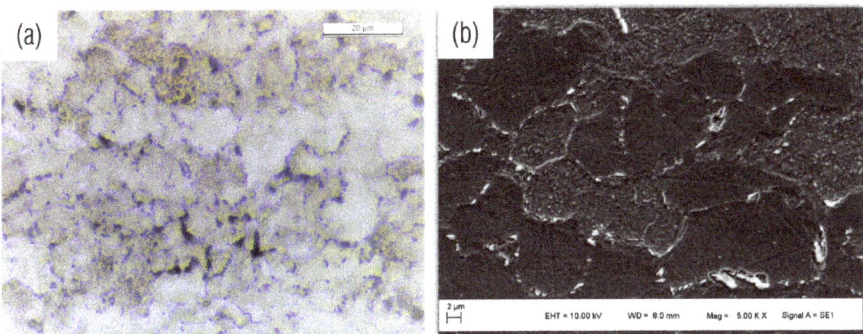

Figure 1. Resulting microstructure of AISI 410 stainless steel after thermal treatments by (**a**) LM 1000×, and (**b**) SEM 5000×, after Vilella acid etching.

2.2. HVOF Coating

Spherical fine powders WOKA-3653 (86WC-10Co-4Cr) from Oerlikon Metco (Pfaffikon, Switzerland) were used with a particle size distribution of nominal range $-45 + 11$ μm and apparent density 4.8–5.8 g/cm^3 [19]. The approximate chemical composition of the WOKA-3653 powder is given in Table 1. HVOF spraying parameters are briefly listed in Table 2.

Table 1. Chemical composition of WOKA-3653 (wt %) [19].

Elements	Composition (%)
W	Balance
Co	8.5–11.5
Cr	3.4–4.6
C (Total)	4.8–5.6
Fe (Max)	0.2

Table 2. Coating parameters of high-velocity oxygen fuel (HVOF) equipment.

Spray Parameters	Value
Propylene flow rate	4620 (L/h)
Oxygen flow rate	15,180 (L/h)
Powder feed rate	42 (g/min)
Spray distance	230 (mm)

2.3. Erodent Particles

The particle size distribution of the alumina erodent (Al_2O_3) was measured using the Cilas 1180 laser diffraction particle size analyzer (Cilas Ariane Group, Le Barp, France) according to the ISO 13320:2009 standard [20]. The abrasive particle size distribution is shown in Figure 2, showing results of the average abrasive particle size of approximately 98.55 μm. Furthermore, irregular angular-shaped particles are depicted by the SEM micrograph in Figure 3.

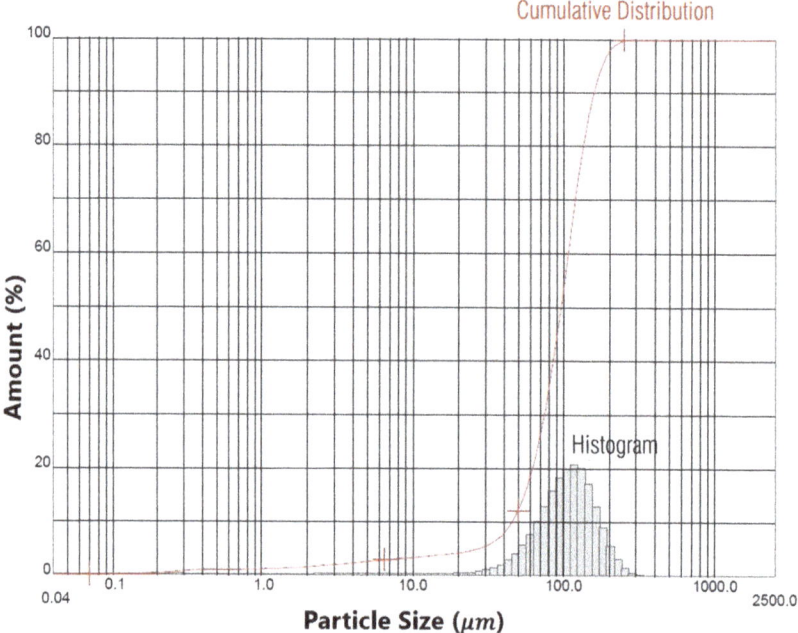

Figure 2. Particle size distribution of aluminium oxide and cumulative distribution.

Figure 3. SEM image showing the abrasive particles' shape and size (56×).

2.4. Microstructural and Qualitative Phase Characterization

The microstructure characterization was done in an EVO MA10 SEM (ZEISS, Oberkochen, Germany) and using a light microscope, LEICA DM2700 M. The total porosity of the coatings was measured through digital image analysis following the ASTM E2109-01 standard [21]. The surface roughness of the steel and as-sprayed samples were measured before jet slurry erosion testing, and after it using a surface roughness tester Mitutoyo SJ-400 (Mitutoyo America Corporation, Aurora, IL, USA). Then, cross-sectional Vickers microhardness measurements were performed using a Buehler Micromet 2001 microhardness tester (HV300 g, 30 s, BUEHLER Ltd., Lake Bluff, IL, USA) according to the ASTM E-384-11 standard [22]. To identify the phases in the coatings and to understand the phase formation during coatings' build-up, surface XRD analysis of the as-sprayed samples was carried out using the PHILIPS Xpert MDP X-ray diffractrometer (PHILIPS, Amsterdam, The Netherlands) with Cu-Kα radiation. Lastly, localized compositional analyses of the coated specimens were done with a SHIMADZU SSX-550 EDS spectrometer (SHIMADZU CORPORATION, Nishinokyo Kuwabara-cho, Nakagyo-ku, Kyoto, Japan) coupled to the EVO MA10 SEM.

2.5. Jet Slurry Erosion Tests

The jet slurry erosion tests were carried out in a modified commercial high-pressure washer Electrolux UWS10 (Electrolux Group, Stockholm, Sweden) based on the ASTM G-76 standard [23–27]. A feeding system for the erodent particles was adapted in the nozzle of the gun using an internal Venturi accelerator inside the test chamber. This equipment allows for control over the angle of impact, the speed of impact, the concentration of erodent particles in the suspension, and the test temperature. Figure 4 shows the configuration of the testing advice.

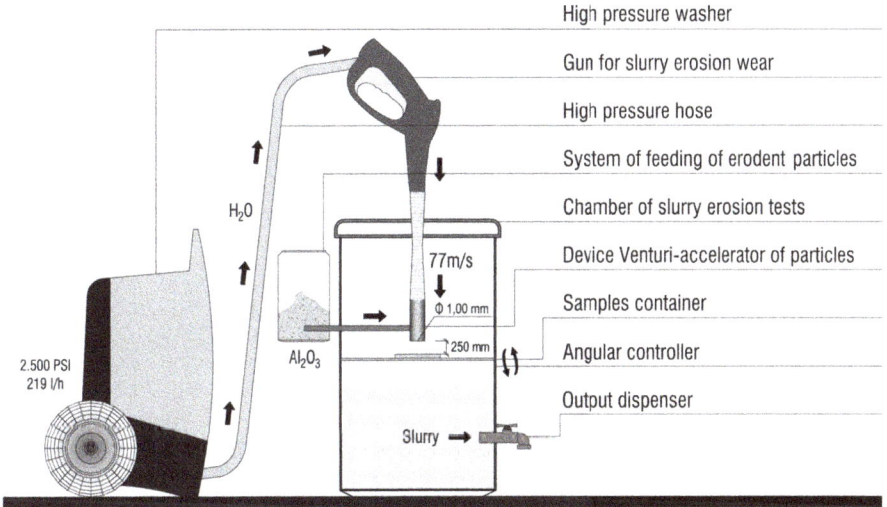

Figure 4. Schematic diagram of the jet slurry erosion tester.

The tests were performed using water between 25 °C and 28 °C with 960 g of erodent. The samples were placed at the nozzle output to guarantee the incidence of impact fluid and mean jet velocity of erosive material in a suspension of 77 m/s. This velocity was calculated using the flow rate, time, and nozzle area measurements of the flow output. The incidence angles studied were 30° and 90° between the axis of symmetry of the fluid flow and the surface of the samples. In all cases, the concentration of particles in the slurry was approximately 7 wt % [28–30]. The erosion resistance was determined from the volume loss results per unit of time, from the difference of the mass loss, considering the relation of the apparent density of the studied materials (adopting substrate as 7.73 g/cm^3 and coating as 12.5 g/cm^3 [31–33]. Mass losses were measured in three samples of coating and three of substrate every 1 min until reaching a total erosion time of 4 min by using a scale with 0.01 mg resolution. The total duration of each test was 4 min. The samples were cleaned in an ultrasonic bath with deionized water before and after each test. Then, after the cleaning operation, they were dried and weighted.

3. Results and Discussion

3.1. X-ray Diffraction Analysis

Figure 5 shows the XRD pattern of the 86WC-10Co-4Cr coating obtained by HVOF with the powder WOKA-3653, where the WC (JCPDS 00-051-0939) and W (JCPDS 01-089-2767) phases can be promptly identified by using the Joint Committee on Powder Diffraction Standards (JCPDS) [34]. The XRD pattern shows that WC is predominant in the powder diffractogram. It is also possible to verify the presence of W and Co as secondary phases formed during the thermal spraying process by the decarburization of WC particles, also reported by other researchers [30,35,36]. The lack of identification of the Co peak by XRD in this figure is justified by the low content of the metal in the coating, and its diffraction peak is most likely situated among the noise range in the graph.

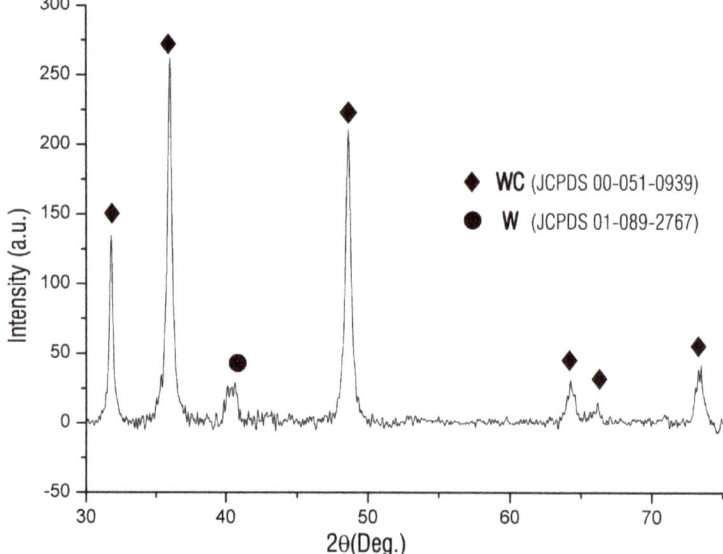

Figure 5. XRD pattern of the HVOF 86WC-10Co-4Cr coating.

3.2. EDS Analysis Coating

Figure 6 shows the elemental microanalysis of one of the regions of the coating cross-section that were scanned using the EDS system, exhibiting the presence of W, Co, Cr, and C in the samples. Figure 7 corroborates the presence of these elements in a quantitative pattern, which agrees with Figure 5, demonstrating a predominance of tungsten. Similar results were found by Thakur et al. [37], considering coatings with the same composition and deposition technique.

3.3. Mechanical Properties of Materials

Figure 8 shows the micrograph of the HVOF 86WC-10Co4Cr coating. It presents a continuous shape and homogeneous microstructure, with a low presence of cracks and pores. It is not possible to distinguish any lamellar structure. This can be explained by a possible melting of the binder metal, which induces a better distribution in the coating material [38]. The average thickness of the HVOF 86WC-10Co-4Cr coating was calculated using micrographs of cross-sections processed using ImageJ software, and among 100 to 200 measurements were performed on different samples for the obtainment of an arithmetic mean of circa 227 µm.

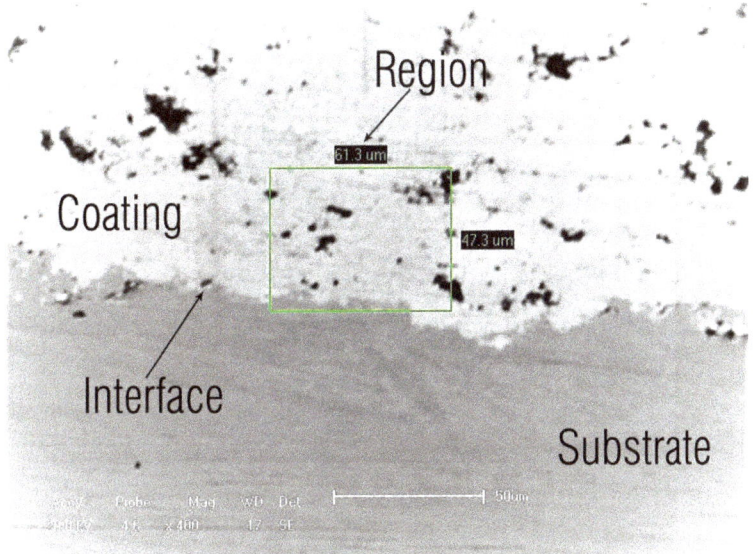

Figure 6. SEM image of the elemental microanalysis region in the cross-section of the HVOF 86WC-10Co4Cr coating using the EDS system (400×).

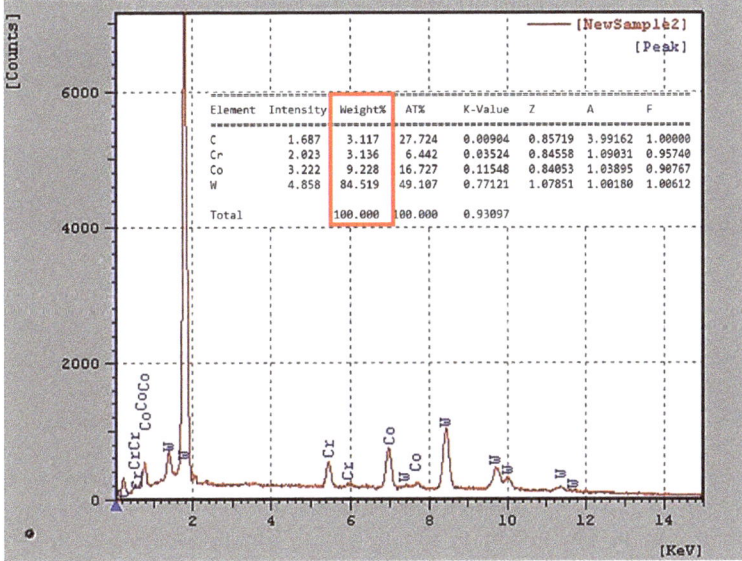

Figure 7. Quantitative pattern of the elemental microanalysis region in the cross-section of the HVOF 86WC-10Co-4Cr coating using the EDS system.

Three analyses of the coating were measured in each one of three samples through digital image analysis according to ASTM E2109-01 [21], yielding an average porosity of 0.7%, which is considered acceptable within the values presented in the literature and according to the criterion specified by the manufacturer Oerlikon Metco of a mean porosity lower than 1.0% [8,10,37].

Microhardness reflects the microstructure and the physical and mechanical properties of both substrate and coating, which in turn are dependent on the materials and processes employed in their manufacture. Hardness is a property that can be considered to be variable throughout the coating in certain zones, due to eventual heterogeneities of the material. For this reason, when measuring this property, one must consider the preparation of the surface, the section analyzed, as well as the number of indentations performed. The microhardness of the coating and substrate are presented in Table 3. The hardness value of the coating is compatible with the literature, and is in agreement with the specification of the coating powder's manufacturer (750 to 1450 HV0.3) [19].

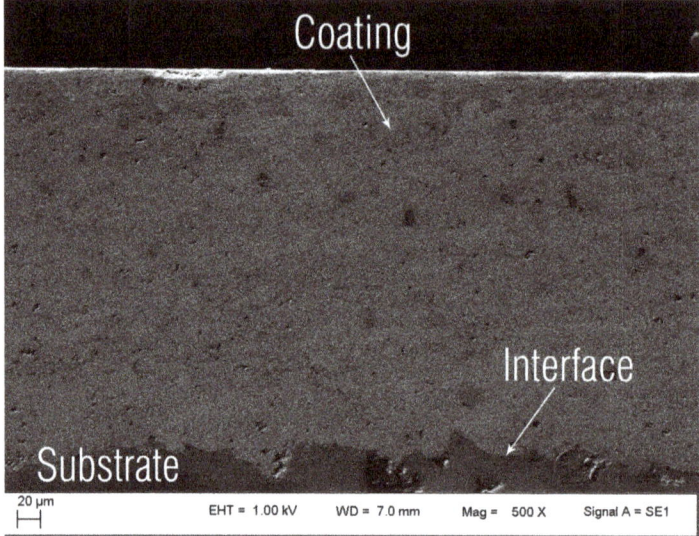

Figure 8. SEM image of the microstructure of the HVOF 86WC-10Co-4Cr coating in the cross-section (500×).

Table 3. Results of Vickers microhardness measurements—cross-section (HV0.3-2.94N).

Material / Coating	86WC-10Co4Cr	Steel AISI 410
Microhardness average	1139	219
Standard Deviation	171	31
Coefficient of Variation (%)	15	14

According to the microhardness values along the cross-section of the spray coating, there were variations due to the distribution of the phases. Castro et al. [39] demonstrated that it is possible to observe this effect when working with sublayers (sprayed coatings), as microhardness values may vary since each indentation point can be placed in different microstructures—in this case, carbides, oxides, inclusions, and the matrix itself.

Trelleborg [40] describes WC-based materials as metal–ceramic composites in which the microhardness values found are associated with each microconstituent. By its turn, the microstructure depends on factors such as the composition of each phase, the morphology of the powder, the spray technique, and the porosity.

The highest value of microhardness performed on the 86WC-10Co-4Cr coating was 1448 HV0.3, being also the closest one to the manufacturer's specification [19]. Thakur et al. [37] and Castro et al. [39] obtained similar results of 1297 ± 45 HV0.3 and 1256 HV, respectively, for WC-CoCr coatings sprayed by HVOF. In the same way, Kumar et al. [8], Maiti et al. [30], Berger et al. [41], and Ahmed et al. [42]

reported microhardness values of 1031 ± 99, 1180 ± 70, 1118 ± 131, and 1106 HV0.3 for similar sprayed coatings applied by the same technique, respectively.

The indentation performed on the substrate, corresponding to thermally treated martensitic stainless steel measured microhardness values, was 219 HV0.3, which is once again very similar to that reported by Maiti et al. [30] of 199 HV0.3.

Table 4 shows the accumulated erosion rate by the mass of erodent in loss of volume of the samples tested as a function of the impact angle for both the substrate and coating in a total time of 4 min.

Table 4. Accumulated erosion rate in volume loss.

Material/Coating	Impact Angle (°)	Volumetric Erosion Rate $(cm^3_{target}/g_{erodent}) \times 10^{-5}$	Standard Deviation $(cm^3_{target}/g_{erodent}) \times 10^{-5}$
Steel AISI 410	30	0.8449	0.0114
	90	0.7025	0.0495
86WC-10Co4Cr	30	0.4490	0.0023
	90	0.6778	0.0283

Figures 9 and 10 show the mass losses measured in three samples of coating and three of substrate every 1 min until a total erosion time of 4 min was reached for the impact angles of 30° and 90°, respectively. Steel exhibits a much higher level of erosive wear at an impact angle of 30° rather than at 90°, demonstrating ductile behavior with a ploughing-predominant material removal mechanism, as is reported in the literature [43,44]. Furthermore, many authors confirm that metals present an enhanced erosion rate at lower angles [45–48]. Figure 10 also shows that, for the impact angle of 90°, the tungsten carbide coating presented an accumulated volumetric erosion rate that is slightly lower than the martensitic stainless steel.

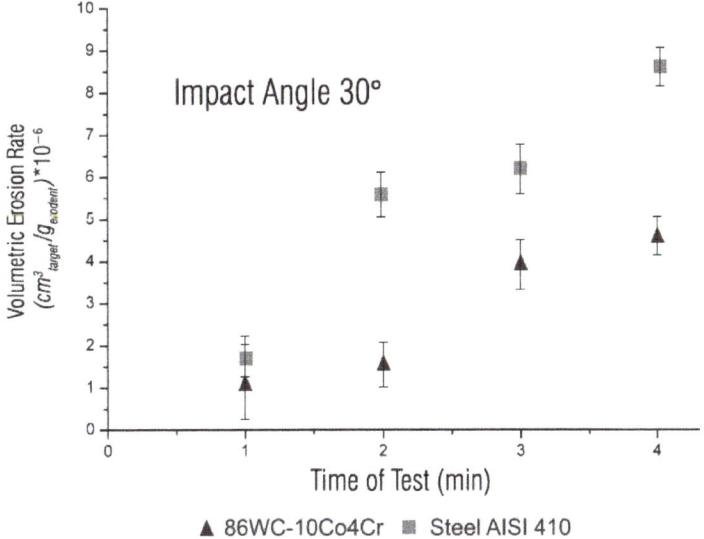

▲ 86WC-10Co4Cr ■ Steel AISI 410

Figure 9. Variation of the accumulated volumetric erosion rate as a function of the impact angle of 30° for the martensitic stainless steel (AISI 410) and tungsten carbide (86WC-10Co-4Cr) coating.

On the other hand, when comparing Figures 9 and 10, it can be seen that the coating presents higher erosive wear at an angle of impact of 90° than at 30°, indicating the coating's fragile behavior. This is due to the repeated action of the erosive perpendicular particles impacted on the surface, producing microfractures that contribute to the wear of the WC hard metal and resulting in its eventual removal, agreeing with the results found in [49,50].

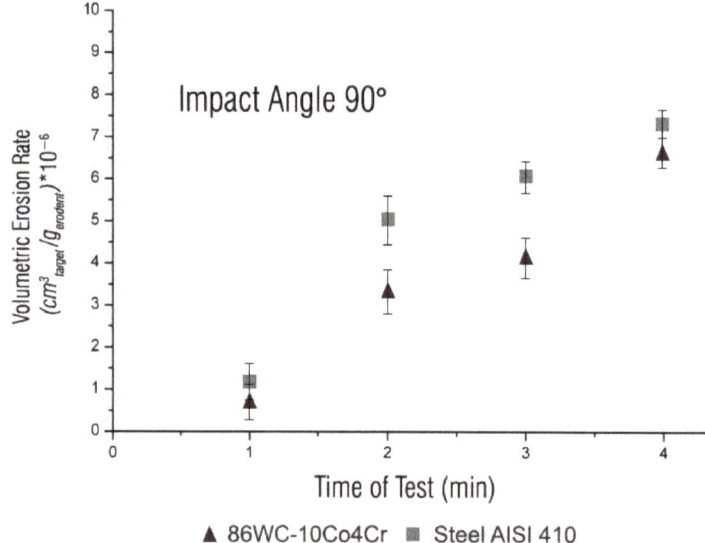

Figure 10. Variation of the accumulated volumetric erosion rate as a function of the impact angle of 90° for the martensitic stainless steel (AISI 410) and tungsten carbide (86WC-10Co-4Cr) coating.

Figure 11 shows, in detail, the eroded areas of the materials subjected to the jet slurry test for the impact angles of 30° and 90°. It is observed that the eroded region of the pieces (highlighted in red) for the 30° angle have an elongated shape and greater eroded area. The erosion of the specimens tested at 90° is more localized, and has a smaller area and a greater depth. The formation of the concentric ring, also known as "halo" erosion, can be observed, which is a mechanical characteristic of the damage due to the development of craters that appear as macro pits on the surface for this angle [51–53]. The roughness of these eroded regions is displayed in Table 5.

Table 5. Average roughness of coating and substrate before and after jet slurry erosion.

Material/Coating	Roughness Ra (µm)				
	Before Erosion	Standard Deviation	Angle (°)	After Erosion	Standard Deviation
86WC-10Co4Cr	0.08	0.02	30	1.60	0.11
			90	1.91	0.13
Steel AISI 410	0.07	0.03	30	2.97	0.47
			90	2.56	0.11

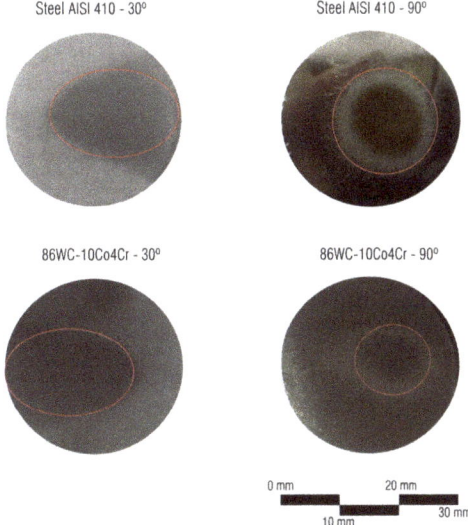

Figure 11. Eroded regions of the materials submitted to the jet slurry test obtained by Image J software (LM, 50×).

Table 6 shows the eroded areas as a function of the impact angle. It is observed that the area is greater for the impact angle of 30°. Compared to the substrate, the coating presented smaller eroded areas for the two impact angles studied.

Table 6. Eroded area as a function of the impact angle for both martensitic stainless steel (AISI 410) and tungsten carbide (86WC-10Co-4Cr) coating.

Material/Coating	Impact Angle (°)	Eroded Area (mm^2)
Steel AISI 410	30	277.7
	90	147.1
86WC-10Co4Cr	30	201.2
	90	94.9

Figure 12 represents the entire surfaces of the specimens' upper cylindrical faces, and elliptical shapes were attained because the coupons were rotated in order for the 3D laser scan to be able to accurately capture the eroded depth. Table 7 shows the respective average erosion depths. As expected, given that 90° tests resulted in more localized worn areas, their depth is superior to those found in samples tested at 30°.

Islam et al. [43] conducted an erosion study in steel tubes, varying both impact angle and velocity, obtaining significantly deeper erosion areas for 90° than for 30°, supporting the results which were hereby found. Similarly, Vite-Torres et al. [53] investigated the performance of the AISI 420 stainless steel subjected to particle erosion tests using two different abrasives at the angles of incidence of 30°, 45°, 60°, and 90°. Their results also demonstrated that the samples evaluated at 90° presented a greater depth profile than those of 30° tested independently of the abrasive which was adopted.

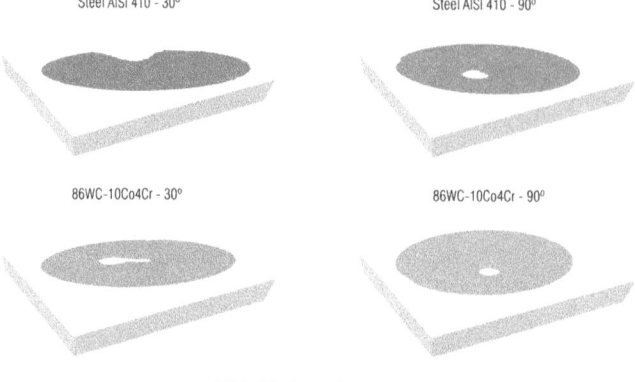

Figure 12. Three-dimensional scanning of the samples submitted to the jet slurry erosion test obtained the eroded depth by using the Geomagic Studio software (Version 2012.1.1, 3D Systems Inc, Rock Hill, SC, USA).

Table 7. Average depth of the eroded regions for substrate and coating.

Material/Coating	Impact Angle (°)	Average Depth (µm)	Standard Deviation (µm)
Steel AISI 410	30	93	±2.88
	90	102	±15.87
86WC-10Co4Cr	30	33	±5.77
	90	62	±3.21

4. Conclusions

In the present investigation, an experimental wear simulation equipment of jet slurry erosion was developed with conditions that allowed for the control of the significant test parameters, such as impact angle, impact velocity, erosive particle concentration in the suspension, and test temperature. The jet slurry erosion behaviors of martensitic stainless steel (AISI 410) and tungsten carbide (86WC-10Co-4Cr) coating were investigated, and from the results obtained in this experimental work, it is possible to infer the following conclusions:

1. The tungsten carbide coating showed higher resistance to jet slurry erosion compared to martensitic stainless steel, possibly due to the low porosity, the hardness of WC particles, and better properties of the CoCr binder material matrix. Thus, the application of such a coating can be considered as beneficial for the structural endurance of the substrate in applications where the part is subjected to similar erosive conditions.
2. The high concentration of WC-Co in the coating blend promotes greater hard-phase connectivity, leading to increased mechanical strength and jet slurry erosion resistance.
3. The martensitic stainless steel presented a higher rate of accumulated volumetric erosion in the impact angle of 30°, denoting ductile behavior. On the other hand, the accumulated volumetric erosion rate for tungsten carbide coating was higher at the impact angle of 90°, indicating a predominantly brittle failure mechanism.
4. For the impact angle of 30°, the tungsten carbide coating presented a cumulative volumetric erosion rate of approximately 50% less than martensitic stainless steel.

5. Patents

For the development of the research a jet slurry erosion tester was designed which is in the patent process by the INPI agency of the Brazilian government with Process Number BR 20 2018 074233 2.

Author Contributions: Conceptualization, C.P.B. and A.S.T.; Methodology, G.S., C.P.B. and A.S.T.; Software, G.S.; Validation, G.S. and F.V.d.C.; Formal Analysis, G.S. and A.S.T.; Investigation, G.S.; Data Curation, G.S. and F.V.d.C.; Writing-Original Draft Preparation, G.S. and F.V.d.C.; Writing-Review C.P.B., C.F. and F.V.d.C.; Supervision, C.P.B., C.F. and A.S.T.; Project Administration, G.S., C.P.B. and A.S.T.

Funding: This research was funded by the Coordination of Improvement of Higher Education Personnel of the Brazilian Ministry of Education (CAPES/MEC), grant numbers 88882.345871/2019-01 and 88882.345831/2019-01.

Acknowledgments: The authors express their gratitude to the following laboratories: Laboratory of Ceramic Materials (LACER), Laboratory of Physical Metallurgy (LAMEF), Laboratory of Design and Selection of Materials (LdSM) and the Laboratory of Conventional Machining (LUC) of the Engineering Institute of the Federal University of Rio Grande do Sul (UFRGS) for the investigation support and to the Research and Technology Center of Rijeza Metallurgy. G.S. and F.V.d.C. thank the Coordination of Improvement of Higher Education Personnel of the Brazilian Ministry of Education (CAPES/MEC) for the doctorate scholarship.

Conflicts of Interest: The authors declare no conflict of interest.

References

1. Popov, V.L. Adhesive wear: Generalized Rabinowicz'criteria. *Facta Univ. Ser. Mech. Eng.* **2018**, *16*, 29–39. [CrossRef]
2. López-Ortega, A.; Bayón, R.; Arana, J.L. Evaluation of Protective Coatings for High-Corrosivity Category Atmospheres in Offshore Applications. *Materials* **2019**, *12*, 1325. [CrossRef]
3. Mohrbacher, H. Martensitic automotive steel sheet-fundamentals and metallurgical optimization strategies. Advanced Materials Research. *Trans. Tech. Publ.* **2015**, *1063*, 130–142.
4. Klueh, R.; Nelson, A.T. Ferritic/martensitic steels for next-generation reactors. *J. Nucl. Mater.* **2007**, *371*, 37–52. [CrossRef]
5. Abe, F. Precipitate design for creep strengthening of 9% Cr tempered martensitic steel for ultra-supercritical power plants. *Sci. Technol. Adv. Mater.* **2008**, *9*, 013002. [CrossRef]
6. Krauss, G. *Steels: Processing, Structure, and Performance*; Asm International: Novelty, OH, USA, 2015.
7. Levy, A.V.; Yau, P. Erosion of steels in liquid slurries. *Wear* **1984**, *98*, 163–182. [CrossRef]
8. Kumar, R.; Kamaraj, M.; Seetharamu, S.; Pramod, T.; Sampathkumaran, P. Effect of Spray Particle Velocity on Cavitation Erosion Resistance Characteristics of HVOF and HVAF Processed 86WC-10Co4Cr Hydro Turbine Coatings. *J. Therm. Spray Technol.* **2016**, *25*, 1217–1230, doi:10.1007/s11666-016-0427-3. [CrossRef]
9. Bergmann, C.P.; Vicenzi, J. Protection against erosive wear using thermal sprayed cermet: A review. In *Protection against Erosive Wear Using Thermal Sprayed Cermet*; Springer: Berlin, Germany, 2011; pp. 1–77.
10. Cunha, M.A.; Guaglianoni, W.C.; Bezerra, B.F.A.; de Camargo, F.V.; Bergmann, C.P. Resistance to Abrasive and Erosive Wear of WCCo/NiCr HVOF Coatings: Comparative Evaluation of Commercial Nanostructured Spray Powders. *Tribol. Ind.* **2019**, in press.
11. Totemeier, T.C.; Wright, R.N.; Swank, W. FeAl and Mo-Si-B intermetallic coatings prepared by thermal spraying. *Intermetallics* **2004**, *12*, 1335–1344, doi:10.1016/j.intermet.2004.04.034. [CrossRef]
12. Ahledel, N.; Schulz, R.; Gariepy, M.; Hermawan, H.; Alamdari, H. Electrochemical Corrosion Behavior of Fe_3Al/TiC and Fe_3Al-Cr/TiC Coatings Prepared by HVOF in NaCl Solution. *Metals* **2019**, *9*, 437. [CrossRef]
13. Zhang, J.; Deng, C.; Song, J.; Deng, C.; Liu, M.; Dai, M. Electrochemical Corrosive Behaviors of Fe-Based Amorphous/Nanocrystalline Coating on Stainless Steel Prepared by HVOF-Sprayed. *Coatings* **2019**, *9*, 226. [CrossRef]
14. Liu, W.H.; Shieu, F.S.; Hsiao, W.T. Enhancement of wear and corrosion resistance of iron-based hard coatings deposited by high-velocity oxygen fuel (HVOF) thermal spraying. *Surf. Coat. Technol.* **2014**, *249*, 24–41. [CrossRef]
15. Hou, G.; An, Y.; Zhao, X.; Zhou, H.; Chen, J. Effect of alumina dispersion on oxidation behavior as well as friction and wear behavior of HVOF-sprayed CoCrAlYTaCSi coating at elevated temperature up to 1000 °C. *Acta Mater.* **2015**, *95*, 164–175. [CrossRef]
16. Nahvi, S.; Jafari, M. Microstructural and mechanical properties of advanced HVOF-sprayed WC-based cermet coatings. *Surf. Coat. Technol.* **2016**, *286*, 95–102. [CrossRef]
17. Hong, S.; Wu, Y.; Zhang, J.; Zheng, Y.; Zheng, Y.; Lin, J. Synergistic effect of ultrasonic cavitation erosion and corrosion of WC–CoCr and FeCrSiBMn coatings prepared by HVOF spraying. *Ultrason. Sonochem.* **2016**, *31*, 563–569. [CrossRef]

18. FEITAL Group. 2019. Available online: www.feital.com.br/produtos/ (accessed on 23 May 2019).
19. Oerlikon Metco Company. 2019. Available online: https://www.oerlikon.com/metco/en/products-services/coating-materials/coating-materials-thermal-spray/cermets/ (accessed on 23 May 2019).
20. International Organization for Standardization. *Particle Size Analysis—Laser Diffraction Methods*; ISO Standards Authority: Geneva, Switzerland, 2009.
21. ASTM International. *E2109-01 (2014) Standard Test Methods for Determining Area Percentage Porosity in Thermal Sprayed Coatings*; ASTM International: West Conshohocken, PA, USA, 2014.
22. ASTM International. *Standard Test Method for Knoop and Vickers Hardness of Materials*; ASTM International: West Conshohocken, PA, USA, 2011.
23. ASTM International. *ASTM G76-13 (2013) Standard Test Method for Conducting Erosion Tests by Solid Particle Impingement Using Gas Jets*; ASTM International: West Conshohocken, PA, USA, 2013.
24. Zu, J.; Hutchings, I.; Burstein, G. Design of a slurry erosion test rig. *Wear* **1990**, *140*, 331–344. [CrossRef]
25. Wentzel, E.; Allen, C. The erosion-corrosion resistance of tungsten-carbide hard metals. *Int. J. Refract. Met. Hard Mater.* **1997**, *15*, 81–87. [CrossRef]
26. Santa, J.; Baena, J.; Toro, A. Slurry erosion of thermal spray coatings and stainless steels for hydraulic machinery. *Wear* **2007**, *263*, 258–264. [CrossRef]
27. Santa, J.; Espitia, L.; Blanco, J.; Romo, S.; Toro, A. Slurry and cavitation erosion resistance of thermal spray coatings. *Wear* **2009**, *267*, 160–167. [CrossRef]
28. Hutchings, I. A model for the erosion of metals by spherical particles at normal incidence. *Wear* **1981**, *70*, 269–281. [CrossRef]
29. Finnie, I. Some reflections on the past and future of erosion. *Wear* **1995**, *186*, 1–10. [CrossRef]
30. Maiti, A.; Mukhopadhyay, N.; Raman, R. Improving the wear behavior of WC-CoCr-based HVOF coating by surface grinding. *J. Mater. Eng. Perform.* **2009**, *18*, 1060. [CrossRef]
31. Madsen, B.W. Measurement of erosion-corrosion synergism with a slurry wear test apparatus. *Wear* **1988**, *123*, 127–142. [CrossRef]
32. Murthy, J.; Rao, D.; Venkataraman, B. Effect of grinding on the erosion behaviour of a WC-Co-Cr coating deposited by HVOF and detonation gun spray processes. *Wear* **2001**, *249*, 592–600. [CrossRef]
33. Liu, X.; Zhang, B. Effects of grinding process on residual stresses in nanostructured ceramic coatings. *J. Mater. Sci.* **2002**, *37*, 3229–3239. [CrossRef]
34. File, P.D. *Joint Committee on Powder Diffraction Standards*; ASTM International: West Conshohocken, PA, USA, 1967; pp. 9–185.
35. Chivavibul, P.; Watanabe, M.; Kuroda, S.; Kawakita, J.; Komatsu, M.; Sato, K.; Kitamura, J. Effects of particle strength of feedstock powders on properties of warm-sprayed WC-Co coatings. *J. Therm. Spray Technol.* **2011**, *20*, 1098–1109. [CrossRef]
36. Guilemany, J.; De Paco, J.; Miguel, J.; Nutting, J. Characterization of the W2C phase formed during the high velocity oxygen fuel spraying of a WC + 12pct Co powder. *Metall. Mater. Trans. A* **1999**, *30*, 1913–1921. [CrossRef]
37. Thakur, L.; Arora, N.; Jayaganthan, R.; Sood, R. An investigation on erosion behavior of HVOF sprayed WC-CoCr coatings. *Appl. Surf. Sci.* **2011**, *258*, 1225–1234. [CrossRef]
38. Castro, G.; Arenas, F.; Rodriguez, M.; Scagni, A. Microestructura De materiales compuestos WC-Co/Ni-W-Cr recubiertos Por HVOF. *Rev. Latinoam. Metal. Mater.* **2001**, *21*, 50–55.
39. Castro, R.; Rocha, A.; Cavaler, L.; Kejelin, N.; Marques, F. Avaliação das características de superfície e do desgaste abrasivo de revestimentos aplicados em hastes de cilindros hidráulicos pelas técnicas de aspersão (HVOF) e eletrodeposição. Master's Thesis, Federal University of Rio Grande do Sul, Porto Alegre, Brazil, 2012.
40. Solution, T.S. Hydraulic Seals/Rod Seals. Available online: http://tk-ines.ru/upload/iblock/26e/26e36169a1738d5faa13585f921374e1.PDF (accessed on 23 May 2019).
41. Berger, L.M.; Saaro, S.; Naumann, T.; Wiener, M.; Weihnacht, V.; Thiele, S.; Suchánek, J. Microstructure and properties of HVOF-sprayed chromium alloyed WC-Co and WC–Ni coatings. *Surf. Coat. Technol.* **2008**, *202*, 4417–4421. [CrossRef]
42. Ahmed, R.; Ali, O.; Faisal, N.H.; Al-Anazi, N.; Al-Mutairi, S.; Toma, F.L.; Berger, L.M.; Potthoff, A.; Goosen, M. Sliding wear investigation of suspension sprayed WC-Co nanocomposite coatings. *Wear* **2015**, *322*, 133–150. [CrossRef]

43. Islam, M.A.; Farhat, Z.N. Effect of impact angle and velocity on erosion of API X42 pipeline steel under high abrasive feed rate. *Wear* **2014**, *311*, 180–190. [CrossRef]
44. Okonkwo, P.C.; Shakoor, R.A.; Zagho, M.M.; Mohamed, A.M.A. Erosion Behaviour of API X100 Pipeline Steel at Various Impact Angles and Particle Speeds. *Metals* **2016**, *6*, 232. [CrossRef]
45. Bayer, R.G. *Mechanical Wear Prediction and Prevention*; Marcel Dekker, Inc.: Monticello, NY, USA, 1994.
46. Papini, M.; Spelt, J. The plowing erosion of organic coatings by spherical particles. *Wear* **1998**, *222*, 38–48. [CrossRef]
47. O'Flynn, D.; Bingley, M.; Bradley, M.; Burnett, A. A model to predict the solid particle erosion rate of metals and its assessment using heat-treated steels. *Wear* **2001**, *248*, 162–177. [CrossRef]
48. Biswas, S.; Satapathy, A.; Patnaik, A. Erosion wear behavior of polymer composites: A review. *J. Reinf. Plast. Compos.* **2010**, *29*, 2898–2924. [CrossRef]
49. Barber, J.; Mellor, B.; Wood, R. The development of sub-surface damage during high energy solid particle erosion of a thermally sprayed WC-Co-Cr coating. *Wear* **2005**, *259*, 125–134. [CrossRef]
50. Javaheri, V.; Porter, D.; Kuokkala, V.T. Slurry erosion of steel—Review of tests, mechanisms and materials. *Wear* **2018**, *408*, 248–273. [CrossRef]
51. Wood, F.W. Erosion by solid-particle impacts: A testing update. *J. Test. Eval.* **1986**, *14*, 23–27.
52. De Souza, V.; Neville, A. Corrosion and erosion damage mechanisms during erosion–corrosion of WC-Co-Cr cermet coatings. *Wear* **2003**, *255*, 146–156. [CrossRef]
53. Vite-Torres, M.; Laguna-Camacho, J.; Baldenebro-Castillo, R.; Gallardo-Hernandez, E.A.; Vera-Cárdenas, E.; Vite-Torres, J. Study of solid particle erosion on AISI 420 stainless steel using angular silicon carbide and steel round grit particles. *Wear* **2013**, *301*, 383–389. [CrossRef]

© 2019 by the authors. Licensee MDPI, Basel, Switzerland. This article is an open access article distributed under the terms and conditions of the Creative Commons Attribution (CC BY) license (http://creativecommons.org/licenses/by/4.0/).

Article

Predicting the Tensile Behaviour of Cast Alloys by a Pattern Recognition Analysis on Experimental Data

Cristiano Fragassa [1,*], Matej Babic [2], Carlos Perez Bergmann [3] and Giangiacomo Minak [1]

1. Department of Industrial Engineering, University of Bologna, IT-40136 Bologna, Italy; giangiacomo.minak@unibo.it
2. Jožef Stefan Institute, SL-1000 Ljubljana, Slovenia; babicster@gmail.com
3. Department of Material Engineering, Federal University of Rio Grande do Sul, Porto Alegre BR-90040-060, Brazil; bergmann@ufrgs.br
* Correspondence: cristiano.fragassa@unibo.it; Tel.: +39-347-6974-046

Received: 18 April 2019; Accepted: 9 May 2019; Published: 13 May 2019

Abstract: The ability to accurately predict the mechanical properties of metals is essential for their correct use in the design of structures and components. This is even more important in the presence of materials, such as metal cast alloys, whose properties can vary significantly in relation to their constituent elements, microstructures, process parameters or treatments. This study shows how a machine learning approach, based on pattern recognition analysis on experimental data, is able to offer acceptable precision predictions with respect to the main mechanical properties of metals, as in the case of ductile cast iron and compact graphite cast iron. The metallographic properties, such as graphite, ferrite and perlite content, extrapolated through macro indicators from micrographs by image analysis, are used as inputs for the machine learning algorithms, while the mechanical properties, such as yield strength, ultimate strength, ultimate strain and Young's modulus, are derived as output. In particular, 3 different machine learning algorithms are trained starting from a dataset of 20–30 data for each material and the results offer high accuracy, often better than other predictive techniques. Concerns regarding the applicability of these predictive techniques in material design and product/process quality control are also discussed.

Keywords: material properties prediction; experimental data analysis; ductile/spheroidal cast iron (SGI); compact graphite cast iron (CGI); Machine Learning (RF); pattern recognition; Random Forest (RF); Artificial Neural Network (NN); k-nearest neighbours (kNN)

1. Introduction

An accurate knowledge of the mechanical properties of materials represents the first step towards their correct use in any field of engineering and in everyday life. Thanks to this information, for instance, it is possible to design structures and components in order to optimize their functionality according to technical parameters of specific interest such as strength, weight, safety, costs and so forth [1–4].

In the case of metals and, especially, of rather common cast irons [5–7], the shared opinion is that their properties are quite predictable (e.g., fatigue [8], abrasion [9], fracture [10], strength [11]).

This is certainly the case when compared to other families, such as organic or composite materials. It is also possible to say that everything depends on the perspective. Since metals are widely known and used, low unpredictability in their properties is expected but actually quite common metal alloys, such as cast irons, can be affected by a not negligible variability in their essential properties [12,13].

From the constituent elements to the process parameters, a large list of factors can interfere with the microstructures of an alloy and, as a consequence, on its properties [14,15].

Although some uncertainties could be easily eliminated (e.g., better control on stoichiometry), it is not so simple in reality. In a traditional foundry, for instance, up to 30% of the metal originates from reused materials, with great advantages in terms of costs and environmental protection [16] but risks in controlling the chemical composition and a certain variability in the cast alloy properties [17].

This variety is the reason for the prevalence of cast iron throughout history and in the present [18]. Grey, white, malleable or ductile—cast alloys can provide largely different properties, offering a valid choice of materials in dissimilar situations. With minimal changes to the composition (e.g., from 3% to 4% in the carbon content) or the use of additives in marginal quantities (<1%), properties such as tensile strength move from 170 to 930 MPa, elongation from 0.5 to 18% and hardness from 130 to 450 HB [19].

This versatility is a weakness when constancy in the material's property is necessary. In Reference [20] it is reported, for instance, that a relevant change in the microstructure and in the mechanical properties of a grey cast iron extracted from identical sand-cast parts produced by different foundries.

In this study, which starts from an extensive experimental test session (with data published in References [21–23]), it can be observed that two successive metal fusions, carried out a few hours later and without voluntary changes in the process parameters, lead to slightly different metallographic profiles between them. Furthermore, specimens extracted from the same casting, although they come from almost identical metal casting conditions, then take on a slight difference in metallographic terms, probably linked to different cooling conditions. These are all rather common knowledge in foundries that, however, lead to an intrinsic variability in the properties of the materials to be considered in the product design and, even, in the process designs.

This study intends to take a further step. Given that it is not possible to reduce the variability in mechanical properties as much as desired, a valid way of predicting them is sought.

Commonly, the mechanical properties of cast irons are related to their metallurgical ones, as reported in the case of Reference [24], which is especially focused on the latest research in solidification and melt treatments. It means, in practice, that micrographs are analysed with the scope to develop models and predictive formulae. A valid example of the current situation is represented by Reference [25], research concerned with the ability of the main theories to predict the fatigue limit of nodular cast iron. It reintroduces several previous studies dealing with the effect of small defects, such as micro-shrinkage cavities or pores and graphite nodules (shape and size) and with the characteristic of a microstructure. At the same time, it also highlights scarce applicability of the existing models and proposes a new correlation in the prediction of the fatigue limit, which involves additional parameters.

Other investigations proposed passing by the overall aspects, and representing the materials that are detectable from micrographs, such as content of graphite, ferrite, perlite, the grade of nodularity, vermicularity and so on (e.g., Reference [26]). Unfortunately, there is no direct connection between one of those characteristics and its mechanical properties and their relationship is hidden behind mutual interactions.

For this reason, recently studies have been successfully proposed to predict the mechanical characteristics of cast alloys by exploiting artificial intelligence approaches (as in References [27–29]).

The list of methods and tools to make concrete a general concept for Artificial Intelligence (AI) is enormous, rapidly evolving and cannot be disclosed in a few words [30]. However, one of the fundamental principles underlying the application of AI to material engineering, including prediction, is to use its ability for 'pattern recognition' [31].

An AI algorithm can be built to recognize patterns (such as recurrences, schemes, similarities) that the human mind cannot. This happens when, as in our case, the patterns are hidden on multiple levels. For instance, the ductility of cast iron—able to double the yield strength in the alloy, an interesting property for its real uses—can be provided by graphite in the form of very tiny nodules [32], together with a specific range of ferrite and pearlite content [33]. This metallurgical situation is obtained only by merging different conditions as (between the others) a specific range in magnesium content and a restricted addition of cerium and different temperatures [34].

While it is objectively complex to develop mathematical models to predict the properties of materials taking into account the combination of several factors, it is relatively simple to use one of the many AI algorithms, accepting a given approximation in its predictions [35].

This is certainly determined by the fact that AI algorithms, even rather complex ones, are now available to everyone and, at the same time, they offer high quality results. It depends on the positive combination of the mentioned pattern recognition with the concept of 'machine learning' (ML), an application of AI that provides systems the ability to automatically learn and improve from experience without being explicitly programmed [36].

The process of learning begins with observations or data, such as examples, direct experience or instruction, in order to look for patterns in data and make better decisions in the future based on this information [37].

This process can be implemented according to one of several different AI strategies, commonly categorized as "supervised," "semi-supervised," "unsupervised" or "reinforcement" machine learning.

In the first case, the dataset of information (input) is provided together with the results (output). The algorithm uses these data to learn the relations existing between inputs and outputs. When finished, it is ready to propose an output per each new set of input values [38].

In the second category—"unsupervised"—the ML process means that there is no distinction between input and output: each set of values is considered as an input and the algorithm is forced to find patterns without a specific external base of information [38].

Semi-supervised learning methods are something in between the other two categories: there is some form of feedback available for each step or action but there is no label or error message [38].

Finally, the reinforcement learning method interacts with its environment by producing actions and discovers errors or rewards [38]. Trial and error search and delayed reward are the most relevant characteristics of reinforcement learning.

Since ML enables the analysis of massive quantities of data, these learning methods often benefit from a data analysis process called "clustering," introduced with the scope of grouping similar entities. It helps to profile the attributes of different groups, giving insight into their different underlying patterns [39,40]. The clustering analysis is also used to reduce the dimensionality of information with respect to the data characterized by a large number of variables. Between other algorithms for clustering, probably the most popular in ML are K-mean Clustering and Hierarchical Clustering.

Even if the unsupervised algorithms are quite uncommon in general and even less common in the fields of mechanical and material engineering, they are presented here since they seem ready to revolutionise investigations of the real world [36–40]. These advanced methods, in fact, can learn without passing for a conventional procedure of ML, based on specific datasets. They simply apply rules and learn by mistakes, following a system of evaluation of potential solutions based on scores [41].

Regardless of the ML technique, the success in prediction depends on the capability of each method to structure information acting on different levels. Supervised learning is useful in cases where a property (label) is available for a particular dataset (training set) but is missing and should be predicted for other cases. Unsupervised learning is used to detect implicit relationships in a given unmarked dataset.

In the case of material properties, as mentioned, the complexity can be related to the fact that these properties are connected with each other and to other chemical-physical ones, in a skein of relations that has to be untied before predictions. Machine Learning comes to the rescue because it allows the unhooking of the intermediate levels ("hidden") from the superficial levels so that these first ones can be configured in full freedom in the search for correlations (pattern recognition). This process allows the collection of the input of a first level of knowledge to transform it into the output of a second level in an operation that always leads to abstracting of concepts. Learning takes the form of a pyramid with the highest concepts learned starting from the lowest levels [42].

It also means that, in the near future, an expert system can probably be installed in a foundry, monitoring the material/process situation but, at the moment, these systems are limited to a few very interesting research experiences. For instance, in Reference [43] an Artificial Neural Network (ANN) was used to classify nodular, grey or malleable cast irons on the basis of their microstructure. After a training session made by 60 samples, the AI showed an accuracy in selection very similar to that obtained by visual human tests. Reference [44] also deals with ductile cast iron and ANN methods; in particular, different networks were built and trained by 700 melts and used to predict tensile strength, elongation and hardness. In addition, in the same article, useful considerations are proposed regarding the ANN modelling and its input parameters, together with an assessment of their significance and availability in an industrial environment.

In brief, it is possible to say that an approach, based on AI and ML, can be conveniently used in material engineering [45,46]. Furthermore, the accuracy in prediction depends on several aspects, including the consistency and the quality of the data used for training.

2. Aims and Scope

This research has three main objectives, as follows:

- Predict the mechanical properties of metals and, in particular, the tensile properties such as, yield strength, ultimate strength, ultimate strain and Young's modulus, starting from experimental data. In addition, it would be possible to investigate the relationship between these properties and fundamental aspects of metallurgy, as in the cases of constituent elements, microstructures, process parameters or treatments.
- Use information directly taken from micrographs by a conventional process of image analysis but globally converted into macro-indicators related to the content of graphite, ferrite, perlite, nodularity and vermicularity. In addition, it would be possible to discuss the interrelations existing between all these features—mechanical and metallurgical—with the scope to recognize essential and overabundant information. This investigation will involve two different families of cast alloys, a nodular cast iron (SGI) and a less common compact graphite cast iron (CGI).
- Select and use these essential data inside a Machine Learning (ML) approach, based on pattern recognition, with the scope to perform an 'intelligent analysis' of experimental measures. In this task, some of the most common methods of ML will be applied, specifically the Random Forest (RF), the Artificial Neural Network (NN) and the k-nearest neighbours (kNN). These classifiers will be implemented by the use of conventional codes and accessible platforms, comparing them in terms of functionality and accuracy in prediction, especially in association with the consistency and quality of the dataset used for training but also considering the overall variability of the phenomena under investigation.
- Introduce essential concerns regarding the real applicability of these techniques for scopes related to the material design, product/process quality control and so on, including practical suggestions on the way to simplify the procedure towards an industrially-oriented application.

3. Materials and Methods

3.1. Experimental Data

Measures were derived from a large experimental campaign undertaken in the past by some authors on foundry alloys, where samples in nodular iron and compact graphite cast iron were manufactured and mechanically characterized. These phases are detailed in References [21–23] and are here briefly summarized.

3.1.1. Casting

Samples were made by a traditional process of green sand moulding, using a hot blast long campaign cupola furnace [47], filled with layers of coke and ignited with torches. When the coke was very hot, solid pieces of metal were charged and alternated with additional layers of fresh coke. The high temperature, together with the other chemical conditions, transformed the solid metal into molten iron. Sand shapes were placed on the pouring line and, then, filled with molten iron.

According to the process parameters, including the use of inoculants, it was possible to produce two different cast alloys:

1) Spheroidal Graphite Iron (SGI), a material also called nodular or ductile iron with respect to its high ductility, offered by the spheroidal shape of graphite;
2) Compacted Graphite Iron (CGI), a material with intermediate properties between grey and nodular iron, thanks to a more compact form of graphite that is becoming quite popular, particularly in the automotive sector [48].

The castings were finally shaped by tool machining to extract samples in accordance with the experimental standards. In particular, 4 metal castings, manufactured on two different days, were created.

Extraordinary care was taken to minimize the risk of unexpected changes in the casting conditions, especially in terms of chemical composition, temperature and other process parameters that could interfere with the metallurgical stability (also in accordance with Reference [49]). For instance, the castings used for extracting samples were produced after a long time of conventional production in order to stabilize both temperature and metallurgy profiles. In addition, the chemical composition of the alloys was verified several times by off-line tests.

3.1.2. Metallurgical and Mechanical Properties

The ML algorithms were trained by experimental data from 48 samples, 27 in SGI and 21 in CGI. In particular, the following metallographic factors were used:

- Quantity of Graphite
- Quantity of Ferrite
- Quantity of Perlite
- Grade of Nodularity
- Grade of Vermicularity

with the related values estimated (in%) by considerations of the micrographs. An image analysis was permitted to consider the presence and the geometry of the graphite inside the cast alloy (in terms of area, perimeter, Feret diameter and so on).

In addition, every set of metallographic characteristics was combined with related mechanical properties, as measured in accordance with the EN ISO 6892-1:2016 [50]:

- Ultimate Tensile Strength [UTS],
- Yield Strength [YS],
- Ultimate Strain [ε],
- Young's modulus [E].

In synthesis, each one of these 48 samples provided a specific set of 5 (five) metallurgical and 4 (four) values, used in the machine learning. These data are available in Tables A1 and A2 of Appendix B for SGI and CGI, respectively.

3.2. Machine Learning Algorithms

The current investigation was implemented by using the Orange program and its ML algorithms [51]. It is an open source platform for machine learning and data visualization, powered by 10 different algorithms, which can be used for data analysis and prediction.

In accordance with a quite common interpretation of the 'No Free Lunch' theorem [52], in the area of ML no universal method exists that can aprioristically provide the best results (for example, in terms of predictions) with respect to all sets of potential data [53]. In other terms, it is not possible to select the proper algorithm in advance.

In this research, in line with previous experiences where the ML has been applied to the prediction of metal properties [54], the analysis has been limited to the use of the following methods:

3.2.1. Random Forest (RF)

Random forest [55] is a popular algorithm of supervised learning, consisting of the use of a committee (ensemble) of decision trees. "Ensemble" means that it takes a bunch of "weak students" and combines them to form one strong predictor. "Weak students" are all random implementations of decision trees, which are combined to form a strong predictor—a random forest. The random forest method is often called the "improved implementation" of the decision tree method, because now to get a more accurate prediction and classification, not one tree is used but the tree committee. The method quickly gained popularity due to the comparative simplicity of setting up and running the analysis procedure. In contrast to the classical algorithms for constructing decision trees, in the random forest method, when constructing each tree, at the stages of splitting vertices, only a fixed number of randomly selected signs of the training set (the second parameter of the method) is used and a full tree is constructed (without cutting), that is, each leaf of the tree contains observations of only one class. A 15 tree, 32 fixed seed for a random generator was used, with a growth control based on the condition that no split was implemented for a subset smaller than 5 (Table 1).

Table 1. Parameters of the Random Forest (RF).

Number of trees	15
Fixed seed for random generator	32
Do not split subset smaller than	5

3.2.2. Neural Network (NN)

A neural network [56] is another very popular algorithm of supervised learning. It is used to build an efficient encryption system using a constantly changing key. Neural networks offer a very powerful and general structure for representing a non-linear mapping of several input variables for several output variables. A neural network can be considered a suitable choice for functional forms used for encryption and decryption operations. The NN topology is an important issue, since the application of the system depends on it. Therefore, since the application is a calculation problem, a multi-layered topology was used. Neural networks offer a very powerful and general structure for representing a non-linear mapping of several input variables for several output variables. The process of determining the values of these parameters on the basis of a dataset is referred to as training and therefore the data set is usually referred to as a training set. A neural network can be considered a suitable choice for functional forms used for encryption and decryption operations. A neural network is a structure (network) consisting of a set of interconnected links (artificial neurons). Each link has a characteristic input / output and implements a local calculation or function. The output of any link is determined by the characteristics of its input / output, its relationship with other links, as well as external inputs, if any. A 5 layer neural network in backpropagation was used. The speed of learning was 0.6, the inertial coefficient was 0.5, the test mass tolerance was 0.02 and the tolerance of the learning set was 0.03 (Table 2).

Table 2. Parameters of the Neural Network (NN).

Learning speed	0.6
Inertial coefficient	0.5
Test mass tolerance	0.02
Tolerance of the learning set	0.03
Number of layers	5

3.2.3. K-Nearest Neighbours (kNN)

The K-nearest neighbours method [57] is one of the methods for solving the classification problem. It is assumed that there is already some number of objects with an exact classification (i.e., for each of them it is known exactly which class it belongs to). It is necessary to work out a rule allowing the classification of a new object as one of the possible classes (the classes themselves are known in advance). At the heart of kNN is the following rule: an object is considered to belong to the class to which most of its closest neighbours belong. Under "neighbours" here are objects that are close to the studied in one sense or another. Note that here it is necessary to be able to determine how close objects are to each other, that is, be able to measure the "distance" between objects. This is not necessarily the Euclidean distance. This can be a measure of proximity of objects, for example, in colour, shape, taste, smell, interests, behaviour, and so forth. Consequently, to apply the kNN method in the feature space of objects, a certain metric must be introduced (that is, a distance function). If the nearest neighbours are divided into classes of approximately the same capacity, then, on the contrary, it makes sense to use those rules that take the distances into account to a greater degree (for example, the average rule). The rule of weighted majority takes into account both the number of objects in a class (via the weighting factor) and the distance to these objects. If there are a lot of classes and the weights are approximately the same, then first, such classes are selected so that their total weight is more than 0.5 and then one of the rules is applied. A Chebyshev metric was used with a number of neighbour equal to 2 and a uniform weight (Table 3).

Table 3. Parameters of the k-nearest neighbours (kNN).

Metric	Chebyshev
Number of Neighbours	2
Weight	Uniform

3.3. Correlations

The existence of a link between the various variables under investigation was sought for by calculating the *Pearson* correlation index (as done in Reference [58]). Also called the linear correlation coefficient of Bravais-Pearson, this statistical index shows an eventual linear relationship between variables.

Given two statistical variables X and Y, the correlation index (r_{xy}) is defined as their covariance divided by the product of the standard deviations of the two variables. This index varies between -1 and $+1$, where a > 0 shows the two variables are directly correlated, a value of 0 is for variables are uncorrelated and, finally, <0 for variables are inversely correlated.

Moreover, for the direct correlation (and similarly for the inverse) it is distinguished:

$0 < r_{xy} < 0.3$ there is a weak correlation;
$0.3 < r_{xy} < 0.7$ there is a moderate correlation;
$r_{xy} > 0.7$ there is a strong correlation.

The Pearson correlation was used in two alternative ways. It permitted relation of the:

- experimental measures by way of estimating the influence between different properties;
- experimental and predicted values by way of estimating the accuracy of the ML methods.

The adoption of the correlation coefficient as a way to evaluate the relationships between variables but also the accuracy in predictions was preferred in this study (similar to References [59,60]). Instead of other statistical approaches, it was considered for its extreme simplicity, both in terms of calculation and comprehension: the correlation coefficient directly represents a clear measure of the strength of the linear relationship between two variables. At the same time, it is useful to highlight that unrelated variables are not always independent: this depends on the fact that the correlation coefficient only detects a linear correlation.

4. Results

4.1. Experimental Measures

Tables A3 and A4 in Appendix C report the material tensile properties for SGI and CGI, respectively, as estimated by the three ML algorithms: Random Forest (RF); Neural Network (NN); k-Nearest Neighbours (kNN). Specifically, for each sample used in the experimental tests (48 in total), the expert system provided 3 different predictions (RF, NN, kNN) for each of the 4 mechanical properties under investigation (Ultimate Tensile Strength, Yield Strength, Ultimate Strain and Young's modulus). These values are later considered and discussed in terms of Mean Values (μ), Variability (σ, $\sigma_\%$) and Correlations (r_{xy}).

4.2. Spheroidal Graphite Cast Iron

In particular, the four diagrams in Figure 1 show these tensile properties in the case of SGI where the values from the measures are reported together with error bars representing the related variability. In addition, all the diagrams are displayed with a *y-scale* ranging from 0 to (approx.) the maximum value. These provide a visual representation of the ability of the ML methods to fit the experimental values. For instance, in the case of UTS, it is possible to see how the largest part of the predictions (specifically 69 on 27 × 3 = 81) falls inside the error bars (±9%). These error bars were evaluated starting from the experimental measures and passing by their relative standard deviation ($\sigma_\%$). Per each property and each material, a specific $\sigma_\%$ can be defined and this percentage can be used for dimensioning the error bar. The related bars are intended to give an idea of the variability of the measures, showing at the same time how a large part of the predictions falls inside this variability.

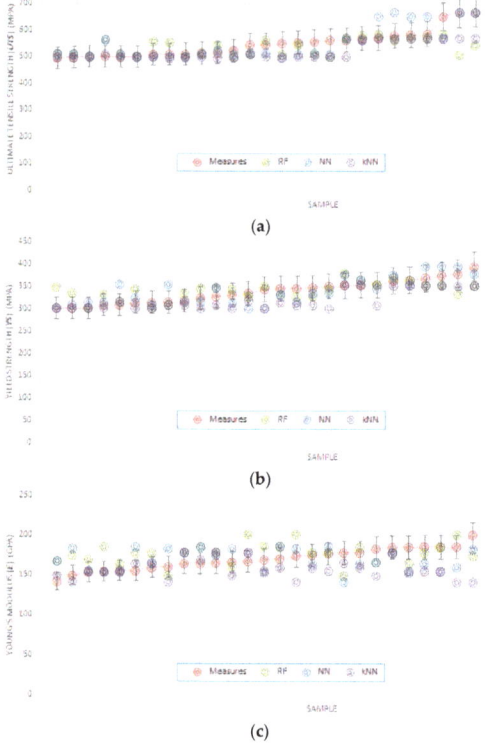

Figure 1. *Cont.*

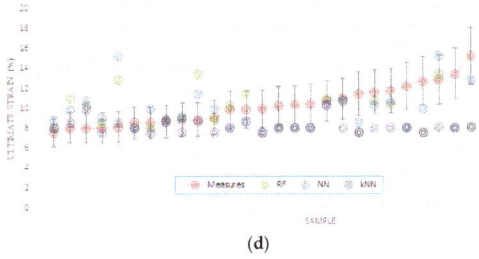

(d)

Figure 1. Metallurgical properties ((**a**) Ultimate Tensile Strength (UTS), (**b**) Yield Strength (YS), (**c**) Ultimate Strain (ε), (**d**) Young's modulus (*E*)) in the case of Spheroidal Graphite Iron (SGI), as measured and predicted.

Comparing these properties, it is also possible to recognize, in general, how prediction on UTS and YS seems better than those on ε and *E*.

4.3. Compacted Graphite Cast Iron

The analogous diagrams in Figure 2 show the tensile properties for CGI. This analogy is also respected for considerations of data and evidence. For instance, also in this case for UTS, the major part of the predictions (specifically 16 values on 21 × 3 = 63) falls inside the error bars (±6%).

Finally, the diagrams are similarly scaled for Figures 1 and 2, permitting a direct comparison between the properties of the two materials. It is evident, for instance, the lower properties of CGI respect to SGI.

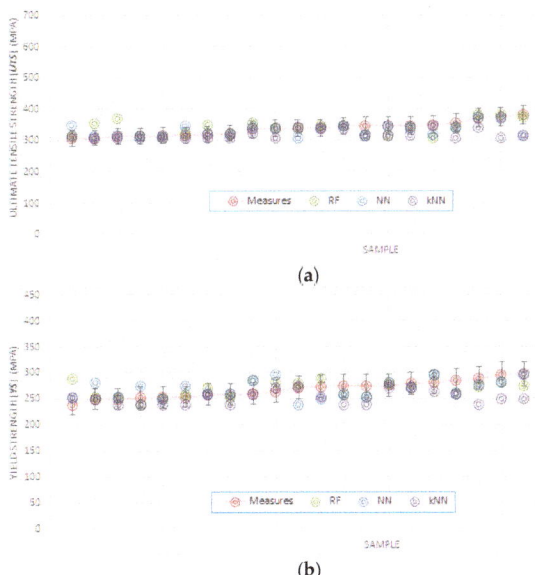

Figure 2. *Cont.*

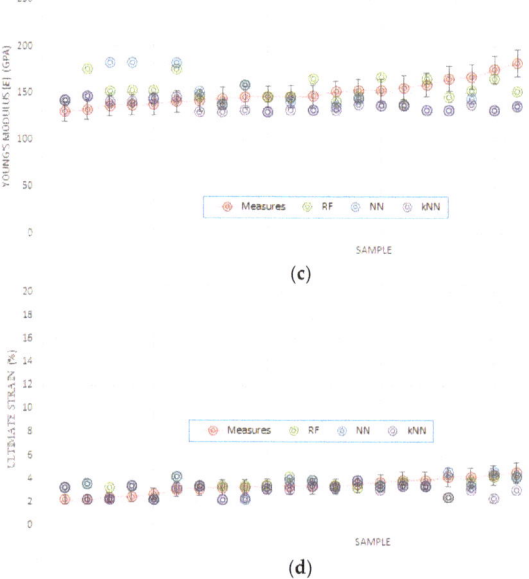

Figure 2. Metallurgical properties ((**a**) Ultimate Tensile Strength (UTS), (**b**) Yield Strength (YS), (**c**) Ultimate Strain (ε), (**d**) Young's modulus (E)) in the case of Compacted Graphite Iron (CGI), as measured and predicted.

5. Discussion

5.1. Prediction Model Validation

Before proceeding with discussing the results, it is appropriate to specify some additional concepts on the use of the learning and validation process within this work and justify them. Usually studies involving ML approaches divide the available data into two categories: those used for learning and those used for validating results. This apparently sensible and correct approach brings with it potential errors. For example, while the learning phase should vary slightly if information from a limited number of data is subtracted by the complete dataset, the validation phase is very sensitive to how many and which specific data are taken into account. Furthermore, whatever the result is, the doubt remains that the particular choice of data modified the general assessment on the validity of the entire prediction procedure.

This situation is frustrated when, as in the present case, external data are themselves subject to a high variability. In other words, it can be accepted that the expert system is trained with data subject to a certain intrinsic variability, waiting for its forecasts to be subject to a comparable variability. But, if these predictions are then compared with some specific data extracted from such a variable sample, the risk of making mistakes is rather high.

To avoid this risk, the simplest way is to increase the dataset of external data, to be used both for training and validating, as done in Reference [44]—where 700 samples were considered—all of them representing a homogeneous situation (same alloy, process conditions and so on). In the meantime, also considering the hypothesis where additional tests are not imaginable in short, a different approach was here proposed for data analysis. It is in line with another recent paper by some of the authors [54] where ML algorithms were used for the evaluation of the surface roughness evaluation in steel after thermal hardening by laser radiation. In brief, the validation is performed on the full dataset of values, not on part of it.

5.2. Data Preliminary Analysis

Considering the experimental measures, as reported in Appendix B in Tables A1 and A2, it is clear there is a not marginal variability in values, in the cases of both metallurgical and mechanical properties.

The relative standard deviation ($\sigma_\%$), expressed as the ratio between the standard deviation (σ) and the mean value (μ) permits a homogeneous comparison among parameters, providing useful information. For instance, in the case of SGI, the metallographic factors, expressed by $\sigma_\%$, spread between 14% (nodularity) and 44% (vermicularity). This variability is lower but also relevant, in the case of CGI, limited between 5% (vermicularity) and 24% (nodularity). Also, in terms of mechanical properties, the same variability of measures is confirmed. In particular, it ranges from 5%–6% in the case of the Ultimate Tensile Strength and the Yield Strength of CGI, up to 19%–21% in the case of the Ultimate Strain of SGI for both alloys.

As a consequence, the complexity of using these data for predictions on material properties is evident. At the same time, in Table 4 the relationships between parameters, estimated by the Pearson correlation coefficient, is reported in the case of the metallurgical properties, showing that several parameters are, as weighted, in medium-strong relations (e.g., perlite vs ferrite or perlite vs graphite).

Table 4. Correlation matrix between the metallurgical properties in the case of SGI and CGI.

SGI	Graphite	Ferrite	Perlite	Nodularity	Vermicularity
Graphite	1.00	0.04	−0.29	0.34	−0.30
Ferrite	0.04	1.00	**−0.79**	0.13	−0.19
Perlite	−0.29	**−0.79**	1.00	0.03	0.05
Nodularity	0.34	0.13	0.03	1.00	**−0.99**
Vermicularity	−0.30	−0.19	0.05	**−0.99**	1.00
CGI	Graphite	Ferrite	Perlite	Nodularity	Vermicularity
Graphite	1.00	−0.20	−0.45	−0.24	0.20
Ferrite	−0.20	1.00	**−0.79**	0.13	−0.19
Perlite	−0.45	**−0.79**	1.00	0.03	0.05
Nodularity	−0.24	0.13	0.03	1.00	**−0.99**
Vermicularity	0.20	−0.19	0.05	**−0.99**	1.00

But the relations of real interest are those that can connect mechanical properties to the microstructural ones, in order to predict the overall behaviour of an alloy knowing its compound and microstructure. Several articles, as mentioned, move in that direction, relating micro- and macro-scales, while the current investigation limits its focus on data analysis.

In particular, Table 5 exhibits, in terms of Pearson correlation coefficients, the relationships between the metallurgical parameters and the mechanical properties.

Table 5. Correlation matrix between the metallurgical and mechanical properties.

SGI	Graphite	Ferrite	Perlite	Nodularity	Vermicularity
Ultimate Tensile Strength (UTS)	−0.25	−0.87	0.90	0.63	−0.65
Yield Strength (YS)	−0.19	−0.83	0.84	0.67	−0.69
Ultimate Strain (ε)	−0.03	0.34	−0.32	0.05	−0.06
Young's Modulus (E)	−0.03	−0.09	0.09	0.18	−0.21
CGI	Graphite	Ferrite	Perlite	Nodularity	Vermicularity
Ultimate Tensile Strength (UTS)	−0.44	−0.46	0.69	0.23	−0.17
Yield Strength (YS)	−0.46	−0.35	0.61	0.25	−0.17
Ultimate Strain (ε)	−0.47	0.00	0.29	0.28	−0.26
Young's Modulus (E)	−0.11	0.08	0.00	−0.39	0.38

Several preliminary considerations emerge from these data. According to the experimental values, as correlated by the Pearson coefficient, it seems that:

- the SGI is more affected than the CGI with respect to changes in the metallurgical properties;
- the content of graphite is not so relevant for the definition of mechanical properties, especially in the case of SGI, while it has a light negative effect on CGI;
- the contents of ferrite and perlite, more than others, directly influence the material strength, especially in the case of SGI;
- the ductility is directly related to ferrite and perlite content but not to graphite in the case of SGI, while the graphite shows a light negative effect on CGI;
- the Young's modulus, evaluated as standard, is practically uncorrelated, except for a slight dependency to the nodularity and to the vermicularity, more relevant in the case of CGI.

5.3. Expert Algorithms

This preliminary data analysis permits the presentation of a general situation, quite common in the strength of materials, where each material property seems to be related to many of the others without strong/predominant dependencies. The consequent network of weak correlations between properties makes the material data analysis complex. Thus, every approach searching for a linear dependency is almost useless, unable to "untangle the skein" and artificial intelligence can help with "cracking" this complexity.

An expert algorithm, in fact, can search for patterns to be recognized without considering their physical significance. It will simply scan data search for hidden analogies and structures, moving the vision between different levels of the knowledge abstraction.

5.4. Mean Values and Variability

In Table 6, the tensile properties for SGI and CGI are reported as predicted by the expert system, in terms of mean values and standard deviations. It is evident how the variability in predictions is in line with the initial variability of measures. For instance, the experimental data on SGI, measured on 27 specimens, show an average value of the UTS equal to $\mu = 549$ MPa with variability a $\sigma = \pm 51$, equal to ±9%. Using these values, the three ML algorithms (RF, NN, kNN) propose three different predictions, equal to, in the mentioned case, respectively, 536 ± 31 (6%), 550 ± 64 (12%), 520 ± 33 (6%). The variability of those predictions (±6%; ±12%; 6%) is almost equivalent to the initial one (±9%). The same concept is also evident in the cases of all the other properties, demonstrating that the selected ML algorithms do not interfere with the base of data increasing its variability.

Table 6. Mechanical properties prediction in terms of mean values and standard deviations.

SGI	Unit	Data	RF	NN	kNN
Ultimate Tensile Strength (UTS)	MPa	549 ± 51 (9%)	536 ± 31 (6%)	550 ± 64 (12%)	520 ± 33 (6%)
Yield Strength (YS)	MPa	340 ± 27 (8%)	341 ± 20 (6%)	341 ± 31 (9%)	320 ± 22 (7%)
Ultimate Strain (ε)	%	10.2 ± 1.9 (19%)	9.4 ± 1.8 (19%)	9.7 ± 2.0 (21%)	8.1 ± 0.1 (8%)
Young's Modulus (E)	GPa	170 ± 14 (8%)	177 ± 13 (7%)	170 ± 13 (8%)	157 ± 12 (7%)
CGI	Unit	Data	RF	NN	kNN
Ultimate Tensile Strength (UTS)	MPa	337 ± 22 (6%)	340 ± 24 (7%)	332 ± 19 (6%)	317 ± 5 (4%)
Yield Strength (YS)	MPa	268 ± 16 (6%)	268 ± 16 (6%)	268 ± 17 (6%)	251 ± 13 (5%)
Ultimate Strain (ε)	%	3.4 ± 0.7 (21%)	3.5 ± 0.5 (14%)	3.4 ± 0.7 (21%)	3.0 ± 0.5 (18%)
Young's Modulus (E)	GPa	150 ± 14 (9%)	154 ± 12 (8%)	146 ± 17 (12%)	136 ± 20 (4%)

5.5. Mean Values and Error Estimation

In Table 7, the distance between the measured and predicted values are reported in terms of percentage errors with respect to the average values of tensile properties. It is evident how, except in rare cases, all the different algorithms can provide valid predictions. Also, in the present case, the confidence placed by Reference [59] in the possibility of accelerating material property predictions by the use of machine learning is confirmed.

Table 7. Error estimation in the prediction of mean values.

Property	SGI			CGI		
	RF	NN	kNN	RF	NN	kNN
Ultimate Tensile Strength (UTS)	−2.4%	0.2%	−5.3%	0.9%	−1.5%	−5.9%
Yield Strength (YS)	0.3%	0.3%	−5.9%	0.0%	0.0%	−6.3%
Ultimate Strain (ε)	−7.8%	−4.9%	−20.6%	2.9%	−11.8%	−11.8%
Young's modulus (E)	4.1%	0.0%	−7.6%	2.7%	−2.7%	−9.3%

In particular, of 24 values under investigation, 7 of them (almost 1/3) are predicted with an error lower than 1%; 11 lower than 3%; 16 (equal to 2/3) lower than 6%. In addition, it can also be noted that the uncorrected values (with percentage errors > 10%) are essentially related to the prediction of the Ultimate Strain (ε).

Even in this extreme case, the correspondence is quite acceptable, as demonstrated in Figure 3 where the comparison between measures and predictions is shown in terms of values frequency. It is possible to see how the density functions for measures and predictions overlap in several situations as, for example, in the cases of RF and NN methods for the Ultimate Strain (ε). Actually, it is possible to say that the NN method can closely retrace the density functions for all the situations under investigation. It means, in practice, that the NN method can predict the real values not limited to the average values but also through its range of variability. This good accuracy in predicting the mechanical properties of cast iron offered by the NN method was also reported by Reference [61] with respect to an investigation involving 24 process parameters and 800 external data.

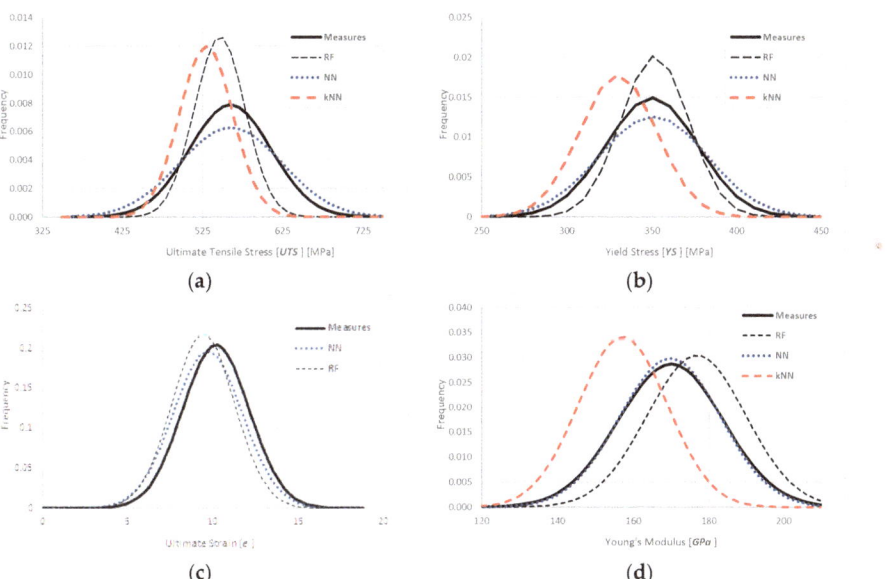

Figure 3. Comparison between measured and predicted values in terms of density functions (case of SGI) for: (a) Ultimate Tensile Stress; (b) Yield Stress; (c) Ultimate Strain; (d) Young's Modulus.

It is not the case that the kNN method shows a tendency to systematically underestimate reality (shown by narrow curves, shifted on the left, towards lower average values). These graphs are related to SGI but similar considerations emerge from data analysis in the case of CGI.

It is also worth noting that the validity of the predictions by ML does not seem to be affected by the specific values assumed by the properties. This contrasts with Reference [28], where it was reported that the prediction by the expert system ceased to be accurate for UTS higher than 100 MPa—in this analysis, the estimation methods proved to be equally valid for two cast iron families with very different properties, that is, SGI and CGI with an UTS of, respectively, 539 and 337 MPa.

Furthermore, it is confirmed that the potentiality of CGI in replacing SGI in particular applications where ductivity may be preferred to high stiffness, as expressed by References [4,62,63].

5.6. Correlations

Even if a prediction is accurate with respect to the average values and to the variability of measures, their distribution could be significantly different. Thus, the linear correlation is also investigated. In Table 8, the linear relations between measured and predicted values are reported in terms of Pearson correlation coefficients: as mentioned, the correlation is much better than the value closer to 1.

Table 8. Correlation between measures and predictions in terms of Pearson correlation coefficient.

Pearson Correlation Coefficient (r_{xy})	SGI			CGI		
	RF	NN	k-NN	RF	NN	kNN
Ultimate Tensile Strength (UTS)	0.39	**0.76**	**0.81**	0.48	0.41	0.37
Yield Strength (YS)	0.56	**0.74**	**0.79**	0.33	0.33	0.22
Ultimate Strain (ε)	−0.12	0.17	−0.14	0.29	**0.50**	0.19
Young's modulus (E)	0.17	−0.07	−0.05	−0.02	−0.48	−0.41

Figure 4 reports a graphical representation of the meaning of this linear correlation proposing a comparison between data with difference Pearson correlation coefficients (r_{xy}). It is the case, in particular, of the values of Yield Strength (YS) and the Young's modulus (E) for SGI, characterized by Pearson correlation coefficients of 0.74 (predicted by NN) and 0.17 (predicted by RF), respectively. The same diagram shows, simultaneously, properties with different units—namely (MPa) and (GPa)—and each point represents a single prediction, positioned by the experimental (x-axis) and predicted (y-axis) values. In this chart, points that are distributed along the bisector ($x = y$) or very close to it—as the values of Yield Strength (YS)—demonstrate a good correlation between experimental measurements and numerical predictions. On the contrary, a distribution of points like clouds, as in the case of Young's modulus (E) values, demonstrates a lack of correlation between measures and prediction.

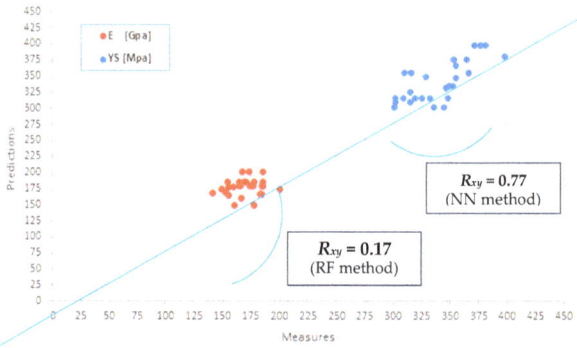

Figure 4. A graphical representation of data with different Pearson correlation coefficients. It is the case of YS (MPa) and E (GPa) for SGI with coefficients of, respectively, 0.74 (NN) and 0.17 (RF).

Instead, Figure 5 reports a graphical representation of data characterized by (substantially) equal Pearson correlation coefficients.

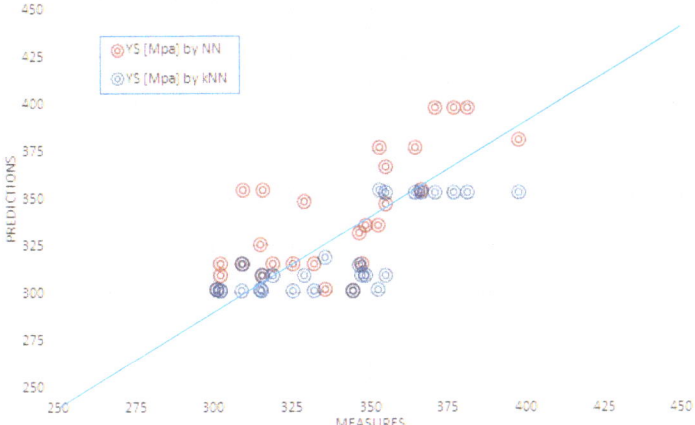

Figure 5. A graphical representation of data with equal Pearson correlation coefficients. It is the case of YS (MPa) for SGI with coefficients of, respectively, 0.74 (NN) and 0.79 (kNN).

It is the case, for instance, of the values of Yield Strength (YS) for SGI where those coefficients are 0.74 or 0.76, depending on the specific ML methods used (NN and kNN, respectively). The correspondence between the coefficients suggests that the two forecasting techniques are similarly effective. At the same time, the diagram shows how some aspects can escape attention if the analysis remains limited to the coefficients. In the figure, it is observed, for instance, how the values predicted by kNN are not randomly distributed around the bisector but are all shifted to the right. This means, in practice, that the kNN method proposes predictions subject to a slight, but systematic, error of underestimation, confirming the similar considerations on the kNN method previously emerged.

5.7. Results Summary

In brief, the accuracy of ML methods in predicting the mechanical properties starting from the metallurgical properties, as estimated by images analysis samples and macro-indicators, was evaluated comparing measures and predictions with respect to their mean values (μ)/standard deviations (σ) (Table 7) and by the overall trend of the Pearson coefficients (r_{xy}) (Table 8). The comparison is also shown by representing, respectively, the way the density functions, defined by μ and σ, overlap (Figure 3) and how these values are distributed, point to point, in correlation graphs (Figures 4 and 5)

Thanks to this analysis it is possible to propose, as results, the following considerations:

- ML methods confirm their general validity in predicting the mechanical properties of metals;
- this seems true, even in the presence of a quite limited dataset to be used for training;
- information can be directly taken from micrographs by a conventional process of image analysis and macro-indicators without the need to go through deeper metallurgical investigations;
- in particular, the NN method seems the most appropriate of those considered;
- the kNN method, although it has good accuracy, also shows a tendency to systematic errors;
- the accuracy in prediction is different for each specific property under investigation, achieving the best results for UTS and YS but also offering acceptable indications in the other cases;
- the average values of experiments and predictions (measured by μ) often coincide in practice;
- the deviation with respect to the average values (measured by σ) shows a variability in prediction in line with the intrinsic variability as revealed by the experimental measurements;
- the Pearson correlation (r_{xy}) can be conveniently adopted for a quick evaluation of data but also for the validation of predictions.

6. Conclusions

There is no doubt that Artificial Intelligence (AI) and Machine Learning (ML) have become widely known over the past few years thanks to their applicability. As Big Data technologies retain the status of the most discussed IT trend of modernity, so ML algorithms can be considered the most powerful tool focused on the predictive application of large amounts of data. It is the same in the material sciences.

This article proposed the use of three between the most common ML algorithms, available in an open source platform, in supporting the material data analysis. In particular, the Random Forest, the Artificial Neural Network and the k-nearest neighbours methods have been preferred for pattern recognition on two datasets of tensile properties of two different foundry alloys—ductile and compact graphite cast irons—as measured in previous experiments. The Expert System, trained on these (extremely limited number of) data, provided predictions in line with the measures, both in terms of mean values and variability, in large part of the situations under investigation. In particular, an extremely accurate correspondence between experimental and predicted data along the full range of values emerges in the case of Ultimate Tensile Strength (UTS) and Yield Strength (YS), with errors lower than 1% (when considered in terms of mean/expected values). But even in the other situations under examination, related to the Ultimate Strain (ε) and Young's modulus (E), the use of AI as an investigation system could provide valid support. As a consequence, it is possible to confirm the benefit of the ML techniques in predicting the mechanical properties of cast alloys. Furthermore, they could help with investigating the strict relationship between these ultimate properties and other fundamental aspects of metallurgy, as constituent elements, microstructures, process parameters or treatments.

Finally, it is worth considering how the validity in prediction could be reasonably improved: selecting additional Machine Learning classifiers, optimizing their parameters and enlarging the dataset adopted for training. All these improvements in the investigation approach are relatively simple without introducing any additional complexity to the predicting procedure.

Author Contributions: Conceptualization, C.F.; methodology, C.F. and C.P.B.; software, M.B.; validation, C.F.; formal analysis, C.F. and G.M.; investigation, C.F. and M.B.; resources, C.F. and M.B.; data curation, C.F.; writing—original draft preparation, C.F. and M.B.; writing—review and editing, C.F. and G.M.; visualization, C.F. and C.P.B.; supervision, C.P.B. and G.M.; project administration, C.P.B.

Funding: This research received no external funding.

Acknowledgments: The experimental section was carried out at the Interdepartmental Centre for Industrial Research on Advanced Applications in Mechanical Engineering and Materials Technology of the University of Bologna, Italy, under the coordination of Lorella Ceschini and Giangiacomo Minak. Casting processes were performed at the CRIF Research Centre of SCM Foundries of Rimini, Italy, under the coordination of Stefano Cucchetti. Special thanks to Andrea Morri.

Conflicts of Interest: The authors declare no conflict of interest. In particular, the Foundry that provided the materials had no role in the design of the study; in the collection, analyses or interpretation of data; in the writing of the manuscript or in the decision to publish the results.

Appendix A

For a better comprehension, the adopted nomenclature is here reported:

SGI	Spheroidal cast iron
CGI	Compact graphite cast iron
GR	Graphite
FE	Ferrite
PE	Perlite
NO	Grade of Nodularity
VE	Grade of Vermicularity

HB	Brinell hardness	
AI	Artificial Intelligence	
ANN	Artificial Neural Network	
ML	Machine Learning	
RF	Random Forest method	
NN	Neural Network method	
kNN	k-Nearest Neighbours method	
μ	Mean Value	
σ	Standard Deviation	
$\sigma_\%$	Relative Standard Deviation	
r_{xy}	Pearson Correlation Coeff.	
UTS	Ultimate Tensile Strength	
YS	Yield Strength	
ε	Ultimate Strain/Ductility	
E	Young's/Elasticity Modulus	

Appendix B

Table A1. Metallographic and mechanical properties of specimens in Spheroidal Graphite Iron (SGI). Data from [21–23].

Specimen	GR %	FE %	PE %	NO %	VE %	UTS MPa	YS MPa	ε %	E GPa
1	9.1	47.5	43.4	53.9	36.8	500.0	315.3	10.2	154.5
2	12.2	47.1	40.8	63.6	27.2	501.0	302.3	10.3	164.6
3	13.6	42.5	43.9	75.2	17.0	508.7	315.7	8.6	184.9
4	8.6	48.6	42.8	62.6	30.4	496.8	301.2	11.6	184.9
5	12.1	48.5	39.5	67.1	26.2	494.8	325.4	8.5	170.5
6	11.2	42.8	46.0	68.8	23.7	508.8	314.8	8.0	185.7
7	8.3	43.6	48.1	50.9	40.9	501.4	309.2	9.8	153.0
8	12.6	43.6	43.8	79.0	15.2	500.5	309.4	8.8	178.2
9	6.3	52.8	40.9	56.4	34.4	510.2	302.1	8.0	155.0
10	8.6	43.7	47.8	65.7	24.3	549.9	344.7	11.7	168.9
11	12.1	44.8	43.1	75.5	17.1	561.5	347.5	13.4	178.2
12	8.1	49.0	42.9	75.7	17.3	545.4	329.1	12.8	165.4
13	9.2	40.8	50.0	66.9	23.6	554.4	352.4	10.4	155.3
14	7.1	44.6	48.3	68.6	22.3	544.8	346.4	10.9	176.7
15	9.4	47.3	43.4	75.1	17.1	557.4	348.7	12.2	174.7
16	13.2	34.2	52.7	86.1	9.4	570.4	354.8	11.4	141.7
17	11.3	30.5	58.2	85.7	9.4	586.4	366.5	7.5	186.0
18	13.7	39.2	47.1	84.4	10.8	564.4	354.9	9.8	167.0
19	9.1	32.1	58.8	78.2	16.3	582.9	370.9	8.0	173.7
20	10.2	30.8	59.1	80.8	14.1	572.5	353.0	8.5	149.2
21	7.6	33.5	58.8	84.6	10.2	581.9	364.4	12.7	200.6
22	9.3	24.6	66.1	89.6	5.9	651.7	376.8	9.9	160.4
23	7.0	22.7	70.3	81.6	11.9	668.7	397.5	9.0	183.3
24	6.5	24.8	68.7	74.2	17.7	666.6	381.2	8.8	166.4
25	10.2	55.7	34.1	77.7	16.6	514.2	319.0	15.2	164.6
26	7.0	51.6	41.4	72.5	19.7	515.7	335.7	8.1	159.6
27	7.1	45.2	47.7	61.9	27.6	523.9	332.0	10.7	185.9
Mean (μ)	9.7	41.2	49.2	72.7	20.1	549.4	339.7	10.2	170.0
St. Dev. (σ)	2.3	9.0	9.4	10.2	8.8	50.6	26.7	1.9	13.9
R. St. Dev. ($\sigma_\%$)	24%	22%	19%	14%	44%	9%	8%	19%	8%

Table A2. Metallographic and mechanical properties of specimens in Compacted Graphite Iron (CGI). Data from [21–23].

Specimen	GR %	FE %	PE %	NO %	VE %	UTS MPa	YS MPa	ε %	E GPa
1	21.0	60.2	18.8	16.1	81.2	318.8	253.1	3.3	136.1
2	17.3	62.3	20.4	12.1	85.1	350.1	274.3	2.2	182.5
3	13.9	62.9	23.2	15.4	82.2	307.5	237.8	3.3	146.3
4	14.5	61.6	23.9	9.4	88.5	314.1	252.4	2.2	137.1
5	14.5	62.6	22.9	23.4	74.3	316.2	259.1	2.5	142.1
6	13.2	64.8	22.0	13.0	84.4	308.4	252.3	2.4	140.8
7	15.7	64.3	20.0	17.5	79.7	321.7	258.5	3.4	151.4
8	12.6	61.9	25.6	11.7	86.4	315.0	249.7	2.7	152.9
9	16.7	53.5	29.8	9.0	88.9	312.4	249.5	3.6	156.3
11	11.4	64.9	23.7	15.1	82.7	338.3	273.0	4.4	175.6
12	9.2	67.6	23.3	21.6	74.6	338.8	257.3	4.2	146.4
14	11.2	65.8	22.9	17.9	80.0	336.8	274.0	4.6	132.1
15	10.3	63.0	26.8	16.7	81.4	339.2	270.9	4.2	145.7
16	14.6	56.6	28.9	19.5	78.5	345.8	263.9	3.4	145.8
17	10.1	62.9	27.0	16.7	81.7	346.4	278.4	3.7	159.2
18	11.1	63.5	25.5	16.2	81.8	354.6	288.3	3.1	165.6
19	9.8	59.9	30.3	17.8	80.5	345.0	274.9	3.4	152.6
20	12.8	58.3	28.9	24.0	74.0	345.7	284.1	3.2	129.8
22	12.9	52.9	34.2	18.5	79.5	370.7	281.0	3.9	144.3
23	12.6	53.4	34.0	16.3	81.9	374.0	295.8	3.4	137.9
24	9.7	55.2	35.2	13.3	85.4	380.8	296.7	3.9	167.9
Mean (μ)	13.0	61.2	25.8	16.6	81.2	337.2	267.9	3.4	149.9
St. Dev. (σ)	3.0	4.3	4.7	4.0	4.2	21.8	16.3	0.7	14.0
R. St. Dev. (σ%)	23%	7%	18%	24%	5%	6%	6%	21%	9%

Appendix C

Table A3. Prediction of mechanical properties of Spheroidal Graphite Iron (SGI).

UTS				YS			ε				E			
MPa	RF	NN	kNN	MPa	RF	NN	kNN	%	RF	NN	kNN	GPa	RF	NN	kNN
495	510	509	497	325	348	316	301	8.5	8.0	8.0	8.0	171	185	186	160
497	510	510	495	301	349	302	302	11.6	10.7	10.2	8.0	185	165	153	155
500	501	501	497	315	302	309	301	10.2	8.0	8.0	8.0	155	185	153	153
501	510	509	497	302	301	315	301	10.3	8.0	8.0	8.0	165	185	185	169
501	562	564	509	309	316	355	316	8.8	9.0	9.0	7.5	178	149	142	165
501	500	500	497	309	332	315	301	9.8	11.4	8.6	8.6	153	169	155	155
509	557	501	501	316	309	355	309	8.6	8.8	8.5	8.8	185	178	178	178
509	554	501	495	315	345	325	302	8.0	10.2	10.7	9.8	186	178	165	155
510	500	500	497	302	336	309	301	8.0	10.9	9.8	8.5	155	165	155	153
514	510	509	501	319	336	316	309	15.2	8.1	12.8	8.1	165	178	178	178
516	545	524	495	336	329	302	319	8.1	12.8	15.2	8.5	160	177	165	165
524	501	510	497	332	345	315	301	10.7	10.9	10.2	10.3	186	185	155	155
545	562	516	501	329	349	349	309	12.8	13.4	15.2	8.1	165	178	175	178
545	509	516	509	346	332	332	315	10.9	10.7	10.7	8.0	177	178	186	155
550	509	497	497	345	352	301	301	11.7	10.4	10.7	8.0	169	186	153	155
554	545	510	501	352	345	336	302	10.4	8.0	8.0	8.0	155	177	185	165
557	509	514	501	349	329	336	309	12.2	8.0	8.0	8.1	175	178	165	160
562	501	509	501	348	329	316	309	13.4	8.0	8.0	8.0	178	185	165	160
564	573	570	501	355	355	348	309	9.8	10.2	8.0	8.0	167	201	178	178
570	582	564	564	355	367	367	353	11.4	7.5	8.5	7.5	142	167	167	149
573	583	652	570	353	381	377	355	8.5	8.0	9.9	7.5	149	174	183	142
582	564	669	570	364	371	377	353	12.7	7.5	9.9	7.5	201	174	183	142
583	573	652	570	371	353	398	353	8.0	8.5	9.0	7.5	174	201	183	142
586	573	652	570	367	364	355	353	7.5	8.0	8.8	8.0	186	201	160	142
652	586	573	570	377	355	398	353	9.9	7.5	7.5	7.5	160	149	183	142
667	510	669	573	381	336	398	353	8.8	13.4	11.4	8.6	166	160	183	149
669	545	667	573	398	353	381	353	9.0	8.8	9.9	7.5	183	166	166	149

Table A4. Prediction of mechanical properties of Compacted Graphite Iron (CGI).

UTS				YS				ε				E			
MPa	RF	NN	kNN	MPa	RF	NN	kNN	%	RF	NN	kNN	GPa	RF	NN	kNN
308	316	350	308	238	288	252	252	3.3	3.4	2.2	2.4	146	146	141	132
308	355	314	308	252	238	274	238	2.4	3.3	2.2	2.2	141	176	183	146
312	371	315	312	250	252	281	250	3.6	3.3	3.9	3.9	156	138	137	137
314	312	314	252	250	250	238	2.2	3.6	3.6	2.2	137	153	183	141
315	308	314	315	250	252	252	238	2.7	2.2	2.2	2.2	153	168	137	137
316	322	346	316	259	284	284	257	2.5	3.4	3.4	3.4	142	146	151	130
319	350	322	319	253	259	274	238	3.3	3.4	2.2	2.2	136	151	183	141
322	319	319	322	259	252	259	238	3.4	3.4	3.2	3.2	151	141	136	132
337	355	339	337	274	257	257	238	4.6	4.2	4.2	3.1	132	176	146	146
338	339	337	338	273	288	250	252	4.4	4.2	4.6	2.4	176	166	132	132
339	346	337	339	271	278	238	273	4.2	2.5	4.6	2.5	146	159	159	132
339	337	308	339	257	271	259	259	4.2	3.7	3.4	3.1	146	166	132	132
345	346	346	345	275	278	281	271	3.4	3.4	3.3	3.3	153	146	144	138
346	316	346	346	264	281	296	271	3.4	4.2	3.9	3.1	146	146	130	130
346	339	339	346	278	271	271	271	3.7	3.4	3.4	3.1	159	166	132	132
346	316	316	346	284	259	259	257	3.2	3.4	3.4	3.4	130	142	142	142
350	308	319	350	274	252	253	238	2.2	3.3	3.3	3.3	183	151	137	136
355	338	339	355	288	273	275	238	3.1	4.2	4.2	3.3	166	146	132	132
371	381	374	371	281	296	296	264	3.9	3.4	3.4	3.4	144	138	138	130
374	381	371	374	296	281	281	250	3.4	3.9	3.9	3.4	138	153	144	144
381	374	312	381	297	274	296	250	3.9	3.7	3.4	3.4	168	153	144	138

References

1. Ashby, M.F.; Jones, D.R.H. *Engineering Materials 1: An Introduction to Properties, Applications and Design*, 4th ed.; Elsevier: Oxford, UK, 2012.
2. Hans, E.; Koski, J.; Osyczka, A. *Multicriteria Design Optimization: Procedures and Applications*; Springer Science & Business Media: Berlin, Germany, 2012.
3. Boyles, A. *The Structure of Cast Iron: A Series of Three Educational Lectures on the Structure of Cast Iron*; American Society for Metals: Russell Township, OH, USA, 1947.
4. Fragassa, C. Material selection in machine design: The change of cast iron for improving the high-quality in woodworking. *Proc. Inst. Mech. Eng. C J. Mech. Eng. Sci.* **2016**, *231*, 18–30. [CrossRef]
5. Campbell, F.C. *Elements of Metallurgy and Engineering Alloys*; ASM International: Materials Park, OH, USA, 2008; p. 453.
6. Elliott, R. *Cast Iron Technology*; Butterworth-Heinemann: London, UK, 1988.
7. Sinha, A.K. *Physical Metallurgy Handbook*; McGraw-Hill Professional Publishing: New York, NY, USA, 2003.
8. Damir, A.N.; Elkhatib, A.; Nassef, G. Prediction of fatigue life using modal analysis for grey and ductile cast iron. *Int. J. Fatigue* **2007**, *29*, 499–507. [CrossRef]
9. Elkholy, A. Prediction of abrasion wear for slurry pump materials. *Wear* **1983**, *84*, 39–49. [CrossRef]
10. Berdin, C.; Dong, M.J.; Prioul, C. Local approach of damage and fracture toughness for nodular cast iron. *Eng. Fract. Mech.* **2001**, *68*, 1107–1117. [CrossRef]
11. Kim, J.; Bae, C.; Woo, H.; Kim, J.; Hong, S. Assessment of residual tensile strength on cast iron pipes. In *Pipelines 2007: Advances and Experiences with Trenchless Pipeline Projects*; Mohammad Najafi, P.E., Lynn Osborn, P.E., Eds.; ASCE: Reston, VA, USA, 2007; pp. 1–7.
12. Atkinson, K.; Whiter, J.T.; Smith, P.A.; Mulheron, M. Failure of small diameter cast iron pipes. *Urban Water* **2002**, *4*, 263–271. [CrossRef]
13. Fragassa, C.; Zigulic, R.; Pavlovic, A. Push-pull fatigue test on ductile and vermicular cast irons. *Eng. Rev.* **2016**, *36*, 269–280.
14. Álvarez, L.; Luis, C.J.; Puertas, I. Analysis of the influence of chemical composition on the mechanical and metallurgical properties of engine cylinder blocks in grey cast iron. *J. Mater. Process. Technol.* **2004**, *153*, 1039–1044. [CrossRef]
15. Fragassa, C.; Minak, G.; Pavlovic, A. Tribological aspects of cast iron investigated via fracture toughness. *Tribol. Ind.* **2016**, *38*, 1–10.

16. Li, Y.; Chen, W.; Huang, D.; Luo, J.; Liu, J.; Chen, Y.; Liu, Q.; Su, S. Energy conservation and emissions reduction strategies in foundry industry. *China Foundry* **2010**, *7*, 392–399.
17. Gonzaga, R.A.; Carrasquilla, J.F. Influence of an appropriate balance of the alloying elements on microstructure and on mechanical properties of nodular cast iron. *J. Mater. Process. Technol.* **2005**, *162*, 293–297. [CrossRef]
18. McNeil, I. *An Encyclopedia of the History of Technology*; Routledge: London, UK, 2002.
19. Angus, H.T. *Cast Iron: Physical and Engineering Properties*; Elsevier: London, UK, 2013.
20. Collini, L.; Nicoletto, G.; Konečná, R. Microstructure and mechanical properties of pearlitic gray cast iron. *Mater. Sci. Eng. A* **2008**, *488*, 529–539. [CrossRef]
21. Radovic, N.; Morri, A.; Fragassa, C. A study on the tensile behaviour of spheroidal and compacted graphite cast irons based on microstructural analysis. In Proceedings of the 11th IMEKO TC15 Youth Symposium on Experimental Solid Mechanics, Brasov, Romania, 30 May–02 June 2012; pp. 185–190.
22. Fragassa, C.; Pavlovic, A. Compacted and spheroidal graphite irons: Experimental evaluation of Poisson's ratio. *FME Trans.* **2016**, *44*, 327–332. [CrossRef]
23. Fragassa, C.; Radovic, N.; Pavlovic, A.; Minak, G. Comparison of mechanical properties in compacted and spheroidal graphite irons. *Tribol. Ind.* **2016**, *38*, 49–59.
24. Tiedje, N. Solidification, processing and properties of ductile cast iron. *Mater. Sci. Technol.* **2010**, *26*, 505–514. [CrossRef]
25. Costa, N.; Machado, N.; Silva, F.S. A new method for prediction of nodular cast iron fatigue limit. *Int. J. Fatigue* **2010**, *32*, 988–995. [CrossRef]
26. Shiraki, N.; Usui, Y.; Kanno, T. Effects of number of graphite nodules on fatigue limit and fracture origins in heavy section spheroidal graphite cast iron. *Mater. Trans.* **2016**, *57*, 379–384. [CrossRef]
27. Santos, C.A.; Spim, J.A., Jr.; Ierardi, M.C.; Garcia, A. The use of artificial intelligence technique for the optimisation of process parameters used in the continuous casting of steel. *Appl. Math. Modell.* **2002**, *26*, 1077–1092. [CrossRef]
28. Calcaterra, S.; Campana, G.; Tomesani, L. Prediction of mechanical properties in spheroidal cast iron by neural networks. *J. Mater. Process. Technol.* **2000**, *104*, 74–80. [CrossRef]
29. Roshan, H.D.; Sudesh, K. Expert system for analysis of casting defects: Cause module. *Trans. Am. Foundrymen's Soc.* **1989**, *97*, 601–606.
30. Artificial Intelligence. Available online: https://en.wikipedia.org/wiki/Artificial_intelligence (accessed on 15 March 2019).
31. Fukunaga, K. *Introduction to Statistical Pattern Recognition*; Elsevier: Amsterdam, The Netherlands, 2013.
32. Dong, M.J.; Hu, G.K.; Diboine, A.; Moulin, D.; Prioul, C. Damage modelling in nodular cast iron. *J. Phys. IV* **1993**, *3*, 643–648. [CrossRef]
33. Gonzaga, R.A. Influence of ferrite and pearlite content on mechanical properties of ductile cast irons. *Mater. Sci. Eng. A* **2013**, *567*, 1–8. [CrossRef]
34. Radisa, R.; Ducic, N.; Manasijevic, S.; Markovic, N.; Cojbasic, Z. Casting improvement based on metaheuristic optimization and numerical simulation. *Facta Univ. Ser. Mech. Eng.* **2017**, *15*, 397–411. [CrossRef]
35. Voracek, J. Prediction of mechanical properties of cast irons. *Appl. Soft Comput.* **2001**, *1*, 119–125. [CrossRef]
36. Kohavi, R.; Provost, F. Glossary of terms. *Mach. Learn.* **1998**, *30*, 271–274.
37. Weiss, S.M.; Kulikowski, C.A. *Computer Systems That Learn: Classification and Prediction Methods from Statistics, Neural Nets, Machine Learning and Expert Systems*; Morgan Kaufmann: San Mateo, CA, USA, 1991; Volume 123.
38. Bishop, C.M. *Pattern Recognition and Machine Learning*; Springer: New York, NY, USA, 2006.
39. Anzai, Y. *Pattern Recognition and Machine Learning*; Elsevier: Amsterdam, The Netherlands, 2012.
40. Maulik, U.; Sanghamitra, B. Genetic algorithm-based clustering technique. *Pattern Recognit.* **2000**, *33*, 1455–1465. [CrossRef]
41. Samuel, A.L. Some studies in machine learning using the game of checkers. *IBM J. Res. Dev.* **2000**, *44*, 206–226. [CrossRef]
42. Schmidhuber, J. Deep learning in neural networks: An overview. *Neural Netw.* **2015**, *61*, 85–117. [CrossRef]
43. Hughes, I.C.H.; Powell, J.; de Albuquerque, V.H.C.; Cortez, P.C.; de Alexandria, A.R.; Tavares, J.M.R. A new solution for automatic microstructures analysis from images based on a backpropagation artificial neural network. *Nondestr. Test. Eval.* **2008**, *23*, 273–283.
44. Perzyk, M.; Kochański, A.W. Prediction of ductile cast iron quality by artificial neural networks. *J. Mater. Process. Technol.* **2001**, *109*, 305–307. [CrossRef]

45. Ward, L.; Agrawal, A.; Choudhary, A.; Wolverton, A. A general-purpose machine learning framework for predicting properties of inorganic materials. *Comput. Mater.* **2016**, *2*, 16028. [CrossRef]
46. Liu, Y.; Zhao, T.; Ju, W.; Shi, S. Materials discovery and design using machine learning. *J. Materiomics* **2017**, *3*, 159–177. [CrossRef]
47. SCM Foundry. Available online: http://www.scmfonderie.it/?l=en&p=azienda (accessed on 15 November 2015).
48. Altstetter, J.D.; Nowicki, R.M. Compacted Graphite Iron—Its properties and automotive applications. *AFS Trans.* **1982**, *82*, 959–970.
49. BS EN ISO 1563. *Founding. Spheroidal Graphite Cast Iron*; BSI: London, UK, 2012.
50. EN ISO 6892-1. *Metallic Materials—Tensile Testing—Part 1: Method of Test at Room Temperature*; ISO: Geneva, Switzerland, 2016.
51. Orange Platform. Available online: https://orange.biolab.si/ (accessed on 10 April 2019).
52. Wolpert, D.H.; Macready, W.G. No free lunch theorems for optimization. *IEEE Trans. Evol. Comput.* **1997**, *1*, 67. [CrossRef]
53. Wolpert, D. The lack of a priori distinctions between learning algorithms. *Neural Comput.* **1996**, *8*, 1341–1390. [CrossRef]
54. Babic, M.; Calì, M.; Nazarenko, I.; Fragassa, C.; Ekinovic, S.; Mihaliková, M.; Janjić, M.; Belič, I. Surface roughness evaluation in hardened materials by pattern recognition using network theory. *Int. J. Interact. Des. Manuf.* **2018**, *13*, 211–219. [CrossRef]
55. Lin, Y.; Yongho, J. Random forests and adaptive nearest neighbours. *J. Am. Stat. Assoc.* **2006**, *101*, 578–590. [CrossRef]
56. Gurney, K. *An Introduction to Neural Networks*; CRC Press: London, UK, 2014.
57. Garcia, S.; Derrac, J.; Cano, J.; Herrera, F. Prototype selection for nearest neighbour classification: Taxonomy and empirical study. *IEEE Trans. Pattern Anal. Mach. Intell.* **2012**, *34*, 417–435. [CrossRef] [PubMed]
58. Wagner, N.; Rondinelli, J.M. Theory-guided machine learning in materials science. *Front. Mater.* **2016**, *3*, 28. [CrossRef]
59. Pilania, G.; Wang, C.; Jiang, X.; Rajasekaran, S.; Ramprasad, R. Accelerating materials property predictions using machine learning. *Sci. Rep.* **2013**, *3*, 2810. [CrossRef] [PubMed]
60. Zhang, Y.; Ling, C. A strategy to apply machine learning to small datasets in materials science. *Npj Comput. Mater.* **2018**, *4*, 25. [CrossRef]
61. Sata, A. Mechanical property prediction of investment castings using artificial neural network and multivariate regression analysis. In Proceeding of the 63rd Indian Foundry Congress, Greater Noida, India, 27 February–1 March 2015.
62. Lucisano, G.; Stefanovic, M.; Fragassa, C. Advanced design solutions for high-precision woodworking machines. *Inter. J. Qual. Res.* **2016**, *10*, 143–158.
63. Dawson, S.; Schroeder, T. Practical applications for compacted graphite iron. *AFS Trans.* **2004**, *47*, 1–9.

© 2019 by the authors. Licensee MDPI, Basel, Switzerland. This article is an open access article distributed under the terms and conditions of the Creative Commons Attribution (CC BY) license (http://creativecommons.org/licenses/by/4.0/).

MDPI
St. Alban Anlage 66
4052 Basel
Switzerland
Tel. +41 61 683 77 34
Fax +41 61 302 89 18
www.mdpi.com

Metals Editorial Office
E-mail: metals@mdpi.com
www.mdpi.com/journal/metals

www.ingramcontent.com/pod-product-compliance
Lightning Source LLC
LaVergne TN
LVHW070420100526
838202LV00014B/1491